INTRODUCTION TO INDUSTRIAL ROBOTICS

Ramachandran Nagarajan
Former Professor in Robotics
School of Engineering
Universiti Malaysia Perlis
Malaysia

Pearson

Associate Editor—Acquisitions: R. Dheepika
Senior Editor—Production: C. Purushothaman

ISBN 978-93-325-4480-2

First Impression

Published by Pearson India Education Services Pvt. Ltd,
CIN: U72200TN2005PTC057128, formerly known as TutorVista Global Pvt. Ltd,
licensee of Pearson Education in South Asia.

Head Office: A-8 (A), 7th Floor, Knowledge Boulevard, Sector 62, Noida 201 309,
Uttar Pradesh, India.
Registered Office: 4th Floor, Software Block, Elnet Software City, TS-140, Block 2 & 9,
Rajiv Gandhi Salai, Taramani, Chennai 600 113, Tamil Nadu, India.
Fax: 080-30461003, Phone: 080-30461060
www.pearson.co.in, Email: companysecretary.india@pearson.com

Compositor: MAP Systems, Bengaluru
Printed in India at Repro India Limited

BRIEF CONTENTS

CONTENTS

PREFACE

This book is on Industrial Robotics, a kind of robots which would help humans in industries. This book introduces some basic ideas and is elaborately explained for easy reading and understanding. The book covers a wide range of topics that may not be found in other books.

Chapter 1 begins with an "Introduction" which describes the Laws of Industrial Robotics. A detailed listing of Chronological Development of Robots Technology and a detailed features of robots are included. A method of computing pay back period using a well known formula is also introduced in this book.

The next Chapter 2 is on Grippers and Tools of Industrial Robots. Here, definitions of Industrial robots are explained along with configuration of work volume. A typical T^3 Industrial Robot with a description in a relevant block diagram is provided. A conventional structural definition with the help of four basic forms of industrial robots and a detailed listing of seven versions of industrial robots are explained. Next, the precision of movement with the help of some details of degrees of freedom are also given. This makes the introduction to six degrees of freedom amenable. The chapter ends with a detailed set of grippers and tools of industrial robots.

Chapter 3 is on Coordinate Transformation. A 2-D and a 3-D coordinate transformations along with the principles of homogeneous matrices are explained. Four submatrices in each of the transformation are detailed with explanations. Trans(a, b, c), Rot(x,θ), Rot(y,θ) and Rot(z,θ) are detailed with the help of examples. Then, inverse transformation in addition to a simplified derivation of inverting (4×4) matrix are also given. Composite transformation matrix is also explained with an example. Object manipulations and wrist articulations are explained with detailed examples.

Chapter 4 is on Kinematics. A brief discussion on inverse kinematics is also given. The use of homogeneous transformation of (4×4) matrix is illustrated. Link and joint parameters are introduced with an example of using D-H notation and its transformation matrix. Symbolic procedure is explained. Four phases of D-H method are also detailed. Several application examples are also illustrated. These are intended to explain how to get coordinate frames, how to determine the kinematic parameters and how to derive a Table having kinematic parameters along with the robot structures. The outcome is a set of matrices. Finally, Jacobian is introduced. Manipulator Jacobian and Jacobian Singularities are explained with diagrams.

Robot Sensors are discussed in Chapter 5. This chapter starts with Internal and External sensors and applications of sensors. Two Tables are discussed: (i) the desirable features of sensors and (ii) general operating parameters of proximity sensors. A good discussion of proximity sensors in addition to retro-reflective sensors are covered. Electromagnetic and electrostatic sensors are also brought in. Then, the touch sensors and the slip sensor are discussed. A coverage of range sensing (triangulation, short range, ultrasonic and laser range sensing) is illustrated with appropriate diagrams. Force sensings (quarter bridge, half bridge and full bridge) are covered with diagrams and derivation on sensitivity. An introduction to vision system for inspection is also covered.

Chapter 6 is on Robot Control. A general block diagram on robot control is given. Joint motion especially for second order plant is derived. A discussion on the control unit is made. An illustration to second order step responses and a block digram illustrating various blocks especially to robot control are also given. Electric, hydraulic and pneumatic drive systems are given and their respective advantages and disadvantages are also discussed. Then an industrial vision system is illustrated with its required features. The inspection using industrial vision is detailed. Finally, the solid state camera is discussed with relevant requirements.

Two interlinked aspects of robot are detailed in Chapter 7—one is on Programming and the other is on Work Cell. On the programming aspect, the language structure is discussed with a diagram and explanation to robot operating system. Robot programming languages are described with relevant examples. An application example is also illustrated. The important part of sensor integration is discussed followed by robot work cell. Four layouts are explained with proper diagrams: robot centered work cell, in-line work cell, tracking window and mobile robot work cell. Finally, some additional aspects of robot work cell is also given.

Robot Vision has been taken up in Chapter 8. Various components of robot vision as a block diagram are discussed. The features of lighting devices and Analog to Digital conversion with an example are explained. Image storage and illumination are also brought in. Feature extraction and a relevent example are discussed. Template matching and structure techniques for object recognition are also discussed. A comparison of 4-point and 8-point inspection is made. Finally, the features of robot vision are detailed.

Robot Applications in industries are discussed in Chapter 9. Applications like pick and place, assembly, inspection and spray coating are given with diagrams. Robots in handling like machine loading and unloading, material transfer, palletizing and welding are also discussed. Two kinds of welding such as arc welding and spot welding are also detailed. The feature of compliance is discussed with a proper block diagram. The equipment RCC and its working are illustrated. Finally, assembly and injection moulding equipments are also given.

Chapter 10 is on Trajectory Planning. Definition of path, trajectory generation, joint space, Cartesian space and path update rate are listed with illustrations. Steps in trajectory planning are also explained. Importance of p-degree polynomial is explained with details. Linear function with parabolic blends which expands to two catagories are illustrated. A comparison between Cartesian space and joint space is given.

Chapter 11 is for Economic Analysis which illustrates three laws of economics—simple payback period, production rate appraisal on pay back and return on investment payback. Each is explained with an example. A Table on investment cost on robot project is produced. Another Table on costs that can be saved is explained. Problems on robot installations are explained which are useful to financial managers. Attitudes of humans on robot working are also given. In addition, the current capabilities and the future expected capabilities are also discussed.

Artificial Intelligence (AI) is discussed in Chapter 12. Two of the general softwares (i) LISP and (ii) PROLOG are brought in with a view of adapting LISP in the future discussion. AI research goals are also detailed. Knowledge representation is considered first. Semantic network concept is explained. Travelling Salesman problem is also considered subsequently. Search techniques are explained further. A set of LISP statements are adapted in discussing example problems.

Robot Dynamics is considered in Chapter 13. Some questions on current developments in AI is followed by relevant answers. A discussion on generalised coordinates and generalised forces are explained. Lagrangian function as the difference between total potential energy and total kinetic energy is considered next. The problem on inverted pendulum is taken subsequently. Slender rod which has a thickness tending to zero and is full of other dimensions is considered as the next example. Lagrangian function is used to derive the energy equations.

Fuzzy Logic Control (FLC) of Robot Joints is considered in Chapter 14. This chapter starts with a description of FLC in successfully parking my car. The advantages and drawbacks of FLC are discussed next. Two important words normally mis-spelt in the concept of FLC are discussed next. Possibility distribution function is considered. An example of fuzzy set is also discussed. Operations on fuzzy set are explained next. Fuzzy relation is also discussed. A methodology in designing FLC is given. A general scheme of FLC is given in a block diagram. The relevent set of designing and adopting FLC are also explained. A detailed view of bringing FLC in control problems is explained. Development of Fuzzy Associative Memory (FAM) is also explained with an example for students better understanding.

Medical Applications of Robots is discussed in Chapter 15. This chapter introduces the classification of medical robots and why there has been a slow growth in recent designing and current applications are explained. Guide robot has been introduced for blind people navigation. Prosthetic arms and exoskeleton after hip surgery are also discussed. The applications of hospital pharmacy robot are given. The clinical robots follows next. Robots for hip surgery are also given. Tumour extracting robot and its functions are explained.

The final Chapter 16 explains on Helping the Visually Impaired for their Autonomous Navigation. Blind people has been covered under the visually impaired Section. The term visually impaired, autonomous navigation and how to help them in their navigation has been deliberated upon. A statistics on blind people in the world has also been presented. Classification of electronic travel aids is given. The earlier classifications such as Guide Cane, Polaron, NavBelt and Sonic Guide are explained. These include Navchair and vOICe. The details of Dobelle Eye are brought in. Then, four research projects and their out comes are explained. They are NAVI-1, NAVI-2, SVETA-1 and SVETA-2. Computation of object distance is also explained.

Each Chapter ends with Review Questions. Exercises, References and Bibliographies are also given at the end of the book.

Acknowledgements

While having written and completion of this book I wish to express my sincere gratitude to my esteemed co-researcher Prof. Dr Sazali Yaacob, who has inspired me during several stages of research. He was the Dean of Engineering at the time of writing this book. I was teaching about 30 students in Robotics each year of Undergraduate and Post graduates. Among the Undergraduates students, some of whom I trusted to supervise a few in their final year project works.

Among the postgraduate Research Students, some of them were ready to take up research works with me. To site a few, among them, Farrah Wong, Sainarayanan, Balakrishnan and Rosalyn Porle were working in NAVI and SVETA projects and availed their Post Graduate Degrees. I am also thankful to En Maraban (the blind) for assisting in the experiments.

I would also like to acknowledge IRPA grants from MOSTI and the facilities provided by Universiti Malaysia Sabah and Universiti Malaysia Perlis.

I would like to thank the following experts for providing financial help for me to travel: Prof. Dr Richard Weston, Loughborough University of Technology, Loughborough, UK, Prof. Dr Sigeru Omatu, Osaka Prefecture University (normally called by locals as Osaka Fu) and Prof. Dr Junzow Watada, Waseda University.

I would like to indicate the following persons from whom I received encouragements for completing the book in time. A few but not the least are mentioned here.

M. Karthikeyan and K. Vijayalakshmi family and R. Chandraprakasam and C. Kanammal family are my inspirations and at times they have pursued with me on my completion of this book.

N. Ganeshkumar, N. Suganthi and G. Premkumar family, K. Sundarraj and N. Jayanthi family were similarly keen on completion of this book. A. Narayana Murthy and C. Vani family, G. Thiruvenkita Prasad and C. Tharini family and S. Kumarasamy and K. Santhanam family were also very keen in the completion of this book. R. Indrani, S. Senthilnathan and V. Kalaimani are much concerned in the completion of this book. The author acknowledges with thanks the directives from R. Dheepika, Associate Editor, Pearson India Education Services Pvt. Ltd, to write the final chapters.

In addition, the contribution from P. Pradeep and P. Praveena were very much helpful in typing this book.

Mrs Mangai Nagarajan, my wife, who has been always with me during my writing the book and she was encouraging me in all my activities.

My heartfelt thanks to all the above.

Ramachandran Nagarajan

ABOUT THE AUTHOR

Professor Dr Ramachandran Nagarajan received the following degrees: BE (Hons), University of Madras, India (Electrical and Electronics Engineering); M. Tech. Indian Institute of Technology, Kanpur, India (Control Engineering); PhD, University of Madras, India (Adaptive Control Systems); and Post Doctoral, Loughborough University of Technology, UK (Robot Control) under the main field of Electrical and Electronics Engineering. His main research fields are Robotics and Robotic Control.

His teaching and research assignments are at PSG College of Technology, University of Madras, India; Loughborough University of Technology (LUT), UK; University of Basrah, Iraq; University Sains Malaysia (USM); Osaka Prefecture University, Japan; University Malaysia Sabah (UMS); and University Malaysia Perils (UniMAP). He has published more than 200 research papers in Journals and Conferences. His additional contributions are of books, book chapters, special issues, monographs and edited proceedings. He has two years industrial experience in India where he was in-charge of developing Computer Numerical Control Systems. His memberships are as Life Fellow IE (India) and as Life Senior Member IEEE (USA).

His assignments were in teaching and research supervising to Under Graduate, Master, Doctoral and Post-doctoral students, reviewer of Research Books, Journal and Conference papers, Examiner for Master and Doctoral Research, Member in Faculty Promotion Committees, Funded Research Evaluation Committees, Organizing Committees of International Conferences and Member in Academic Evaluation Committees.

He has offered several Professorial Lecture programmes, to site a few are Present and Future Trends in Robots (USM, 1996), Robots for Hospital and Healthcare Applications (UMS, 2002), Helping the Visually Impaired for their Autonomous Navigation (UniMAP, 2008) and Intelligent Control of Exo-thermal Process in Japan, Malaysia, Taiwan and India.

He has received prestigious National and International Medals of Honor for Research Products, Memorial Prizes for best papers, University and National Research Fundings, Post-doctoral Fellowship in UK and in JSPS (Japan).

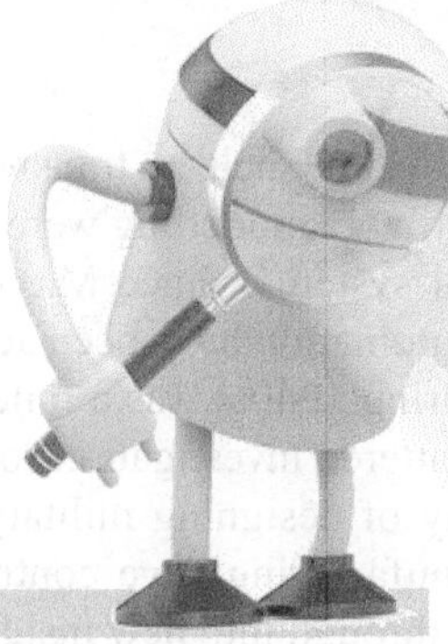

INTRODUCTION 1

Humanity has the stars in the future, and that future is too important to be lost under the burden of juvenile folly and ignorant superstition

Sir Isaac Asimov

This chapter covers the following key topics:

Industrial Automation and Computers – Industrial Robot – The Laws of Robotics – Robot Populations and Applications – Do Robots Create More Unemployment? – Payback Period of a Robot – Robot Applications in Manufacturing

---

1.1 INDUSTRIAL AUTOMATION AND COMPUTERS

The term '*Industrial automation*' is defined as the technology concerned with control of systems in the process of achieving an end product. The process of achieving the end product has to be with minimum or no human intervention. In this book, the term 'automation' is used in place of 'industrial automation' unless otherwise required. Automation first became popular in automobile industries. Automation helps the industrial manufacturers in achieving high productivity, high level of accuracy, consistent quality and increased *labour saving*. Automation is based on foundations of automatic control and obviously on feedback theory. It is applicable to any branch of science, engineering and technology wherever and whenever a feedback is required.

The use of *feedback* in order to control a system has a fascinating history. The first application of feedback control is believed to be a float regulator for level control in Greece during the period starting somewhere between 300 and 1 BC. This control mechanism is used in maintaining water level in the water clocks and oil level in oil lamps. Pressure regulators for steam boilers (similar to pressure cooker valve) were in use from 17 AD. James Watt's fly ball governor, invented in 1769, is still in use in many industrial speed control systems.

Intuitive inventions were contributing to the development of automatic control and hence automation till 1868 when Maxwell formulated the mathematical model (in terms of differential equations) to describe a system. He demonstrated the effect of parameters on system performance. The concepts of *accuracy* and *stability* were understood. The mathematical models in different forms such as transfer function, pulse transfer function, describing function and state variable modelling were considered as inevitable tools for analysing and designing of control systems. Since Maxwell's formulation of modelling, the automatic control and hence automation have been accepted to be a science rather than an art.

World War II (during 1940s) was a milestone in the field of automation. In this period, the whole area of scattered investigations of automation consolidated into a discipline. The compelling necessity of designing military equipment, prototype production and testing and mass scale manufacturing have contributed to a tremendous growth in the field of automation. A push-through into new insights and methods was achieved. Many automatic control theories were formulated. The pioneering contributions of H. Nyquist, H. W. Bode, V. M. Popov and many others have strengthened the discipline of automatic control and are still being used in developing new technologies of designing, controlling and perfecting automation systems.

The idea of using computers in automation emerged during 1950s. Analogue computers were used as on-line controllers in continuous processes such as steel and paper industries. Almost during the same time, the power of rapid decision-making properties of digital computers was found feasible in automation. A dramatic first application of digital computer control was established when a polymerization unit went on-line. This pioneering work in *computer control* was noticed by industrialists, manufacturers and research experts; the industrialists realized a potential tool for increased automation, the manufacturers recognized a new market and the research experts saw diversified research fields. This has resulted in a tremendous growth in modern automation methods. Automation without a computer is now hard to imagine.

The cost of computers used as controllers was the major concern in implementing automation. The cost of analogue controllers increased linearly with increased control loops. On the other hand, even though initial cost of digital computer was large, the cost of adding additional loops was small. Additional advantages such as digital instrumentation panels, implementation of complex control strategies, easy human interface, reduced human interaction and high reliability were the forcing functions of increased use of digital computers in automation. The first microprocessor developed by Intel Corporation, USA, during 1971, has initiated the microcomputer age. Since then, the research efforts in microelectronics and development of computer software have been offering powerful computers for automation at reduced cost. Since the necessities of computer control are unlimited, one can safely guess that there will be a sustained growth in industrial automation during many years to come.

1.2 INDUSTRIAL ROBOT

Automation and *industrial robots* are two closely related technologies. According to definition of automation, an industrial robot can be considered itself as a form of automation. The term 'robots' is used in place of 'industrial robots' in this book. A robot is a general purpose programmable machine which possesses the characteristics of a human arm. The robot can be programmed by its computer to move its arm through sequences of motion in order to perform some useful tasks. It repeats the motions over and over until it is reprogrammed to perform some other task. Many industrial operations involve robots working together with other equipment.

The word 'robot' was introduced by Karel Čapek a Czechoslovakian play writer in 1921. Robot was derived from the Czech word 'robota' which means forced worker. Early writings and plays during 1920s and 1930s pictured the robot as ferocious humanoid machine which was intended for killing human beings.

Sir Isaac Asimov (Figure 1.1) through his prodigious imagination contributed a number of stories about robots starting from 1939. He coined the word 'robotics' which means the science of robots. Later, Sir Isaac Asimov and Dr George Devol started a robotics industry named as Unimation Inc. They started developing fully hydraulic powered UNIMATE robot. In later years, Ford Motors used UNIMATE successfully for die casting.

1.2.1 The Laws of Robotics

In 1942, Sir Isaac Asimov developed the famous three laws (Law one, Law two and Law three) of robotics which still remain as worthy industrial design standard. However, the laws have been extended and revised by him and others since 1985 to accommodate his creations, his attitude to robotics and the modern requirements of humanity. The extended set of laws is as follows:

The Meta-Law
A robot may not act unless its actions are subject to the laws of robotics.

Law Zero
A robot may not injure humanity, or through inaction, allow a humanity to come to harm (humanity is the family of all human beings and other biologically living things).

Law One
A robot may not injure a human being, or through inaction, allow a human being to come to harm, unless this would violate a higher order (an earlier stated) law.

Law Two
A robot must obey orders given by human being, except where such orders would conflict with a higher order law.

A robot must obey orders given by subordinate robots, except where such orders would conflict with a higher order law.

Figure 1.1 Sir Isaac Asimov

Law Three
A robot must protect the existence of a subordinate robot as long as such protection does not conflict with a higher order law.

A robot must protect its own existence as long as such protection does not conflict with a higher order law.

Law Four
A robot must perform the duties for which it has been programmed, except where that would conflict with a higher order law.

The Procreation Law
The robot may not take any part in the design or manufacture of a robot unless the new robot's actions are subject to the laws of robotics.

The robots which are strictly manufactured in accordance with the above rules do behave better than human beings.

When the concept of robot was introduced and strengthened, the necessity of industrial automation was also deeply felt. Moreover, the technological progress in thermionic valve (1904), hydraulic and pneumatic systems (1906), logic circuits (1943), digital computer (1946), transistor (1947), microelectronics (1970) and microcomputer (1977) have all made automation and robotics a reality. The first commercial robot, controlled by limit switches and cams, was introduced in 1959. Since then, the development in robot technology has been in constant growth. Nowadays, the service robot within industry and in other areas of applications has made a breakthrough in robot applications.

1.3 ROBOT POPULATION AND APPLICATION

Table 1.1 presents a brief chronological listing of the historical developments in robotics. Population of robots in industrial scene grows steadily year by year. This is due to success in conventional and unconventional applications of robots, new dedicated software technologies, development in microelectronics, new manufacturing techniques, advances in control technologies and in new research findings in artificial intelligence. The growth of robot population in manufacturing applications of the world is reasonably predictable to estimate the number of robots in use in the next decade. It is estimated that the world population of robot exceeded 2.5 million in 2010 itself. Table 1.2 shows the statistics of average price index based on 1990.

Effectively, robot applications in every factory of a country are growing every year. In addition, the cost of robot of same quality is continuously decreasing so that even medium- and small-scale industries can afford to have robots for enhancing their productivity.

A robot need not be replaced very often since the average life of a robot is around 8–12 years. The reprogramming ability of a robot can be advantageously utilized in industries even when the life cycle of a product is small and a new product is taken up for manufacturing. This ability perfectly fits the robot into the concept of flexible manufacturing system.

Robotics is a technology of future, for the future and with a future. Extensive research both in educational and in industrial institutions are continuously striving for developing a robot of higher level of application capabilities. The sustained technology developments in artificial intelligence, nano-electronics, computer interfacing, materials, drives, control and sensors have nurtured the robots for increased requirements of modern factories.

Table 1.1 Chronological Developments of Robot Technology

Year	Development
1921	The word 'Robot' was coined
1939	Early humanoid robot exhibited in 1939, 1940 World Fairs by Westinghouse Electric Corporation
1942	The word 'robotics' appears in Sir Isaac Asimov story 'Runaround'
1952	Numerical control machine demonstrated at Massachusetts Institute of Technology, USA
1954	George Devol designed the programmable, teachable and digitally controlled article transfer robot
1956	First robot company UNIMATION formed
1959	First commercial robot controlled by limit switches and cams
1961	First hydraulic drive robot UNIMATE in die casting machine in Ford Motors
1968	Mobile robot designed by Stanford Research Institute
1971	Electrically powered 'Stanford Arm' developed by Stanford University
1974	ASEA introduced all electric drive IRb6 robot
1974	KAWASAKI installed arc welding robot
1974	Cincinnati Milacron introduced T^3 robot with computer control
1978	PUMA robot introduced by UNIMATION for assembly applications
1978	Cincinnati Milacron's T^3 robot applied in drilling and routing operations
1979	SCARA robot introduced for assembly applications at Yamanashi University
1980	Bin-picking robotic applications demonstrated at the University of Rhode Island
1983	Flexible automated assembly line using robots by Westinghouse Corporation
1986	'ORACLE' robot used in commercial wool harvesting from sheep, Australia
1992	Flexible hydraulic microactuator for robotic mechanism developed in Japan
1992	First Humanoid by Honda Corporation, Japan, recognizes human faces
2000	Humanoid robot, ASIMO, put in service to society
2004	NASA in USA developed RED BALL robot with an intention of protecting the astronaut coming out of space vehicle for repairs
2006	Jumping robot has been developed to investigate the surface of any unknown areas
2010	RED BALL has been manufactured for protecting the astronaut

Table 1.2 Power Index of Industrial Robot for International Comparison (Based on 1990 USD Conversion Rate)

Year	1990	1992	1994	1996	1998	2000	2002	2004	2006	2008
% Robot Not Quality Adjusted	100	82	92	68	58	60	55	58	58	55
% Robot Quality Adjusted	100	65	48	40	28	29	25	25	22	22

Table 1.3 Main Features of Robot

A robot does not become fatigued or distracted
It is capable of producing a job with consistent quality at a steady rate with practically zero rework and wastage
Continuous working throughout the year except during repairs and maintenance
It can take up repeated and tiresome jobs even in unsafe and unhealthy environment
It does not demand wage increase, fringe benefits, pension, holiday, sick pay, etc.
Capital cost of robot is paid only once
Robot's upkeeping cost is increasing at a slower rate compared with labour maintaining cost every year
Robot can lift heavier weights, be more precise, work at higher speeds and can exert larger force than is humanly possible

The main advantages of robot-based manufacturing compared with conventional human-based manufacturing are listed in Table 1.3. It is a certainty that, with all these features, the robot helps to *conserve* the human intelligence for more productive tasks instead of working for the so-named 3D environment (dull, dangerous and difficult).

1.4 DO ROBOTS CREATE MORE UNEMPLOYMENT?

This is a standing question from a person who is introduced to robot technology. Industrial robot is now considered an inevitable piece of an industrial automation system. Robots do displace workers in the same way that technological changes have displaced workers. The introduction of personal computers to offices during 1980s created the same fear in the office workforce. Nowadays, learning about computer and its related subjects is considered as important as any other techniques to learn. Introduction of robots in an already running industry does displace a piece of workforce. However, there is a certainty that robot will create new jobs to the displaced workers. This displaced workforce can be utilized in maintaining, programming and applying the robot in new jobs. This cultivated experience is far better than that gained by doing the work in a 3D environment. In addition, the advantages of introducing the robot in industries have created several industrial set-up such as robot and robot parts manufacturing, robot systems engineering, robot software development and in unbounded areas of robot research and applications.

In order to monitor the growth of robot population in industries, a robot deployment index (RDI) has been constructed. The RDI is the ratio of total number of industrial robots in use to a percentage of manufacturing workforce multiplied by 10. Hence, higher the index value, stronger the deployment of robot in a country. Table 1.4 compares the human unemployment percentage with RDI of some of robots in different countries in 1980s. Table 1.4 also indicates that the countries which employ higher population of robots have lower unemployment levels. It can be concluded that any level of robot population in industries does not cause massive unemployment and, in fact, their applications appear to be indicative of higher levels of employment with a good manufacturing habit. Various statistical figures of robot and their applications are regularly published by robot associations of respective

Table 1.4 RDI and Human Unemployment

Details	UK	Italy	USA	France	Germany	Sweden	Japan
Manufacturing Workforce (millions)	5.4	5.7	19.2	4.8	7.8	1.0	12.3
Unemployment (percentage of population)	11.5	11.0	6.9	10.3	9.0	2.7	2.8
Robot Population (thousands)	3.7	5.5	27.0	7.5	12.4	3.8	90.0
Robot Population (percentage of manufacturing workforce)	0.07	0.10	0.14	0.16	0.16	0.40	0.73
RDI	0.7	1.0	1.4	1.6	1.6	4.0	7.3

countries. India has robot population much less than the countries listed in Table 1.4 and hence it has not been included in the statistics.

1.5 PAYBACK PERIOD OF A ROBOT

The economic viability of an industrial robot can be ascertained if a payback period is computed. The payback period is the length of time required for the net accumulated cash flow to equal the initial investment in the robot project. The usual working life of a robot can be easily 8 years. The robot works faster or slower than human operator depending on applications. A robot is faster than human being in a pick and place task; but, slower when the task has a decision-making function. A simple but effective payback analysis is done by

$$n = \frac{R}{W - M + q(W + D)}, \tag{1.1}$$

where n is the payback period, R is the total capital investment in robot and accessories, W is the wages and other benefits paid to the workers per year, M is the annual expenses of robot upkeep (maintenance, software development, etc.), q is the production rate coefficient; negative if robot is slower than worker, otherwise positive and D is the annual depreciation cost of robot and its peripheral equipment.

As an example, let us have the following values of currency:

$$R = 55{,}000; \qquad W = 24{,}000/\text{shift}; \qquad M = 2{,}600/\text{shift}; \qquad D = 30{,}000.$$

Then the payback periods (in years) for two different combinations of production rate with two different combinations of shift/day, with $q = 0.2$ and $q = -0.2$, are computed as in Table 1.5.

It is convincing that larger the expected life of robot, larger the returns during its life cycle.

This method of computing payback period is least complicated. If the payback period is very short, it is not strictly necessary to apply more complex methods of economics. The techniques such as accounting rate return or discounted cash flow can be used for financial appraisal and for justification of robot installations.

Table 1.5 Payback Periods

Shift/Day	When Robot is Slower	When Robot is Faster
1	5.19	1.71
2	2.02	0.94

1.6 ROBOT APPLICATIONS IN MANUFACTURING

Robot applications are of wide variety in the manufacturing side. These include a large variety of production operations. Here the operations are classified into the following categories.

(i) Materials transfer and machine loading and un-loading
These are applications in which the robot grasps and moves a work part from one location to another. This category includes applications of robots as they transfer parts into and out of a production machine. These include metal machining operations, die casting, plastic moulding and certain forging operations.

(ii) Processing operations
Here the robot uses a tool as an end effector. This tool will accomplish some processing operations on a work part positioned for the robot during the work cycle. Spot welding, arc welding, spray coating and some machining operations fall in this category.

(iii) Assembly and inspection operations
A set of new applications such as robot-based assembly and inspections is the final set of robot applications. The robot is used to put components together into an assembly and/or to perform some automated inspection operations. More robots are equipped with vision capability facilitating the performance of these operations. The applications of robot cameras are to inspect welding, brazing and bonding with glues and other fasteners.

These applications are discussed and detailed in Chapter 9.

REVIEW QUESTIONS

1. When did Maxwell formulate the mathematical model of systems?
2. The cost of digital computers was a major concern, when and why?
3. Who coined the words *Robot* and *Industrial Robot*? What was reaction of people at the time of introducing these words?
4. What are the laws of robotics? Explain these laws?
5. Give the names of three more celebrities of robotics. Why are they called as celebrities?
6. List some of the missing applications of robots till date.
7. Have robots created industrial unemployment? Why or why not?

GRIPPERS AND TOOLS OF INDUSTRIAL ROBOT

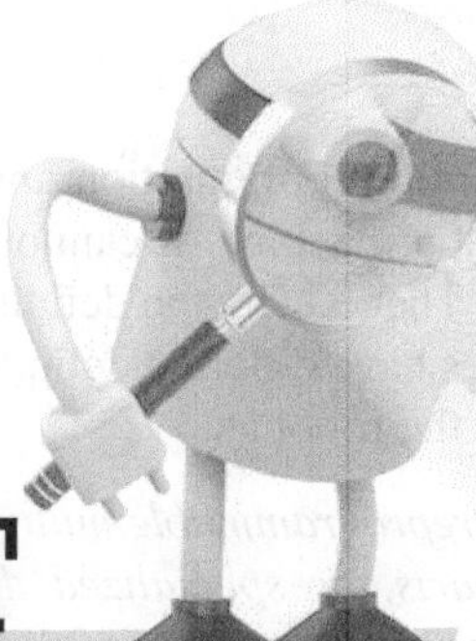

2

I do not fear computers. I fear the lack of them.
Sir Isaac Asimov

This chapter covers the following key topics:

Introduction – Definitions of Industrial Robot – Configuration and Work Volume – Configuration of Human Body – Human Work Volume – Industrial Robot Configuration – Structural Configuration – Robot Work Volume – Precision of Movement – Spatial Resolution – Example – Accuracy – Repeatability – Degrees of Freedom – Examples – End Effectors – Grippers – Tools

2.1 INTRODUCTION

This chapter *imparts* the perception of the industrial robot as an integrated structure consisting of electrical and non-electrical subsystems. These subsystems are interfaced to a computer and controlled by its software. Human structure is considered as the model for robot and articulations of human arm in performing a task are compared with robot arm articulations. Varieties of robots with a number of specific configurations have been introduced in the market. Manufacturers, suppliers and the consumers use various terms when describing an industrial robot and its functions. These terms can be compared with those when a human is standing firmly on the ground and performing a task. The structure, the space in which the human works, various joints and limbs, and the way of performing a task are compared with an industrial robot. In addition, certain well-known robots that are used in industrial applications are also introduced. The importance of *spatial resolution*, *degrees of freedom* (*DoF*), *accuracy* and *repeatability* are illustrated. The use of a subsystem similar to our fingers is also explained.

2.2 DEFINITIONS OF INDUSTRIAL ROBOT

Any automatic machine cannot be considered as a robot. Robot is to have a specific set of characteristics. Interestingly, a 3-axis computer numerical control (CNC) milling machine may have a very similar configuration and control system of a robot arm. However, the CNC machine is just a machine. It cannot do jobs other than milling. But the robot must do something more. That is why the definitions are proposed for a machine to be a robot.

Different countries have different definitions for a robot. The Robot Institute of America (RIA, 1985) defines the robot as

> *A robot is a reprogrammable multi-function manipulator designed to move materials, parts, or specialized devices through variable programmable motions for the performance of a variety of tasks.*

This definition restricts robots in industrial applications. The two important key words are 'reprogrammable' and 'multi-functional'. If the machine is single functional, it cannot be reprogrammable. Reprogrammable means that (i) the robot motion is controlled by a written program and (ii) the program can be modified to change significantly the robot motion. Multi-functional implies that a robot is able to perform many different tasks depending on the program in the memory and tooling at the end of arm. This means that the robot can be programmed for welding with a welding tool at the end of arm and can be reprogrammed if the end of arm has a totally new facility such as for gripping.

Another, a little broader definition is proposed by McKerrow (1986) as

> *A robot is a machine which can be programmed to do a variety of tasks in the same way that a computer is an electronic circuit which can be programmed to do a variety of tasks.*

This definition excludes numerical control machines because they can be programmed for variations within only one task. Teleoperators are also not considered as robots because there is a human in the control system. They provide extended capabilities, not a replacement of a human.

The International Standards Organization (ISO 8373) defines a robot in a similar way as follows:

> *A robot is an automatically controlled, reprogrammable, multi-purpose, manipulative machine with several reprogrammable axes, which may be either fixed in place or mobile for use in industrial automation applications.*

This definition specifically mentions 'reprogrammable axes' for industrial tasks. Such a definition particularly points out that industrial robots are very suitable to modern industries.

There are several such definitions on robots – industrial robots in particular. One way or other, each definition has to be expanded to suit the functioning of the modern industrial robots. In most cases, the definition given by RIA is accepted to be closer to industrial robots of modern times and such a definition is considered worth designing industrial robots.

2.3 CONFIGURATION AND WORK VOLUME

We can compare the configuration of a human body, its work volume and its other features with those of an industrial robot. They are similar in many aspects. Human is natural and the robot is man-made to work for his/her objectives. The objectives can be faster, higher accuracy and/or more powerful compared with those of human being. Both configurations have their own advantages and disadvantages. In the following descriptions, we compare the configuration of a human body with that of an industrial robot.

2.3.1 Configuration of Human Body

A human body with arms (especially one arm) mimics a stationary industrial robot. Every human being has a similar *structure* but with varied sizes and *power*. He (she inclusive) can stand firmly on the ground. He has a *waist, body, shoulder, forearm, elbow, lower arm, wrist and fingers* and each one of these can have its own *articulations*. Articulations are rotations about or translations along several *axes* of the *joints*. Typically, the waist (a joint between body and legs) can rotate about three axes, the shoulder (a joint between body and forearm) about three axes, the elbow (a joint between forearm and lower arm) about two axes and the wrist (a joint between lower arm and the fingers) about two axes.

Each joint connects two *links* (*limbs*). Waist connects the leg and the body. The shoulder connects the body and the forearm, the elbow connects forearm and lower arm, and the wrist connects the lower arm and the fingers. These joint *rotations* and link lengths create a *work space* (*work volume*) around the human. The work volume is a set of all points which the arm can reach. The fingers can give several independent rotations in each of its joints. The human has a number of in-built sensing abilities for *touch, pressure, olfaction, vision, hearing, heat, force, torque* and *compliance* with which the human body does almost unlimited tasks. The compliance is an ability to perform little changes in articulations when unexpected structural disturbance is experienced. A small momentary thrust on the human body shakes the body a little without modifying the rotational angles. Above all, the brain controls all such articulations based on the sensory outputs. A *sensor* creates a level of intelligence. Existence of countless sensors with an analytic brain exclusively differentiates a human as highly intelligent than the machines, especially the robots.

The human, with his vast sensing abilities, can be faster and accurate than a robot in performing several tasks. He cannot be consistent in *speed* and *accuracy* through a longer period of time due to distractions and tiredness. However, a robot can maintain consistency over a long period of time.

2.3.2 Human Work Volume

Let us consider a *task* of a human taking a nail from a table and inserting it in a hole above his head on the wall. In doing so, he has to articulate his various joints. The articulation angles are decided by the brain with an information continuously acquired from its *sensors*, especially the *eyes*. Inserting a nail into the hole without a mistake is the *accuracy*. Repeating the task very many times with the same accuracy is the *repeatability*. He can perform this task with high accuracy and high repeatability several times, of course, with a perfect *vision*. If the vision is poor, the task performed by him has poor accuracy and poor repeatability. If he is blind, he cannot do this job at all without using other sensors.

He has an *intelligence,* immeasurably abundant. His brain is supposed to have this intelligence. Due to his various sensor capabilities, he can *learn* a task by making decisions even in an unpredictable environment. For example, if an object (*obstacle*) comes between the man and the wall, he can take a quick *decision* for what to be done. Either he can wait till the obstacle goes away or bypass this obstacle by performing additional articulations aided by sensors. If the human is a novice, he has to be taught on how to undertake the task

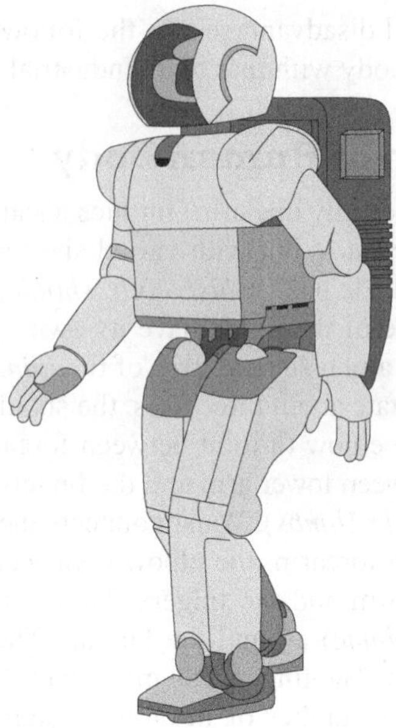

Figure 2.1 Japan's Honda

successfully. Someone is to *teach* him through a set of *programmed motions* of joints and links so that he will perform the task repeatedly.

Figure 2.1 shows a well-known product, Japan's Honda called ASIMO. ASIMO cannot do industrial jobs but can be a good associate to human in many ways. He can bring a newspaper every day, chat with children, help old people and maintain rooms.

2.3.3 Industrial Robot Configurations

An industrial robot is constructed for a specific purpose. It has to be economical and be able to do varieties of industrial jobs. Hence, it has to be of varied structures so as to suit specific requirements. For example, for spray painting, a robot has to carry the weight of spray painting gun. The robot for this job can be of light weight. However, a robot which is intended to stack cement bags in a lorry has to be powerful, strong in structure, with long arms and less articulations. Hence, the configurations of these robots have to be different. The configurations of industrial robots are many to suit several job requirements. To indicate a few, the job requirements can be of picking and placing an object in a machine tool, spray painting a finished product, welding two metal plates, assembling a number of pieces together and palletizing (stacking the products within a specified space).

Figure 2.2 illustrates one of the configurations of an industrial robot manufactured by Cincinnati Milacron, USA. This combines a variety of several robots. This is called T^3 (twist power 3) robot, that is, a twist exerted at the base and another two twists exerted at the body. This has a base firmly attached to the ground or a platform. The robot articulation is possible due to rotations of several joints such as body rotation (about a vertical axis), shoulder swivel (about an horizontal axis), elbow rotation (about another horizontal axis) and wrist rotations named as yaw, pitch and roll in three different axes. An end

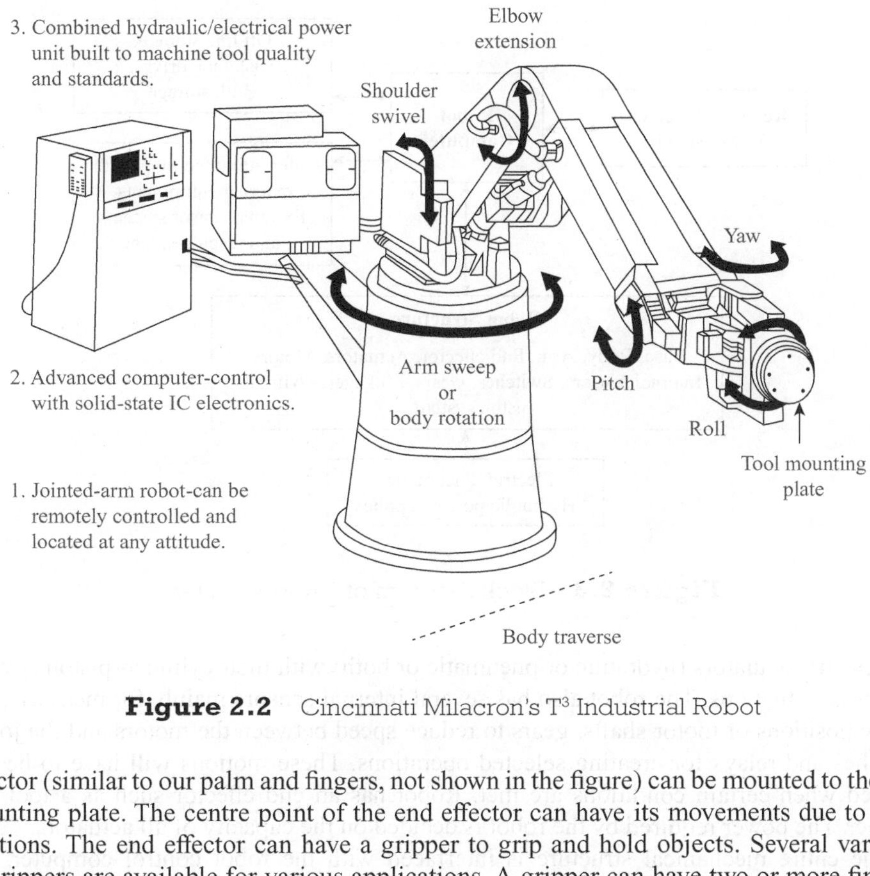

Figure 2.2 Cincinnati Milacron's T³ Industrial Robot

effector (similar to our palm and fingers, not shown in the figure) can be mounted to the tool mounting plate. The centre point of the end effector can have its movements due to wrist rotations. The end effector can have a gripper to grip and hold objects. Several varieties of grippers are available for various applications. A gripper can have two or more fingers. The end effector can be a tool to perform a process on a product. Drilling certain locations is a process. A welding torch is a tool fitted to the tool mounting plate to perform welding processes.

The robot has its brain in terms of a robot computer. The computer has the set of software required for the functioning of the robot. The computer has facilities to interface with number of sensor systems such as a vision camera that can be attached to the robot or placed elsewhere near the robot. The computer can have facilities to interface with other industrial equipments typically a conveyor system which brings the product to the robot so that the robot can perform handling or processing operations.

The robot is also connected to a power supply unit that gives the required power – electrical or hydraulic or pneumatic or combination of these. Electric supply is required to operate the internal motors coupled to various joints. Hydraulic and pneumatic powers are useful in operating cylinder–piston mechanisms built in the robot structure. There is a provision to teach a robot on how to perform a task. This can be done in various ways. The usual method is through a teach pendant.

All these components of a robotic system described above are also illustrated in Figure 2.3. A robotic system basically consists of robot mechanical structure, a robot control computer and power sources. The robot structure has all necessary electro-mechanical, pneumatic and hydraulic elements such as electrical actuators (motors for rotary operations) and

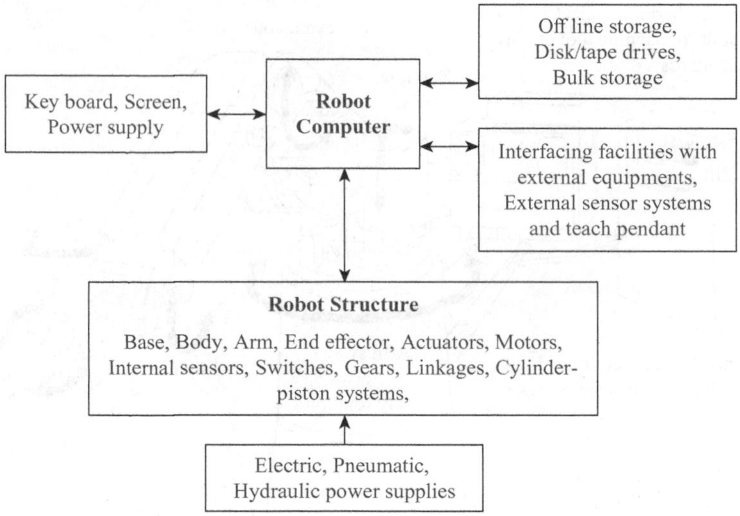

Figure 2.3 Block Diagram of a Robot System

non-electric actuators (hydraulic or pneumatic or both) with their cylinder–piston systems for linear actuations. The robot also has several internal sensors mainly for measuring the rotary positions of motor shafts, gears to reduce speed between the motors and the joints, switches and relays for creating selected operations. These motions will have to be performed when certain conditions are met. Robot has an end effector such as a tool or a gripper. The power required by the robot is decided on the capacity of all actuators.

The entire mechanical structure is interfaced with the robot control computer. The computer has several softwares required for the functioning of the robot structure. The computer has coordinate transformation software (discussed in other chapters), control software to control the speed and position of actuators and interfacing for teaching and learning a specified task and for robot safety to avoid any possible damage to robot structure. In addition, the computer has its own operating system for operating its internal electronics, peripheral devices such as key board and display, disk drives, mass storage and associated equipments.

Figure 2.4 shows another variety of robot, FANUC S-420i, six axes and electrically powered. This has a tool mounting plate to which an end effector has to be fixed. Generally, robots are sold without the end effector and the user has to design or buy the end effector suitable for his applications. This S-420i family consists of five members. This family of robot is normally used in automotive industry for spot welding, body assembly and material handling.

Table 2.1 lists the specifications of three of the S-420i family. Each axis has its own range of rotation and speed. Maximum reach is the farthest point which can be reached by the centre of the tool mounting plate (or by the centre of the end effector, if end effector is mounted). Centre of the end effector is generally known as tool centre point irrespective of gripper or tool at the tool mounting plate. The maximum reach is a function of robot's configuration and its link lengths including obstacles on its articulations. Load capacity (or pay load) is the maximum load (kg) that a robot can lift, hold and position by the

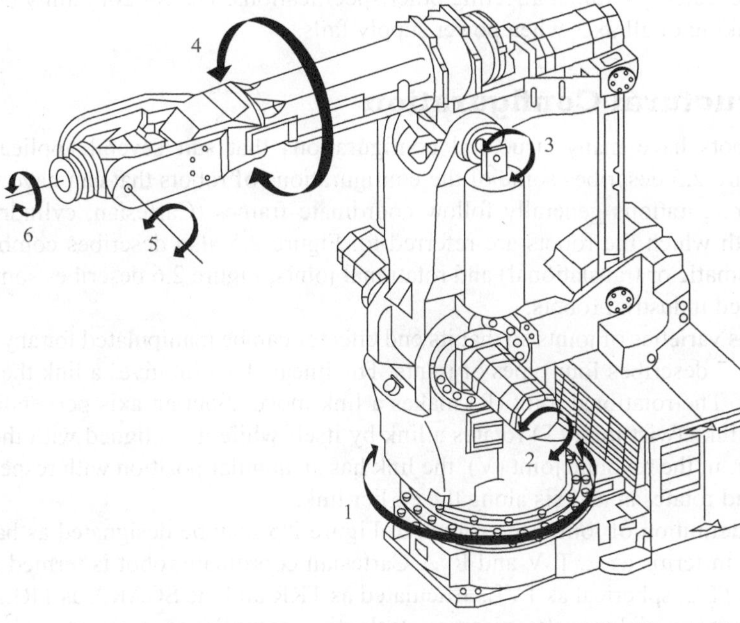

(1) Waist rotation; (2) Shoulder rotation; (3) Elbow rotation; (4) Forearm rotation; (5) Yaw rotation and (6) Pitch rotation

Figure 2.4 FANUC S-420i Electrically Powered Robot

Table 2.1 Specifications of FANUC S-420i Robots

Specification		S-420iF	S-420iL	S-420iS
Maximum reach (m)		2.4	3.00	2.25
Load capacity (kg)		120	120	80
Axis 1	Range (deg.)	360	360	360
	Speed (deg./s)	100	90	70
Axis 2	Range (deg.)	142	142	142
	Speed (deg./s)	110	110	110
Axis 3	Range (deg.)	135	135	135
	Speed (deg./s)	100	90	100
Axis 4	Range (deg.)	600	600	480
	Speed(deg./s)	210	210	210
Axis 5	Range (deg.)	260	260	260
	Speed (deg./s)	150	150	150
Axis 6	Range (deg.)	720	720	720
	Speed (deg./s)	210	210	210
Repeatability (mm)		± 0.4	± 0.4	± 0.4
Mounting method		Upright	Upright	Upright
Mechanical brakes		All axes	All axes	All axes
Weight (kg)		1500	1600	1500

end effector repeatedly without affecting other specifications. The S-420i family does have automatic braking of all axes when power supply fails.

2.3.4 Structural Configuration

Industrial robots have many structural configurations that suit several applications in industry. Figure 2.5 describes some of the configurations of robots that can have varieties of joints. Configurations generally follow coordinate frames (Cartesian, cylindrical and spherical) with which the robots are referred to. Figure 2.5 also describes combinations of linear (prismatic or translational) and rotational joints. Figure 2.6 describes some of the usually adopted industrial robots.

A robot has varieties of joints so that its end effector can be manipulated for any specific task. Figure 2.7 describes four types of joints. The linear (L) joint gives a link the translational motion. The rotational joint (R) makes a link move about an axis perpendicular to the previous link. Twist joint (T) rotates a link by itself while it is aligned with the earlier link. However, in the revolute joint (V), the link has an angular position with respect to the earlier link and rotates at an axis along the earlier link.

With this definition of joints, the robots of Figure 2.5 may be designated as having its configuration in terms as L, T, V and R. A Cartesian coordinate robot is termed as LLL, cylindrical as TLL, spherical as TVL, articulated as TRR and the SCARA as TRL.

SCARA is a special variety, meaning 'selective compliance assembly robot arm'. SCARA is especially useful in electronic component assembly for picking and inserting ICs in their respective sockets much faster and with greater accuracy and repeatability compared with other types of robots. Its specific characteristic is that it is more compliant in the horizontal plane but is stiff in the vertical plane and thus has its selective compliance.

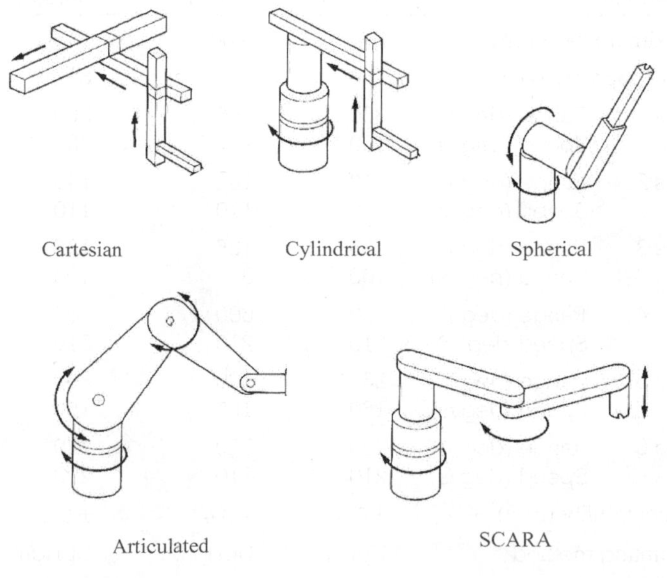

Figure 2.5 Configurations of Industrial Robots

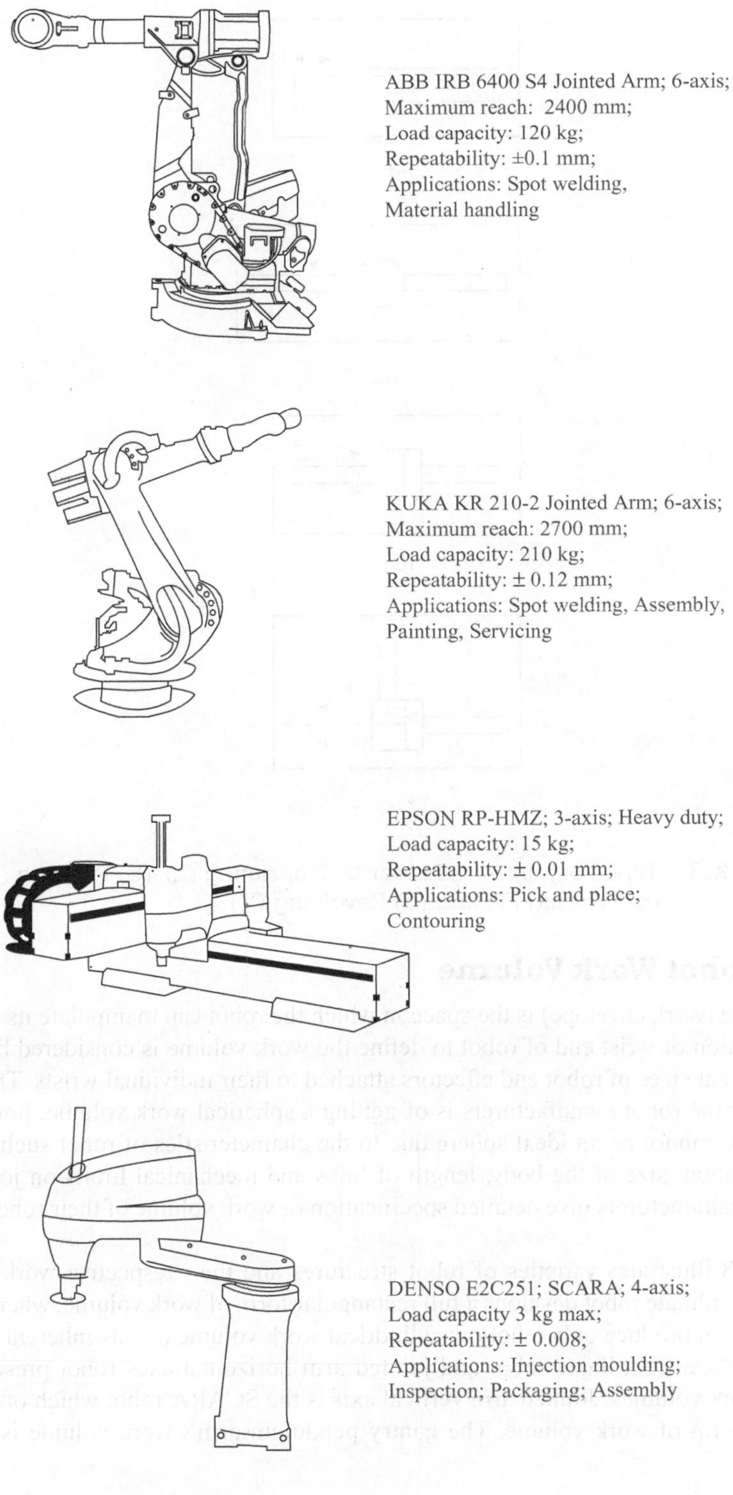

Figure 2.6 Commonly Known Industrial Robots

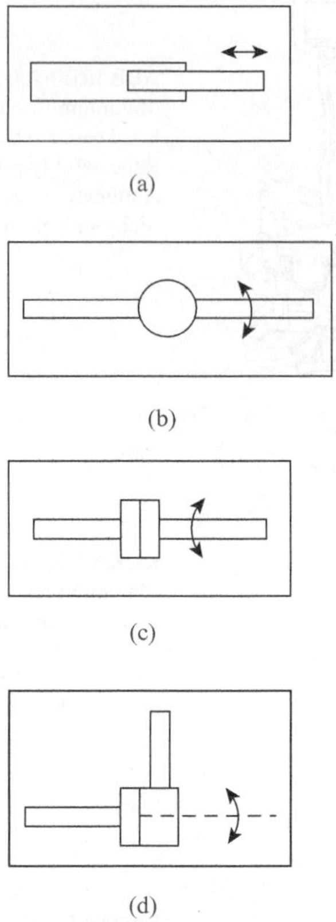

Figure 2.7 Types of Joints: (a) Linear or Translating (L), (b) Rotating (R), (c) Twisting (T) and (d) Revolving (V)

2.3.5 Robot Work Volume

Work volume (work envelope) is the space in which the robot can manipulate its wrist end. This convention of wrist end of robot to define the work volume is considered better than adding different sizes of robot end effectors attached to their individual wrists. The general objective for the robot manufacturers is of getting a spherical work volume; however, the work volume cannot be an ideal sphere due to the characteristics of robot such as physical configuration, size of the body, length of links and mechanical limits on joint movements. All manufacturers give detailed specification of work volume of their robots in their pamphlets.

Figure 2.8 illustrates varieties of robot structures and their respective work volumes. Cartesian coordinate robot develops a full rectangular form of work volume, whereas cylindrical robot can produce only a hollow cylindrical work volume due its inherent structural limitations. Polar coordinate robot and jointed arm horizontal axes robot present nearly spherical work volumes. Jointed arm vertical axis is the SCARA robot which offers a near cylindrical form of work volume. The gantry pendulum arm's work volume is a perfect

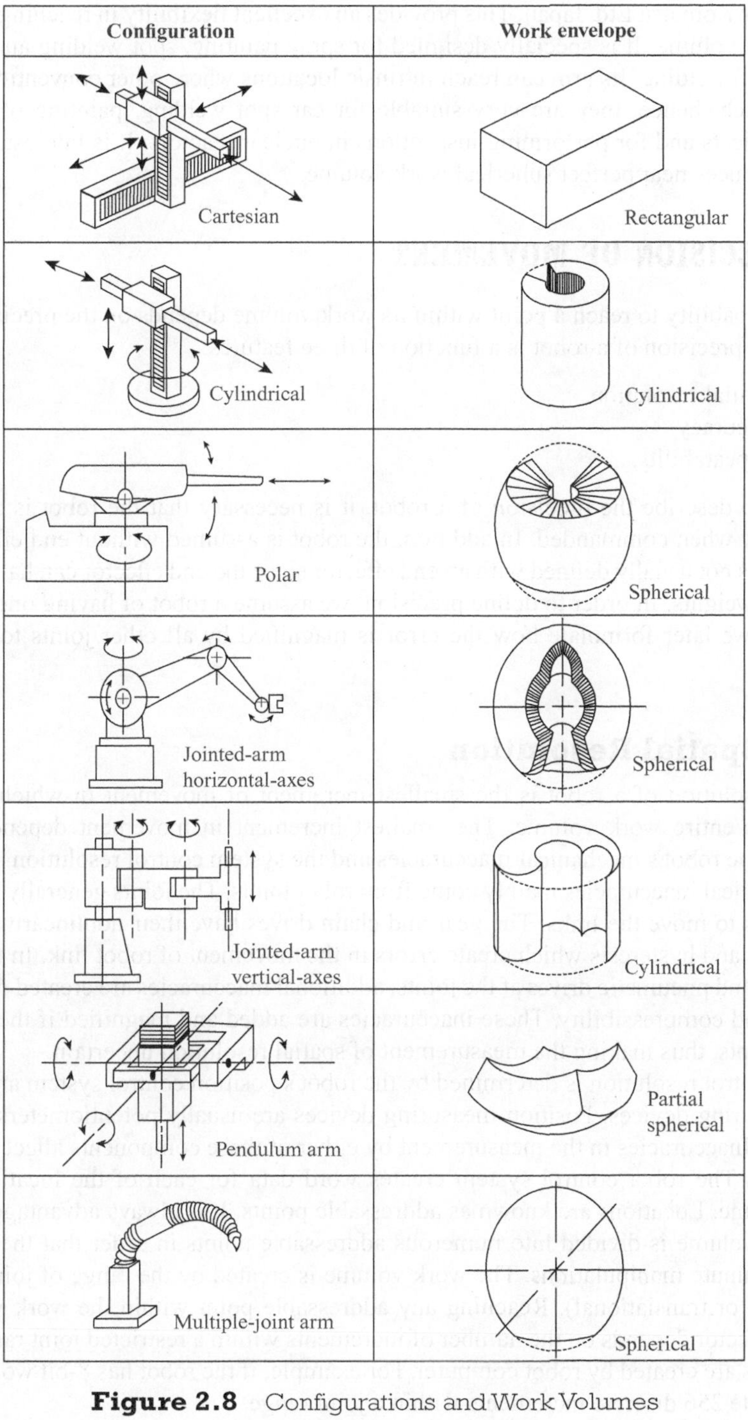

Figure 2.8 Configurations and Work Volumes

section of a hollow sphere. The multiple jointed arm robot is named as spine robot manufactured by Komatsu Ltd, Japan. This provides an excellent flexibility in reaching a location in its work volume. It is specially designed for spray painting, spot welding and seam (or continuous) welding. Its arm can reach intrinsic locations where other conventional robots cannot reach; hence, they are very suitable for car spot welding, painting of irregular-shaped objects and for performing inspections in nuclear reactors. It is interesting to note that it produces near perfect spherical work volume.

2.4 PRECISION OF MOVEMENT

Robot's capability to reach a point within its work volume depends on the precision of the robot. The precision of a robot is a function of three features:

(i) Spatial resolution
(ii) Accuracy
(iii) Repeatability.

In order to describe the precision of a robot, it is necessary that the robot is at rest and works only when commanded. In addition, the robot is assumed without end effector. The precision is not usually defined with an end effector since the end effector can have different sizes and weights. In order to define precision, we assume a robot of having one prismatic joint and we later formulate how the error is magnified by all other joints to affect the precision.

2.4.1 Spatial Resolution

Spatial resolution of a robot is the smallest increment of movement in which the robot divides its entire work volume. The smallest increment in movement depends on two factors – the robot's mechanical inaccuracies and the system control resolution.

Mechanical inaccuracies mainly come from robot joints. The joints generally have gears and chains to move the links. The gear and chain drives have their nonlinearities such as back clash and hysteresis which create errors in the movement of robot link. In the case of hydraulic and pneumatic drives at the joints, additional inaccuracies are created due to fluid leakage and compressibility. These inaccuracies are added and magnified if the robot has several joints, thus making the measurement of spatial resolution uncertain.

The control resolution is determined by the robot's position control system and its position measuring devices. Position measuring devices are usually potentiometers and shaft encoders. Inaccuracies in the measurement by either of these components affect the spatial resolution. The robot control system creates word data for each of the locations in the work volume. Locations are known as addressable points. It is always advantageous when the work volume is divided into numerous addressable points in order that the robot can perform minute manipulations. The work volume is created by the range of joint motions (rotational or translational). Reaching any addressable point within the work volume by the end effector depends on the number of increments within a restricted joint range. These increments are created by robot computer. For example, if the robot has 8-bit word storage, it can create 256 discrete increments within a joint range.

EXAMPLE 2.1

Consider a robot having a sliding joint which can extend the link maximum to 1 m. The robot control memory has 12-bit word storage. What is the control resolution of the axis motion?

Solution:

Maximum number of increments within the joint range is $2^{12} = 4096$.
The minimum increment in sliding joint motion is 1 m/4096 = 0.24414 mm.
Therefore, the control resolution is 0.244 mm (to three digits accuracy).

In Example 2.1, if the robot is required to reach 2.5 mm, a compromise has to be made in the control software on whether to go for 10 increments or 11 increments.

A robot has several joints, some of which are rotational and others translational. Each joint has its own joint range and hence its control resolution. Hence, the total control resolution of the entire robot cannot be obtained by algebraic addition of all individual resolutions. This is a complex procedure in which each resolution has to be considered as a vector in a three-dimensional space; the total resolution is, then, the vector addition of all joint resolutions.

2.4.2 Accuracy

Accuracy refers to a robot's ability to position its wrist end at a desired target point within its work volume. The accuracy is related to spatial resolution because the accuracy depends on how closely the neighbouring points are defined within the work volume and on how accurately the mechanical structure is produced. For example, consider the robot of Example 2.1. Figure 2.9 illustrates the definition of accuracy. *A* and *B* are the nearest points as determined by control resolution. *T* is a point halfway between *A* and *B*. If the mechanical structure is precise, then the robot can reach *A* or *B* exactly, but not at a point in between them. Spatial resolution and the control resolution are the same if the mechanical structure is precise. The robot accuracy is then defined as half of control resolution from either side (±0.122 mm). However, the mechanical structure usually has its own inaccuracies. Due to this, even when the robot is commanded to reach *A* or *B*, it will not reach them exactly, but will reach any point around *A* or *B*.

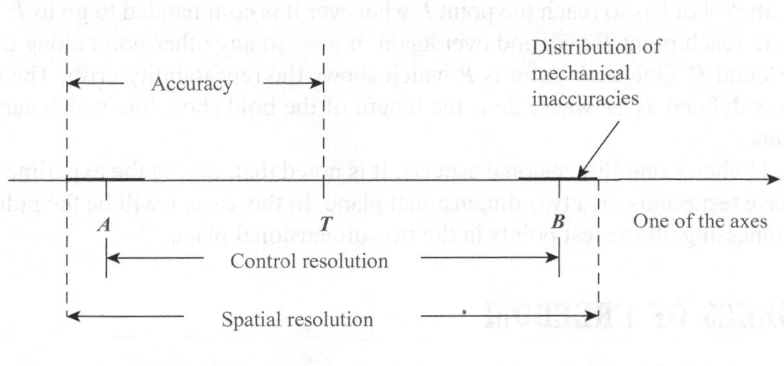

Figure 2.9 Robot Accuracy, Spatial Resolution and Control Resolution

In Figure 2.9, the mechanical inaccuracies are represented as points along the bold short lines around *A* or *B*. They can be regarded as having the statistical normal distribution along the bold lines. With the mechanical inaccuracies, the spatial resolution is defined as the distance between two farthest points that a robot can reach when it is commanded to go to any point between *A* and *B*. Then the accuracy is defined as one-half of the spatial resolution. This definition seems to imply that the accuracies are the same everywhere within the work volume. In fact, the accuracies mainly depend on two factors. One when the robot works at the extreme areas near to the borders of work volume. Here the robot has to extend its arm. The other factor in defining accuracy is the load being carried by the robot end effector when the load is attached to the wrist end. Larger load with extended arm produces greater deflections of the mechanical links, thus reducing the accuracy. The robot accuracy, when arm is fully extended, is worse than when the arm is close to the centre of the work volume or close to the robot's base. This is similar to the accuracy with which we place a 20 kg weight at a location reachable by our extended arm compared with a 20 kg weight placed near to our legs.

2.4.3 Repeatability

The repeatability and accuracy of a human in throwing darts in a dart game are illustrated in Figure 2.10. A person is said to be accurate and repeatable if he is able to throw the darts every time very close to and around the target point *T* (Figure 2.10(d)). If the darts always fall around and close to a point that is not *T*, then the person has a capability of repeating with no accuracy (Figure 2.10(c)). Figure 2.10(b) shows all the darts thrown close to and around *T*, showing accuracy but not repeatable. If the darts fall everywhere on the board (Figure 2.10(a)), the person is understood as having no repeatability and no accuracy.

The same configurations of repeatability and accuracy in dart game will also happen to a robot. Assume a robot holds a pin at its gripper and is commanded to go to the target location on a board, kept horizontal, again and again and prick. By studying the test (pricked) points, we can estimate the robots ability of repeatability and accuracy. In Figure 2.11, *T* is the location to which the robot is commanded to go. However, due to its errors and errors in control resolution, the robot will not reach *T* but it reaches the point *P*. If repeatability is the highest, the robot has to reach the point *P* whenever it is commanded to go to *T*. Due to its inability to reach point *P* over and over again, it goes to any other point along the bold short line around *P*. One such point is *R* which shows the repeatability error. The repeatability is now defined as $\pm r$ where $2r$ is the length of the bold short line which carries all the test points.

Figure 2.11 shows one-dimensional aspects. It is noted that, during the experiments, the robot can have test points on a two-dimensional plane. In this case, r will be the radius of a circle encompassing all the test points in the two-dimensional plane.

2.5 DEGREES OF FREEDOM

We always work in Cartesian coordinate system. In robotics, this is also known as world coordinate system or frame, since it is universally used in locating a point in the robot work space. Even though other coordinate frames such as polar and spherical are also used in robotics, we always prefer a robot to work in Cartesian coordinate frame. Figure 2.12(a)

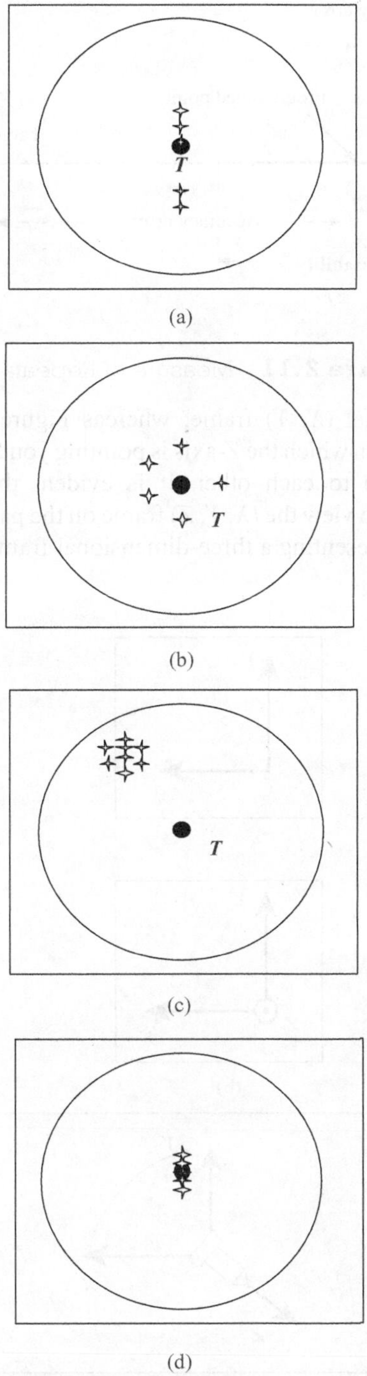

Figure 2.10 Accuracy and Repeatability: (a) Not Accurate Neither Repeatable, (b) Accurate But Not Repeatable, (c) Not Accurate But Repeatable and (d) Accurate and Repeatable

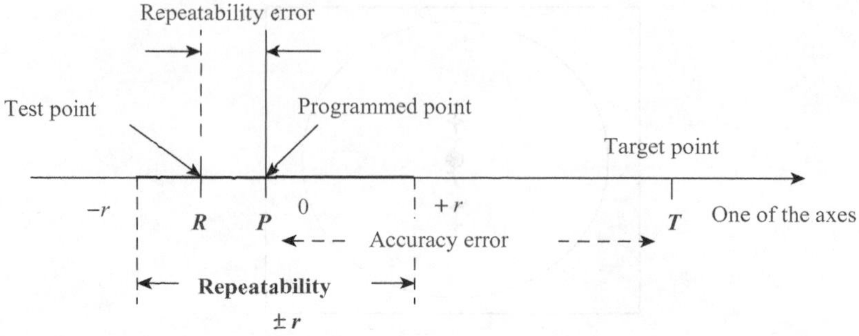

Figure 2.11 Measure of Repeatability

represents a two-dimensional (X, Y) frame; whereas Figure 2.12(b) describes a three-dimensional (X, Y, Z) frame in which the Z-axis is pointing you from the paper. The (X, Y, Z)-axes are orthogonal (+90°) to each other. It is evident that the representation of Z (Figure 2.12(b)) is difficult to view the (X, Y, Z) frame on the paper; however, Figure 2.12(c) is a convenient form of representing a three-dimensional frame. The (X, Y, Z)-axes follow

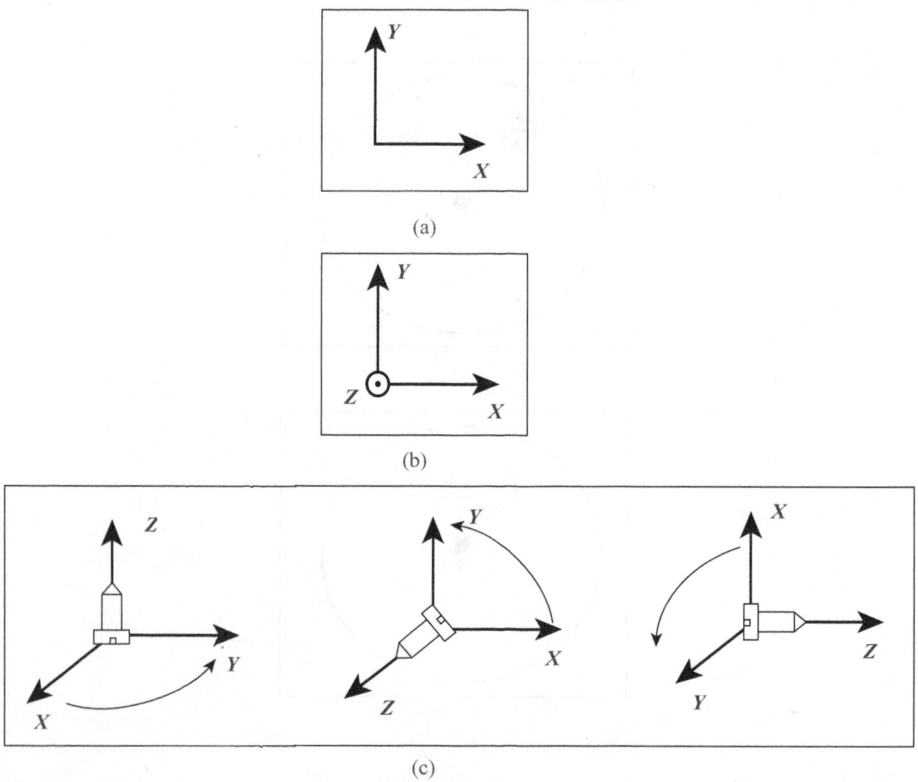

Figure 2.12 Cartesian Coordinate Frames: (a) Two-dimensional and (b) Three-dimensional Frames. (c) Right-hand Screw Method of Representing the Frame

the right-hand screw rule. The rule states that if a right-hand screw is placed with its head at the origin and its tail point directed towards the *Z*-axis, it advances along the *Z*-axis when it is rotated from *X* to *Y*. We shall follow this way of representing a Cartesian coordinate frame in robot work space analysis.

In industrial robotics, the object is the one to be manipulated by the robot. One of the manipulations is placing an object in a machine tool at a particular location with a particular orientation so that the object is machined successfully.

Here at least two sets of world coordinate frames are to be considered – one frame (*x*, *y*, *z*) corresponds to the object and the other (*X*, *Y*, *Z*) to the machine tool table.

For positioning the object, the object can be moved (translated) along any one, or two or three of its coordinate axes to reach a desired position in the machine tool table. For orientation, the object has to be rotated about any one or more of its axes.

Figure 2.13 shows a simple manipulation which involves two translations (30 cm along the object's *y*-axis and 10 cm along the object's *z*-axis) and a rotation (90° about the object's *x*-axis) to match the frames (*x*, *y*, *z*) and (*X*, *Y*, *Z*). Effectively, the origin of object frame is translated and rotated while performing manipulations. Note that if the sequence of manipulation is different, (*x*, *y*, *z*) may or may not coincide with (*X*, *Y*, *Z*).

The number of 'independent' movements that a robot exerts on an object with respect to a coordinate frame is known as the number of degree of freedom (DoF). An unconstrained manipulation can translate the object along all three axes and rotate the object about these axes. This results in six DoF. That is, this manipulation requires translational motions along *x*, *y* and *z* and rotational motions about *x*, *y* and *z* directions. The manipulation of any object can be decomposed into these six independent motions of the mechanical constraints of a robot or the specific positional constraints of the object can limit the DoF.

If an object of Figure 2.13 is to be always on a surface of a flat table even when it is manipulated by a robot, then the object loses three of the six DoF. That is, if it is constrained to be always placed flat on the table, then it can be manipulated only in terms of moving the object along the *x* and *y* axes and rotating it about the *z*-axis.

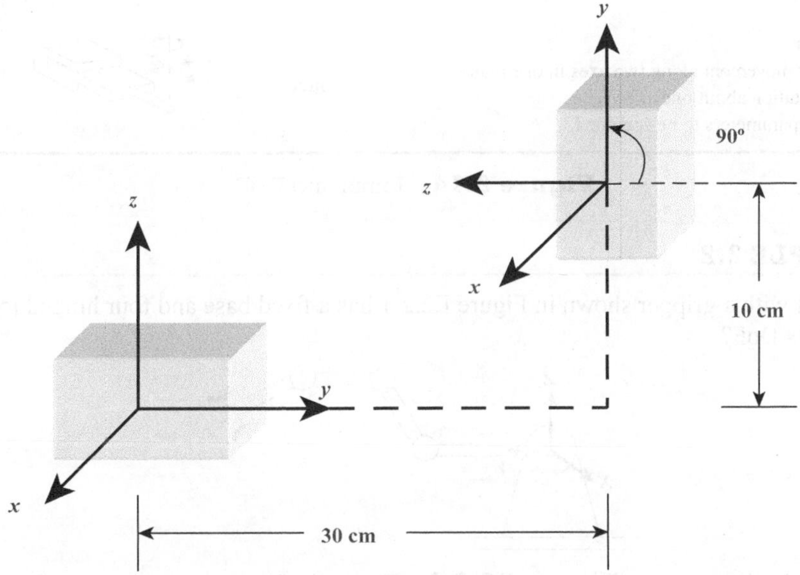

Figure 2.13 A Simple Manipulation of Object

Figure 2.14 illustrates a set of joints and DoF which normally an engineer in industry comes across. The students are to be advised to understand Figure 2.14.

Comments	Number of degrees of freedom	Type
Revolute Rotation about one axis Requires only one parameter to specify position in space	One	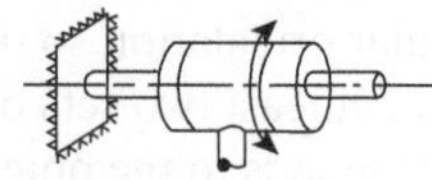
Prismatic Linear movement along one axis Requires only one parameter to specify position in space	One	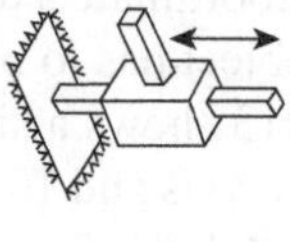
Screw Composite linear and rotational movement along one axis Translation defined by pitch and rotation Only one parameter need be specified	One	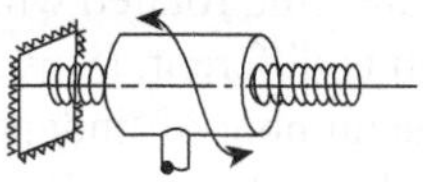
Cylindrical Linear and rotational Independent movement along and around one axis Two parameters to be specified	Two	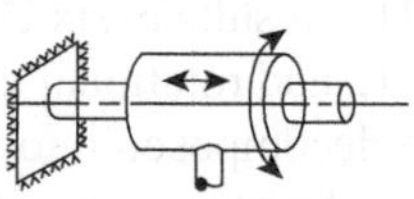
Spherical Can rotate about three axes (precession, nutation and spin) Three parameters to be specified	Three	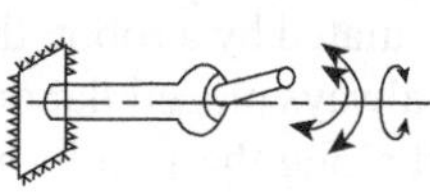
Planar Linear movement along two axes in one plane and rotation about one axis Three parameters to be specified	Three	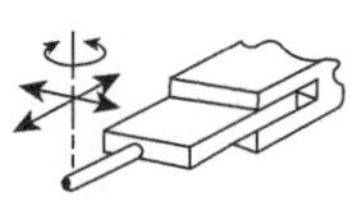

Figure 2.14 Joints and DoF

EXAMPLE 2.2

The robot with a gripper shown in Figure E2.2.1 has a fixed base and four hinged joints. What is its DoF?

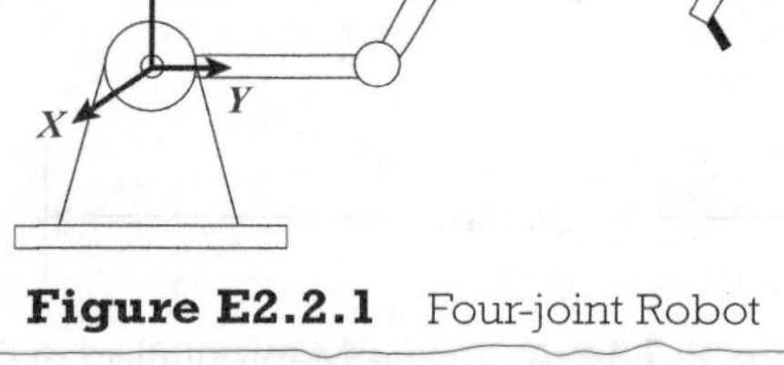

Figure E2.2.1 Four-joint Robot

Solution:

The robot, through its end effector (gripper in this case), can translate an object along the Y-direction and the Z-direction by folding its links. In addition, it can manipulate the object about the X-axis. Since the robot has only three independent manipulations, it has three DoF.

EXAMPLE 2.3

Observe various joints and possible manipulations as shown in Figure E2.3.1 Indicate, with an explanation, the DoF of each joint.

Solution:

R_X, R_Y and R_Z are rotations about the X, Y and Z axes respectively. Figure E2.3.1 shows all six DoF. We compare Figure E2.3.1 with the motions of each joint as listed in Table E2.3.1. Assume that a gripper is attached to movable joints.

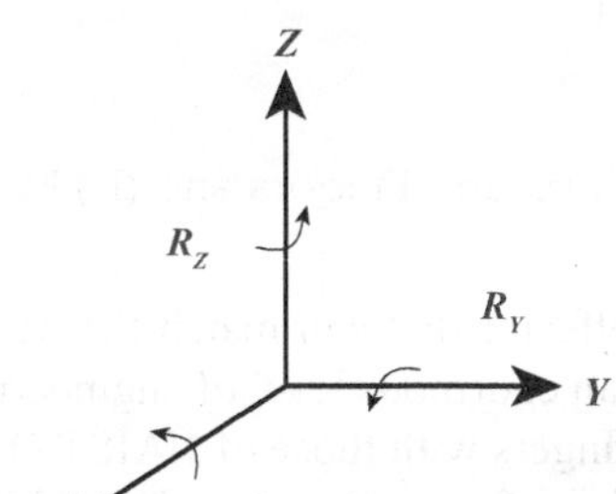

Figure E2.3.1 Possible Six DoF

Table E2.3.1 Solution to Problem

No.	Nature	Rotation/Translation	DoF
1	Revolute	Rotated about the Y-axis	One
2	Prismatic	Translated along the Y-axis	One
3	Screw	Translated along the Y-axis or rotated about the Y-axis	One
4	Cylindrical	Translated along the Y-axis and rotated about the Y-axis	Two
5	Spherical	Rotated about any axis (X, Y, Z)	Three
6	Planar	Translated along two axes (Y and X) and also can be rotated about the Z-axis.	Three

2.6 END EFFECTORS

End effector is an attachment to the wrist of a robot that enables to perform specialized tasks. The ideal form of a general purpose end effector is to do all the jobs that human palm and fingers do. Human fingers have several joints and each joint has two or three DoF that can grasp, hold and manipulate the object irrespective of its shape and size. Robot manufacturers and researchers around the world have been working for achieving an economical,

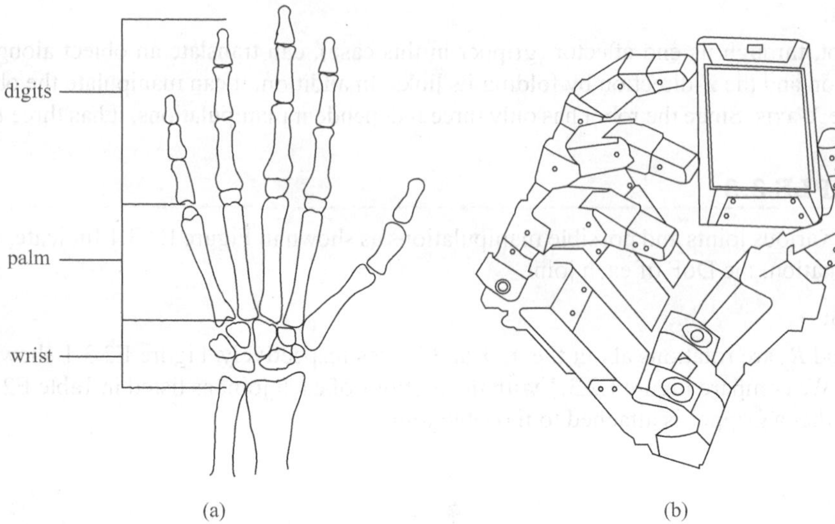

Figure 2.15 (a) Human Palm and Fingers and (b) Man-made Palm (NAIST) and Fingers

powerful, multi-purpose end effector that can match the performance of the human palm and fingers. This task requires an enormous level of engineering feat. Figure 2.15 compares the joints of human palm and fingers with those of NAIST (12 DoF) produced by researchers in Japan. The days are not long for man-made palm and fingers that perform tasks very similar to those of human being.

The end effector has to grasp an object irrespective of its shape, hold it without the objects coming out of the grasp, lift and shift the object to a new location and place it in a specified orientation. The end effector also needs to do some processing functions such as welding and screwing. There are wide varieties of end effectors required to perform different industrial work functions. The various types of end effectors are divided into mainly two categories: (1) grippers and (2) tools.

2.6.1 Grippers

Grippers are used to grip or grasp the objects. The objects (or parts) are generally the work pieces that are to be manipulated by the robot. This aspect of part handling includes machine loading and unloading, palletizing and picking parts from conveyor. Figure 2.16 illustrates some of the generally used grippers in industries.

Interchangeable finger pair is the one which can be selected from several pairs and attached to the wrist to suit the required part size. Constricted fingers are suitable for handling special contoured parts such as cylinders and spheres. Standard angular and parallel grippers are designed to pick up smaller or larger sized parts of rectangular shapes. Inflatable grippers are used to handle cup-shaped parts. The inflatable cylindrical diaphragm is inserted into the cup and inflated so that the inside surface of the cup is held firmly by the diaphragm.

All the above grippers are of ON-OFF type; when it is ON, the gripper grips the part and when it is OFF the part is released. In addition, the ON-OFF grippers are open-loop type. Open loop means that the robot computer commands the fingers to make a

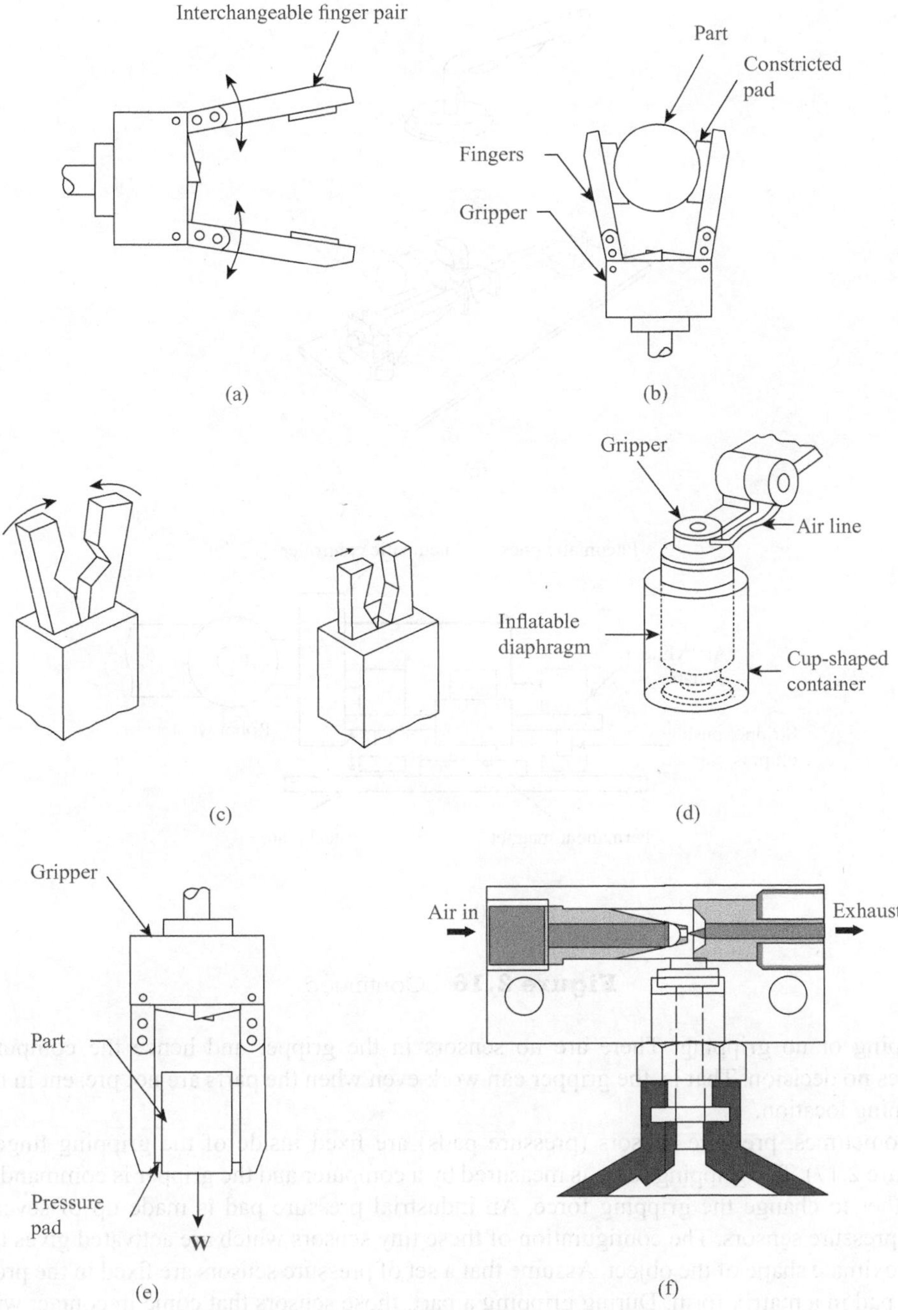

Figure 2.16 Examples of Robot Grippers: (a) Interchangeable Gripper, (b) Constricted Gripper, (c) Standard Angular and Parallel Grippers, (d) Inflatable Gripper, (e) Gripper with Pressure Pad, (f) Suction Cup and Venture Device, (g) Magnetic Grippers (Single and Dual) and (h) Stripper Push off Pins in the Permanent Magnet Gripper

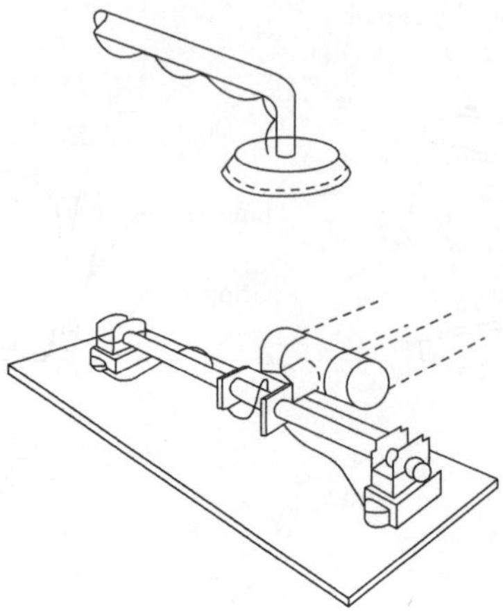

(g)

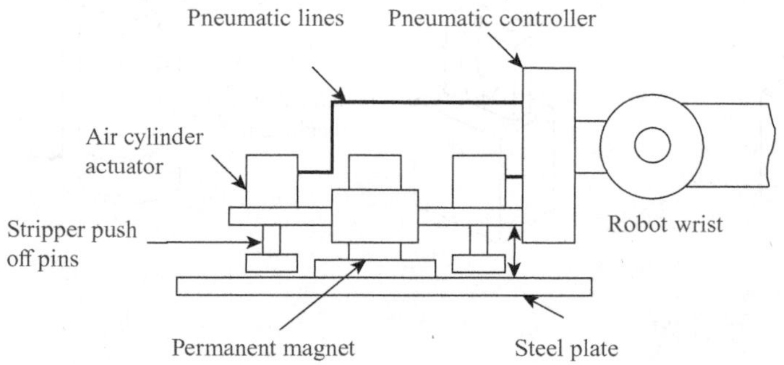

(h)

Figure 2.16 Continued

gripping or no gripping. There are no sensors in the gripper and hence the computer makes no decision. That is, the gripper can work even when the parts are not present in the gripping location.

Sometimes, pressure sensors (pressure pads) are fixed inside of the gripping fingers (Figure 2.17). The gripping force is measured by a computer and the gripper is commanded whether to change the gripping force. An industrial pressure pad is made up of several tiny pressure sensors. The configuration of these tiny sensors which are activated gives the approximate shape of the object. Assume that a set of pressure sensors are fixed in the pressure pad in a matrix form. During gripping a part, those sensors that come in contact with the part are activated and others are not.

Figure 2.17 illustrates such a sensor matrix. The set of activated pressure sensors are seen as dots while grasping the object and the non-activated sensors are not visible in the matrix. The pattern of activated pressure sensors can show the shape of object. The computer continuously measures the pressure subjected by each sensor. A situation may arise when the gripped part is heavy and slips off from the fingers. The computer recognizes this situation

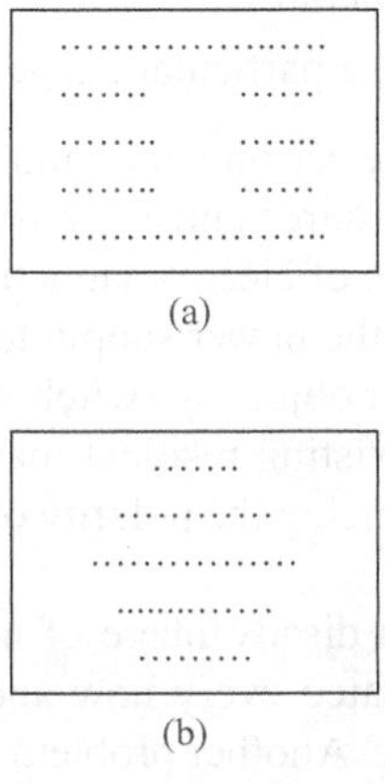

Figure 2.17 Patterns of Activated Sensors in a Pressure Matrix: (a) Rectangular Object and (b) Prismatic Object

by the variations in the pattern of activated sensors and commands for an increased gripper holding pressure to avoid a total slipping off of the object from the gripper.

A heavy metal object has to be gripped with a force larger than that for a thin glass material. Suction cup and venture device are some of the sensors that the industries need. A flat surface is gripped by a suction cup.

Magnetic grippers are of single and dual that can handle smaller or larger flat surfaces. A simple push and pull gripper with permanent magnet is shown in Figure (2.16(h)).

Vacuum cups, also called suction cups, are employed as gripper devices of certain objects. This requires that the surface of the object to be handled to be clean, smooth and flat with no holes so that a sustained vacuum between the object and the cup is created. The vacuum cups are made up of elastic material such as rubber or soft plastic. The vacuum pump can be of a piston–cylinder operated by pneumatic device and powered by an electric motor. If the part to be handled is a flexible material such as rubber, then the vacuum cup has to be made from a hard material. When the object surface is broad in area, multiple suction cups arrangements can be used. Some of the features and advantages of suction cup gripper are as follows:

(a) Requires only one surface for grasping
(b) Relatively light weight
(c) Application of an uniform pressure distribution in the gripping area
(d) Useful even to handle brittle objects such as glass plates
(e) Not useful when the surface is corrugated and having holes.

The magnetic grippers are very useful devices when the object is of ferrous material. In general, the magnetic gripper has the following advantages:

(a) Pick-up time is very short
(b) Variations in part sizes are not a problem since multiple magnetic gripping arrangements are possible
(c) It can handle ferrous metal objects even when there are holes on the surface of the parts and the surface is not necessarily smooth and tidy

(d) Requires only one surface to hold

(e) Its design does not belong to a particular shape of object.

There are two types of magnets – electromagnet and permanent magnet. Electromagnet gripper needs a separate DC power source and a controller unit. Picking an object, holding it and releasing are easier in the case of electromagnetic gripper. If the object is heavy, the releasing is easy by switching OFF the power supply to the electromagnet. However, there will be a problem of releasing lighter object by switching OFF the power source. Releasing may not be successful due to the existing residual magnetism in the magnetic gripper. A quick releasing is performed by reversing the polarity of power source to the gripper before switching OFF the power source.

Permanent magnet gripper has a disadvantage of not having a separate power source. Permanent magnet is to be rejuvenated every now and then since the magnetizing force weakens due to age and repeated use. Another problem is that the object cannot be released unless some gadget is attached to push out the object at the releasing location. This gadget can be a stripper push off pin mechanism operated by a pneumatic controller. At the releasing location, the pneumatic push off pins extend and strip off the object from the permanent magnet. The permanent magnet gripper is highly advantageous for handling tasks in hazardous and explosive environments since this gripper needs no electric ON-OFF devices.

In addition to the above grippers, there are several other varieties such as adhesive grippers (that uses adhesive agents to pick up small pieces of objects), hooks (to pick up objects such as bags) and scoops for handling liquids, molten metals and powders (such as chemical substances). The gripper is application oriented and has to be from a new design for an entirely new application.

2.6.2 Tools

In many applications, the robot has to manipulate a tool to be operated on a work piece rather than manipulating a work piece. Sometimes, a suitable gripper can be used to hold and handle a tool. In some cases, multiple tools can be handled by using a single gripper. However, in most cases the tool is to be attached to the wrist to perform certain process on the work piece. Several applications of tools can be cited in a manufacturing industries. Some applications of tools include spot welding, arc (seam) welding, heating, spray painting, drilling, grinding, brushing, cementing and water jet cutting. Figure 2.18 shows certain types of tools used by industrial robots and their descriptions next to each tool.

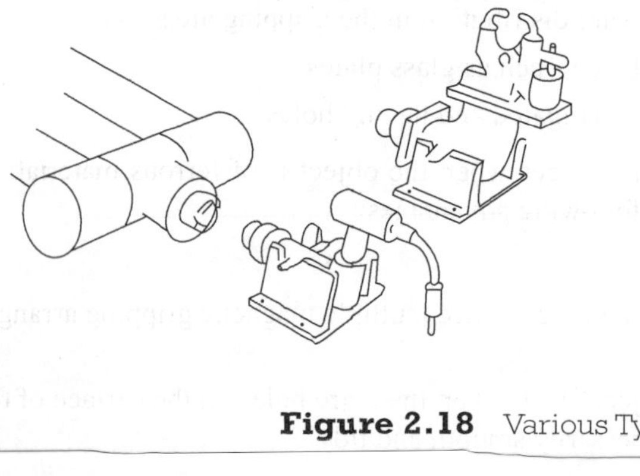

Tool changers are often used to change from one tool to the other in accordance with sequentially programmed operations. The tools can be of different types and sizes. A cradle is used to place the tool. The tool and the robot wrist have a snap-in mechanism. When placing a tool on the cradle, the robot lowers its tool into the cradle and pulls its wrist away. This process is reversed when picking up another tool from its cradle

Figure 2.18 Various Types of Tools

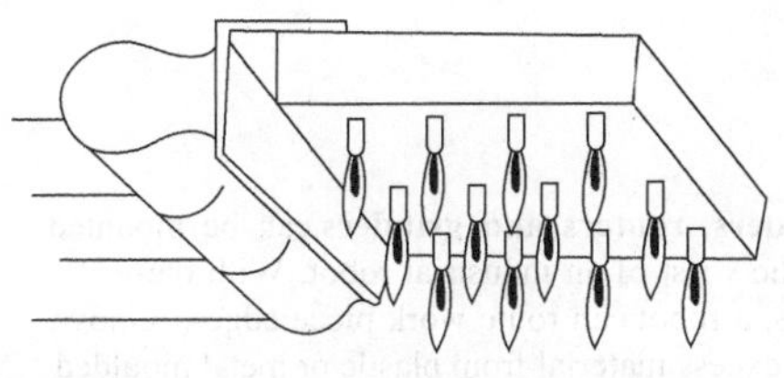

The industrial robot can also manipulate a **heating torch** to break out foundry moulds by playing a torch over the surface. When more heat is required, the flame is lingered over the portion of the surface. This break out operation is faster than sending the mould using a conveyor through a gas-fired oven

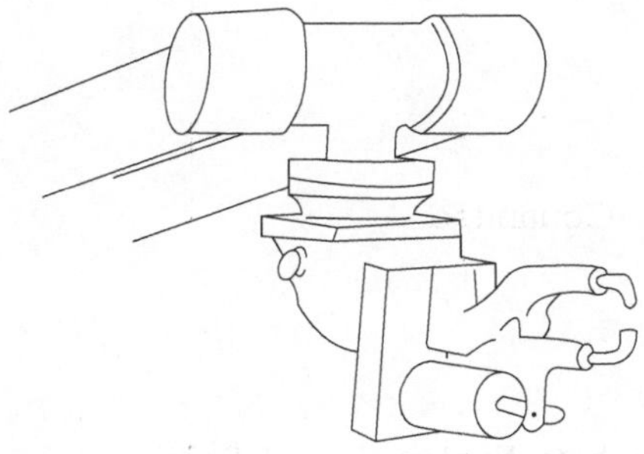

An industrial robot can handle a **spot welding gun** to place a series of spot welds on a flat or a simple curved or compound curved surfaces. This is an inevitable application in an automobile industry

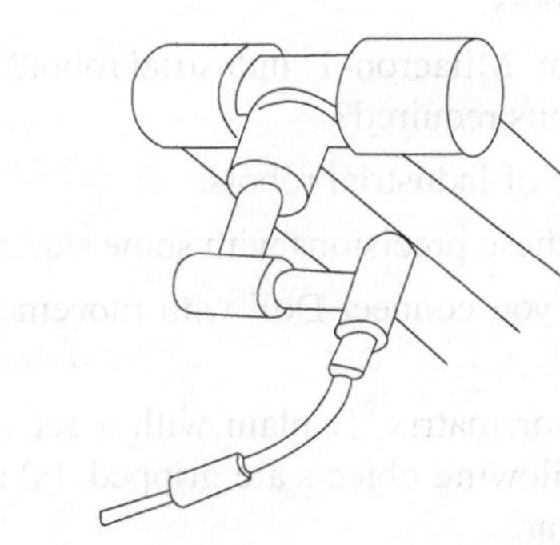

Robot welding by using an **arc welding torch** is an important industrial application. The welding torch dwells briefly on a location of welding two metals till a sufficient current passes through the welding torch. Then the torch is taken along the weld path at a close proximity of the metals. This can work for straight, simple or compound curved weld paths. The robot can relieve the human from this unhealthy and dangerous job environment

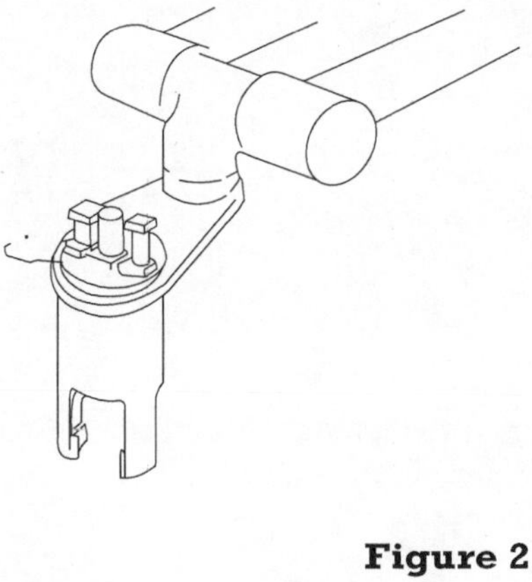

Pneumatic nut-runner drills and impact wrenches are used in a general purpose robot working in hazardous environments. Drilling and counter sinking are done with the aid of positioning guides. Mechanical guides are used to position the tool

Figure 2.18 Continued

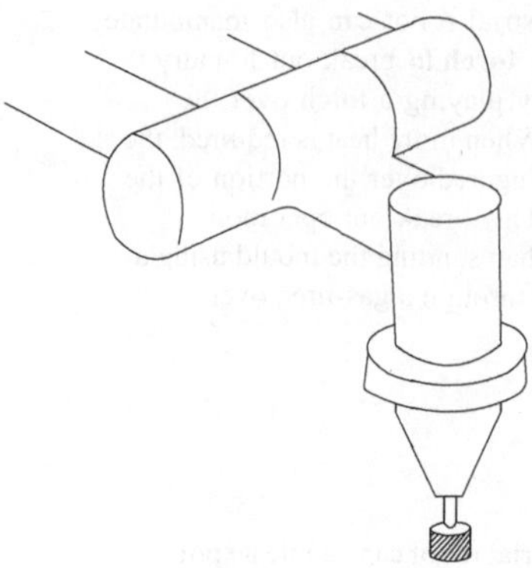

Sanders, routers and grinders can be mounted on the wrist of an industrial robot. With these tools, a robot can route work piece edges, remove any excess material from plastic or metal moulded objects, apply adhesives along a route and create routed polishing

Figure 2.18 Continued

REVIEW QUESTIONS

1. List four more definitions of industrial robots. Explain these definitions.
2. Explain the configuration of human body.
3. What are the equivalent configurations in robots?
4. How many articulations are there in Cincinnati Milacron-T^3 industrial robot? Name these articulations? Why are these articulations required?
5. List and explain the structural configurations of industrial robots.
6. What are precisions of movement? Explain these precisions with some statistics?
7. What do you understand by DoF? How do you connect DoF with movements of robot joints?
8. Discuss the configurations of pressure sensor matrix. Explain with a set of diagrams on pressure-pad patterns when the following objects are gripped: (a) a bolt, (b) a plate with number of holes and (c) a gear.

COORDINATE TRANSFORMATION 3

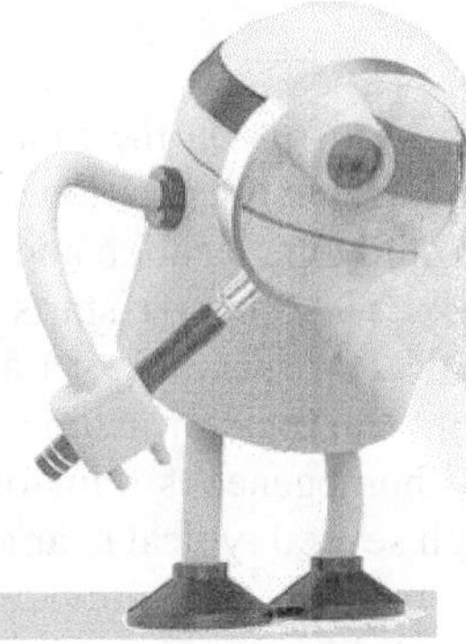

You end up with a tremendous respect for a human being if you're a roboticist

Joseph Engelberger

This chapter covers the following key topics:

Introduction – 2D Coordinate Transformation – Example – 2D Homogeneous Transformation – Example – Description of Object – Example – 3D Coordinate Transformation – Examples – Inverse Transformation – Kinematic Chain – Composite Transformation Matrix – Algorithm for Composite Transformation – Examples – Object Manipulation – Example – The Wrist – Example

---

3.1 INTRODUCTION

Human hand continuously manipulates the objects of interest very skilfully in a three-dimensional (3D) space. Our hand and fingers are controlled by our computing brain to move to locations, picking up objects, placing objects, moving objects, rotating the objects and working on the objects. We have several sensors and the most important is our vision to know where the object is placed, where the fingers are located and what to be done. A conventional industrial robot cannot perform these tasks unless we give sufficient information. The first part of information is about instructions on where the object in 3D space is located, where the robot base is situated and where the gripper is positioned and also various joints and links are located at a particular time. The second part of information is to contain data on what to be done such as moving the gripper on to the object, orientating the gripper in such a way to carefully grasp the object, lift the object, move the object in a certain direction, rotating the object in a certain way, and placing the object in a required orientation. The second part is created by programming the robot so that it performs these smaller efforts towards completing a major task. Several high-level software platforms are available for programming each variety of robot and these make the second part easier. These programs obviously require data we create through the first part of information.

Generating data through the first part requires a careful description of locations of object, gripper, robot base and joints in terms of position and orientation in a 3D space.

This chapter mainly discusses on the first part of information to create required data to make the robot to understand the position of the object and its own position in a 3D space. Through this we can study the location and orientation of the object when the robot begins and completes a task. This chapter starts with basic vectors and their transformations in a two-dimensional (2D) space (X, Y) and extends to three dimensions (X, Y, Z). In robotics, these vectors and transformations are defined in a different way known as homogeneous vectors and homogeneous transformations in three dimensions. These concepts are illustrated through several typical examples.

3.2 2D COORDINATE TRANSFORMATION

We shall begin with a 2D coordinate transformation. Let a point p in the 2D space (X, Y) be described as

$$p^{T} = [a, b] = [2, 1], \tag{3.1}$$

where the superscript, T, is the transpose. The point (also called as vector) has a magnitude m and an angle θ with respect to the X-axis, thus (Figure 3.1)

$$m = \sqrt{(a^2 + b^2)};\ \cos\theta = (a/m);\ \sin\theta = (b/m). \tag{3.2}$$

The point p can be transformed to a new point $p_1^{T} = [a_1, b_1]$ by means of a (2 × 2) transformation matrix P such that $p_1 = P \times p$ where

$$\begin{bmatrix} a_1 \\ b_1 \end{bmatrix} = \begin{bmatrix} p_{11} & p_{12} \\ p_{21} & p_{22} \end{bmatrix} \begin{bmatrix} a \\ b \end{bmatrix}. \tag{3.3}$$

Let us describe a few transformation matrices that transform the vector p to p_1.

(a) No change: $P = \begin{bmatrix} 1 & 0 \\ 0 & 1 \end{bmatrix}$; P is an identity matrix.

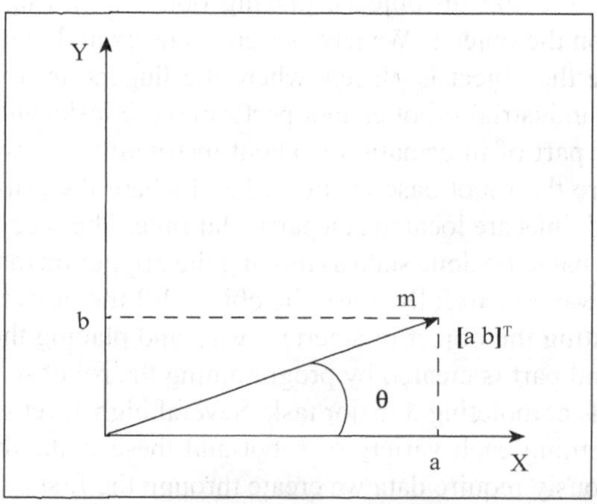

Figure 3.1 Vector p

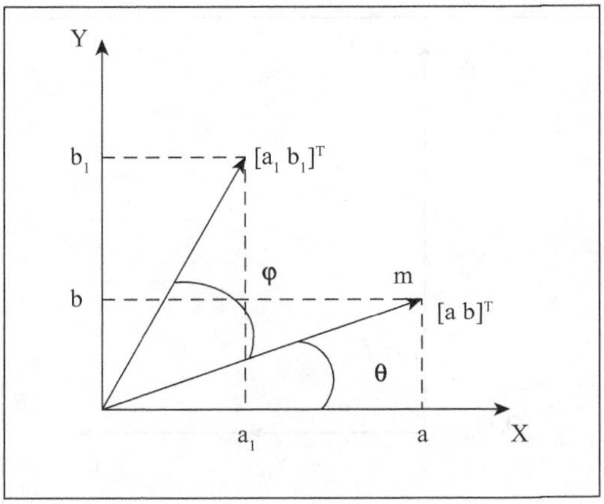

Figure 3.2 Vectors p and p_1

(b) x-scale change: $P = \begin{bmatrix} 3 & 0 \\ 0 & 1 \end{bmatrix}$; x-scale factor is 3; $p_1^T = [3a, b]$

(c) y-scale change: $P = \begin{bmatrix} 1 & 0 \\ 0 & 4 \end{bmatrix}$; y-scale factor is 4; $p_1^T = [a, 4b]$

(d) x- and y-scale changes: $P = \begin{bmatrix} 3 & 0 \\ 0 & 4 \end{bmatrix}$; x, y-scales are independent; $p_1^T = [3a, 4b]$

(e) **Orientation of vectors:** The vector p has a magnitude m and angle θ as represented in Equation (3.2). A new vector p_1 can now be obtained by a rotation of the vector p about the origin to an angle φ. We use the universal way of representing the angle as positive if rotation is counterclockwise. The rotation can be regarded as a compound scale change; that is, x-scale and y-scale are dependent on each other.

Referring to Figure 3.2, vector $p_1 = P \times p$, in which P can be derived as follows:

$$a_1 = m\cos(\theta + \varphi) = m\cos\theta\cos\varphi - m\sin\theta\sin\varphi = a\cos\varphi - b\sin\varphi. \quad (3.4)$$

$$b_1 = m\sin(\theta + \varphi) = m\cos\theta\sin\varphi + m\sin\theta\cos\varphi = a\sin\varphi + b\cos\varphi. \quad (3.5)$$

That is,

$$P = \begin{bmatrix} \cos\varphi & -\sin\varphi \\ \sin\varphi & \cos\varphi \end{bmatrix}. \quad (3.6)$$

The (2×2) transformation matrix P produces a rotation of p, φ°, about the origin of (X, Y) plane. This transformation matrix P defines the orientation of vector p_1 with respect to vector p.

(f) **Orientation of frames**: Let us assume that the magnitude of vector p shrinks to 0; that is, $m = 0$. This imaginary vector may be considered as a point object located at the origin O of (X, Y) frame. In robotics, every object has its own coordinate system and hence has its own origin. Let this origin be represented as O_o. Figure 3.3(a) describes such a situation. At this instant, frame O and frame O_o coincide. Let us call the coordinates of point object as (X_o, Y_o) and the position of the point object as O_o.

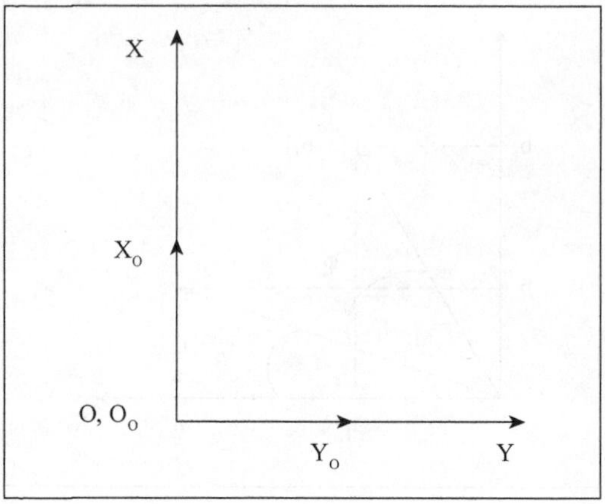

(a)

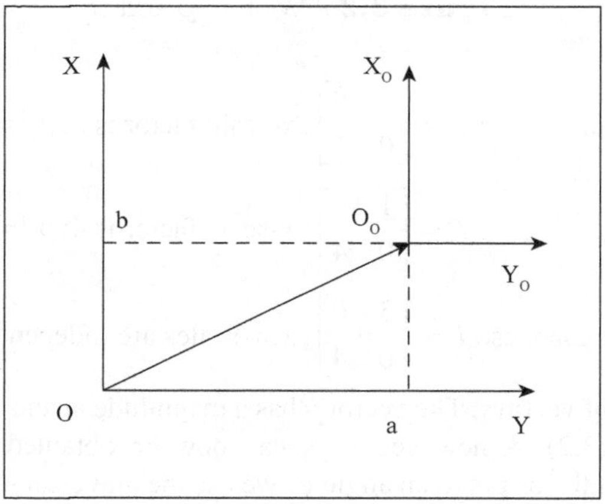

(b)

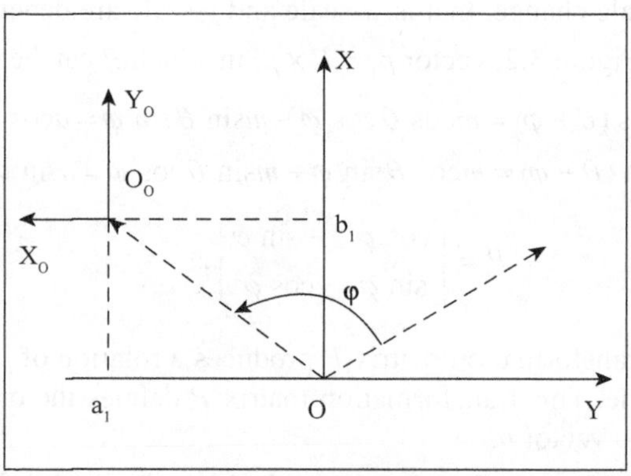

(c)

Figure 3.3 Transformation of Point Object O_o: (a) O and O_o coincide, (b) Translation of O_o and (c) Translation and Rotation of O_o

Now the point object is moved through 'a' units along the X-axis and through 'b' units along the Y-axis. We call the motion along the X-direction and Y-direction as translation. Such a translation results in a transformed position of O_o. The point object carries its coordinate frame (X_o, Y_o) as it translates. This translation is illustrated in Figure 3.3(b).

Further, the point object is rotated about the origin O of (X, Y) by an angle φ to reach (a_1, b_1). During this rotation, the frame (X_o, Y_o) is also rotated to the same angle φ with respect to O (Figure 3.3(c)). Then, the orientation of (X_o, Y_o) with respect to (X, Y) is defined by the transformation matrix which is identical to Equation (3.6) as

$$P = \begin{bmatrix} \cos\varphi & -\sin\varphi \\ \sin\varphi & \cos\varphi \end{bmatrix}. \tag{3.7}$$

EXAMPLE 3.1

Let the origins of (X, Y) and (X_o, Y_o) frames be O and O_o, respectively. Let the origin O_o be shifted 2 units along the X-axis, 1 unit along the Y-axis and rotated 90° about O. What are the new position and orientation of O_o?

Solution:

Due to translation, the origin O_o reaches a point $[2, 1]^T$ in the (X, Y) frame. After rotation, the new position of O_o is obtained using Equation (3.7) as

$$\begin{bmatrix} -1 \\ 2 \end{bmatrix} = \begin{bmatrix} 0 & -1 \\ 1 & 0 \end{bmatrix} \begin{bmatrix} 2 \\ 1 \end{bmatrix} \tag{E3.1.1}$$

and the orientation matrix is $\begin{bmatrix} 0 & -1 \\ 1 & 0 \end{bmatrix}$. (E3.1.2)

Both in the above discussion and in Example 3.1, we realize that the transformation matrix contains the information on rotation and not the information on translation. It will be useful if the transformation matrix has both these information so that with an observation on the transformation matrix, one can easily estimate the final orientation and translation of the object with respect to O if the object undergoes several translations and rotations. For this we define the concept of homogeneous transformation that utilizes a set of new forms of vectors and matrices.

3.2.1 2D Homogeneous Transformation

The concept of homogeneous coordinates was introduced in computer graphics. This coordinate system represents an n-dimensional space in terms of $(n + 1)$ coordinates. For example, in a 2D space, a point $[a, b]$ is represented as $[aw, bw, w]^T$ where w is a selected scale factor

$$\begin{bmatrix} 2 \\ 1 \\ 1 \end{bmatrix} = \begin{bmatrix} 0.2 \\ 0.1 \\ 0.1 \end{bmatrix} = \begin{bmatrix} 20 \\ 10 \\ 10 \end{bmatrix} \tag{3.8}$$

This means, the vectors represent the same vector $[2, 1]^T$. In robotics, w is always considered as unity.

By using these homogeneous (3 × 1) vectors in all our computations, the transformation matrix has to be of (3×3) form. This has the following form with the sub-matrices 1, 2, 3 and 4.

$$P = \left(\begin{array}{c:c} \mathbf{1} & \mathbf{2} \\ \hdashline \mathbf{3} & \mathbf{4} \end{array}\right) \tag{3.9}$$

Sub-matrix 1: This is a (2 × 2) sub-matrix representing the rotation (or orientation) similar to Equation (3.6). Every rotation of object by an angle about the origin of the (*X*, *Y*) frame creates this sub-matrix.

Sub-matrix 2: This is a (2 × 1) sub-matrix representing the translation along the axes of the (*X*, *Y*) frame. After a number of translations, the sub-matrix 2 gives the final position of the object in the (*X*, *Y*) frame.

Sub-matrix 3: This is a null vector of dimension (1 × 2).

Sub-matrix 4: This is a scalar that has the value of scale factor $w = 1$.

EXAMPLE 3.2

Consider the problem given in Example 3.1. The point object O_o with respect to the (*X*, *Y*) frame originally coincides with *O*. Then the point object is represented by the homogeneous vector $[0, 0, 1]^T$. Three transformations are done in three consecutive stages: (i) O_o is moved 2 units in the *X*-direction, then (ii) it is moved 1 unit in the *Y*-direction and then (iii) rotated 90°about the origin *O*. Determine the final position and orientation of the point object.

Solution:

We shall consider these transformations one by one:

(i) Translation along the *X*-axis

$$\begin{bmatrix} 2 \\ 0 \\ 1 \end{bmatrix} = \left[\begin{array}{cc:c} 1 & 0 & 2 \\ 0 & 1 & 0 \\ \hdashline 0 & 0 & 1 \end{array}\right] \begin{bmatrix} 0 \\ 0 \\ 1 \end{bmatrix}. \tag{E3.2.1}$$

We create the transformation matrix with the sub-matrix 1 as an identity matrix indicating no rotation (or orientation) and the sub-matrix 2 indicating 2 unit translation in the *X*-direction and 0 unit in the *Y*-direction. The position of O_o is now $[2, 0]^T$.

(ii) Translation along the *Y*-axis

$$\begin{bmatrix} 2 \\ 1 \\ 1 \end{bmatrix} = \left[\begin{array}{cc:c} 1 & 0 & 0 \\ 0 & 1 & 1 \\ \hdashline 0 & 0 & 1 \end{array}\right] \begin{bmatrix} 2 \\ 0 \\ 1 \end{bmatrix}. \tag{E3.2.2}$$

Transformation matrix indicates the *Y*-axis translation of 1 unit without any rotation. Thus, the position of O_o is $[2, 1]^T$.

(iii) Rotation about origin *O* [by substituting $\phi = 90°$ in Equation (3.6)]

$$\begin{bmatrix} -1 \\ 2 \\ 1 \end{bmatrix} = \left[\begin{array}{cc:c} 0 & -1 & 0 \\ 1 & 0 & 0 \\ \hdashline 0 & 0 & 1 \end{array}\right] \begin{bmatrix} 2 \\ 1 \\ 1 \end{bmatrix}. \tag{E.3.2.3}$$

Transformation matrix indicates a rotation of 90° about origin O with no translation. The final position is $[-1, 2]^T$.

The sequence (i), (ii) and (iii) can be combined without changing the order of transformations as

$$\begin{bmatrix} -1 \\ 2 \\ 1 \end{bmatrix} = \begin{bmatrix} 0 & -1 & 0 \\ 1 & 0 & 0 \\ 0 & 0 & 1 \end{bmatrix} \begin{bmatrix} 1 & 0 & 0 \\ 0 & 1 & 1 \\ 0 & 0 & 1 \end{bmatrix} \begin{bmatrix} 1 & 0 & 2 \\ 0 & 1 & 0 \\ 0 & 0 & 1 \end{bmatrix} \begin{bmatrix} 0 \\ 0 \\ 1 \end{bmatrix}$$

$$= \begin{bmatrix} 0 & -1 & -1 \\ 1 & 0 & 2 \\ 0 & 0 & 1 \end{bmatrix} \begin{bmatrix} 0 \\ 0 \\ 1 \end{bmatrix}, \quad \text{(E3.2.4)}$$

thus forming a compound transformation matrix which has the information on two translations and one rotation.

3.3 DESCRIPTION OF OBJECT

An object in 2D space is described by selected points on the object's periphery. As an example, consider the square object, M, of side 1 unit placed in the (X, Y) plane such that sides AD and AB coincide with the X- and Y-axes, respectively (Figure 3.4). The object is described by its corner points as individual vectors in homogeneous form. The object M is defined as

$$M = \begin{matrix} A & B & C & D \end{matrix} \\ \begin{bmatrix} 0 & 0 & 1 & 1 \\ 0 & 1 & 1 & 0 \\ 1 & 1 & 1 & 1 \end{bmatrix}. \quad (3.10)$$

When the object is translated and rotated, every point on the object is translated and rotated. This is illustrated in the following Example 3.3.

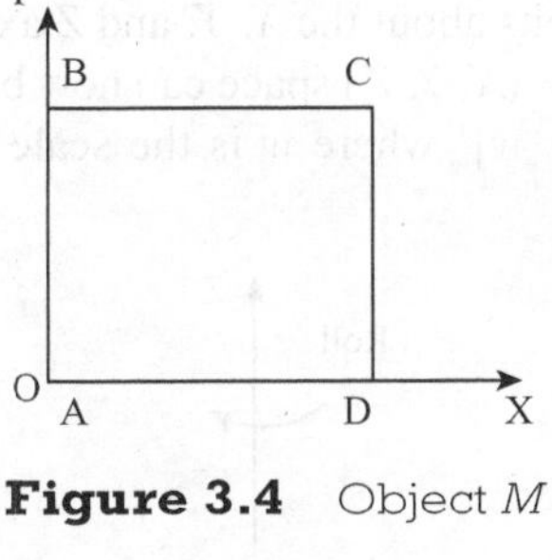

Figure 3.4 Object M

EXAMPLE 3.3

The square object M of 1 unit side is placed in the (X, Y) frame. Describe the position of the object when it is translated −1 unit and 2 units along the X-axis and the Y-axis, respectively and then rotated about the origin O by 90°.

Solution:

When the transformation matrix is H, the new location of the object is described by

$M_{new} = HM$

$$M_{new} = \begin{bmatrix} 0 & -1 & -1 \\ 1 & 0 & 2 \\ 0 & 0 & 1 \end{bmatrix} \begin{bmatrix} 0 & 0 & 1 & 1 \\ 0 & 1 & 1 & 0 \\ 1 & 1 & 1 & 1 \end{bmatrix} = \begin{array}{c} \begin{matrix} A & B & C & D \end{matrix} \\ \begin{bmatrix} -1 & -2 & -2 & -1 \\ 2 & 2 & 3 & 3 \\ 1 & 1 & 1 & 1 \end{bmatrix} \end{array}. \quad \text{(E3.3.1)}$$

The transformed object is given in Figure E3.3.1. Note the new locations of the corner points.

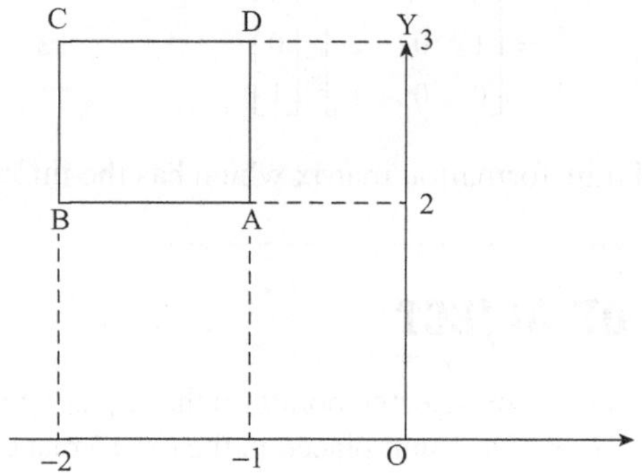

Figure E3.3.1 Transformed Object

3.4 3D COORDINATE TRANSFORMATION

The 3D homogeneous coordinate system is an extension of 2D homogeneous coordinate system to involve all the six degrees of freedom (DoF) of the robot in the 3D space. The six DoF by which an object is manipulated is depicted in Figure 3.5. Manipulation is the skilful handling of objects in terms of its translation and orientation. The set of first three DoF is described by translational motions along the *X*, *Y* and *Z* directions and the next three DoF is described by rotational motions about the *X*, *Y*, and *Z* axes.

The 3D vector $[a, b, c]^T$ in the (X, Y, Z) space can now be represented by a (4×1) homogeneous vector as $[aw, bw, cw, w]^T$ where w is the scale factor; hence denotes the same

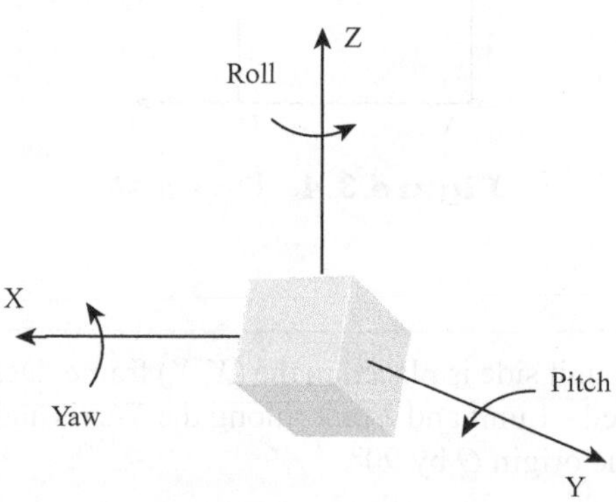

Figure 3.5 Six DoF

3D vector, $[5.0, 2.5, 4.0]^T$. We follow the usual convention of maintaining the scale factor $w = 1$.

$$\begin{bmatrix} 20 \\ 10 \\ 16 \\ 4 \end{bmatrix} = \begin{bmatrix} 10 \\ 5 \\ 8 \\ 2 \end{bmatrix} = \begin{bmatrix} 5.0 \\ 2.5 \\ 4.0 \\ 1 \end{bmatrix} \tag{3.11}$$

The homogeneous transformation matrices, now, are of (4×4) dimensions. The general form of transformation matrix is

$$H = \left[\begin{array}{c:c} \mathbf{1} & \mathbf{2} \\ \hdashline \mathbf{3} & \mathbf{4} \end{array} \right]. \tag{3.12}$$

The homogeneous transformation matrix may be interpreted as defining a relationship between two frames in terms of how far are their origins separated and the way of orientation of one frame to the other.

Sub-matrix 1: This is a (3×3) sub-matrix referred to as the rotation or orientation between two frames–the reference frame and the new frame. Every rotation of the new frame about the X, about the Y, and about the Z axes of reference frame creates this sub-matrix.

Sub-matrix 2: This is a (3×1) sub-matrix representing the translation vector that gives the position of the origin of new frame with respect to the origin of the reference frame.

Sub-matrix 3: This is a null vector of dimension (1×3).

Sub-matrix 4: This is a scalar representing the scale factor $w = 1$.

A vector is a point in a the 3D frame (X, Y, Z); it can be translated parallel to the X, Y and Z axes of the frame and can also be rotated about each of the X, Y and Z axes to get a new vector. This is known as transformation. For instance, a homogeneous vector v is transformed into homogeneous vector u by the following matrix operation:

$$u = H\,v, \tag{3.13}$$

where H is the homogeneous transformation matrix.

Since we always deal with homogeneous transformations, we call a homogeneous transformation matrix as transformation matrix, unless otherwise specifically required. Similarly, a homogeneous vector will be called as vector itself.

The transformation to accomplish a translation of a vector in space by distances a, b and c in the X, Y and Z directions, respectively, is given by the transformation matrix

$$H = \text{Trans}(a, b, c) = \left[\begin{array}{ccc:c} 1 & 0 & 0 & a \\ 0 & 1 & 0 & b \\ 0 & 0 & 1 & c \\ \hdashline 0 & 0 & 0 & 1 \end{array} \right], \tag{3.14}$$

where 'Trans' represents translation. Sub-matrix 1 is an identity matrix indicating no rotation of the vector with respect to axes of the reference frame.

EXAMPLE 3.4

A vector $[25, 10, 20]^T$ is translated by a distance of 8 units in the X-direction, 5 units in the Y-direction and 0 units in the direction of Z. Determine the translated vector.

Solution:
The transformation matrix is

$$H = \text{Trans}\,(8, 5, 0) = \begin{bmatrix} 1 & 0 & 0 & 8 \\ 0 & 1 & 0 & 5 \\ 0 & 0 & 1 & 0 \\ 0 & 0 & 0 & 1 \end{bmatrix} \tag{E3.4.1}$$

and the translated vector is

$$u = Hv = \begin{bmatrix} 1 & 0 & 0 & 8 \\ 0 & 1 & 0 & 5 \\ 0 & 0 & 1 & 0 \\ 0 & 0 & 0 & 1 \end{bmatrix} \begin{bmatrix} 25 \\ 10 \\ 20 \\ 1 \end{bmatrix} = \begin{bmatrix} 33 \\ 15 \\ 20 \\ 1 \end{bmatrix}. \tag{E3.4.2}$$

Rotations of a vector (Rot) about each of the three axes by an angle θ are accomplished by rotation transformation matrices. The rotation transformation matrix about the X-axis is

$$H = \text{Rot}(x, \theta) = \begin{bmatrix} 1 & 0 & 0 & 0 \\ 0 & \cos\theta & -\sin\theta & 0 \\ 0 & \sin\theta & \cos\theta & 0 \\ 0 & 0 & 0 & 1 \end{bmatrix}. \tag{3.15}$$

About the Y-axis, H is given by

$$H = \text{Rot}(y, \theta) = \begin{bmatrix} \cos\theta & 0 & \sin\theta & 0 \\ 0 & 1 & 0 & 0 \\ -\sin\theta & 0 & \cos\theta & 0 \\ 0 & 0 & 0 & 1 \end{bmatrix} \tag{3.16}$$

and about the Z-axis

$$H = \text{Rot}(z, \theta) = \begin{bmatrix} \cos\theta & -\sin\theta & 0 & 0 \\ \sin\theta & \cos\theta & 0 & 0 \\ 0 & 0 & 1 & 0 \\ 0 & 0 & 0 & 1 \end{bmatrix}. \tag{3.17}$$

EXAMPLE 3.5

A vector $v = [1, 1, 0]^T$ is described in a reference frame (X, Y, Z) and is rotated by an angle 45° about the Z-axis; determine the transformed vector, u.

Solution:

The transformation matrix corresponding to the rotation about the Z-axis has been derived. Hence, the required transformed vector, u, is computed as

$$u = H\,v = \text{Rot}(z, 45°)\, v = \begin{bmatrix} 0.707 & -0.707 & 0 & 0 \\ 0.707 & 0.707 & 0 & 0 \\ 0 & 0 & 1 & 0 \\ 0 & 0 & 0 & 1 \end{bmatrix} \begin{bmatrix} 1 \\ 1 \\ 0 \\ 1 \end{bmatrix}$$

$$= \begin{bmatrix} 0 \\ 1.414 \\ 0 \\ 1 \end{bmatrix}. \tag{E3.5.1}$$

The transformed vector, u, occupies the new position $[0, 1.414, 0]^T$ in the reference frame (X, Y, Z).

EXAMPLE 3.6

A frame coincident with a reference frame (X, Y, Z) is rotated by 90° about its Z-axis to obtain a new frame (X_n, Y_n, Z_n). What is the orientation of this new frame with respect to the reference frame?

Solution:

The transformation matrix after the rotation is given in Equation (E3.5.1). The (3×3) partition matrix is the orientation of the new frame (X_n, Y_n, Z_n) with respect to reference frame (X, Y, Z).

$$H = \text{Rot}(Z, 90) = \begin{bmatrix} \cos 90 & -\sin 90 & 0 & 0 \\ \sin 90 & \cos 90 & 0 & 0 \\ 0 & 0 & 1 & 0 \\ 0 & 0 & 0 & 1 \end{bmatrix} = \begin{array}{c} \begin{matrix} X_n & Y_n & Z_n & \end{matrix} \\ \begin{bmatrix} 0 & -1 & 0 & 0 \\ 1 & 0 & 0 & 0 \\ 0 & 0 & 1 & 0 \\ 0 & 0 & 0 & 1 \end{bmatrix} \end{array} \begin{matrix} X \\ Y \\ Z \\ \end{matrix}. \tag{E3.6.1}$$

Equation (E3.6.1) depicts the reference frame and the new frame after the rotation about the Z–axis.

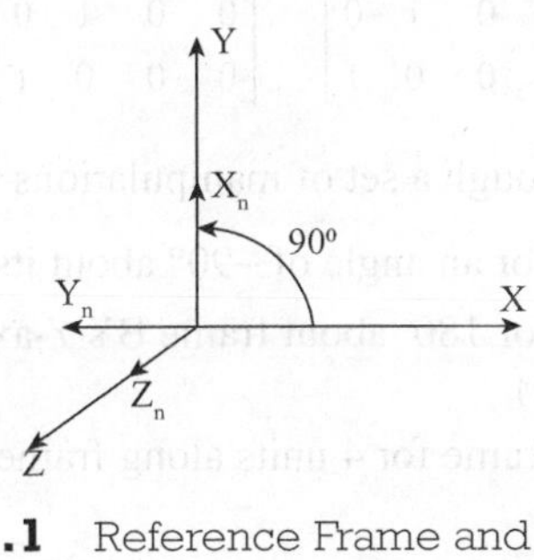

Figure E3.6.1 Reference Frame and the New Frame

Let us compare Figure E3.6.1 and Equation (E3.6.1). The H matrix in Equation (E3.6.1) shows that X and Y_n are in opposite directions (as indicated by the (X, Y_n) element = −1). H also shows that the Y and X_n and Z and Z_n are in the same directions.

EXAMPLE 3.7

Figure E3.7.1 describes three frames A, B, and C displaced and rotated from one to the other. Determine

(a) transformation matrix H_A^B which transforms A frame to B frame,

(b) transformation matrix H_B^C which transforms B frame to C frame and

(c) transformation matrix H_A^C which transforms A frame to C frame.

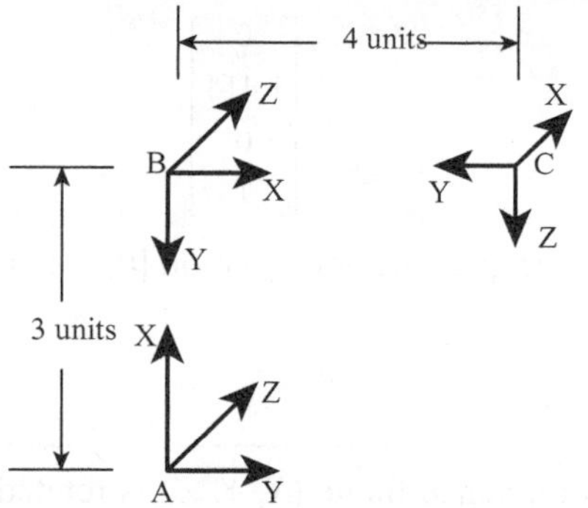

Figure E3.7.1 A, B, C Frames

Solution:

(a) Frame B is obtained through a set of manipulations with respect to frame A's axes:

(i) Rotating frame A for an angle of 90° about its Z-axis to get an intermediate frame I_1 (let us call it as an intermediate frame I_1)

(ii) Moving the I_1 frame along frame A's X-axis for a distance of 3 units to get frame B

Hence

$H_A^B = \text{Trans}(3, 0, 0)\ \text{Rot}(Z, 90),$

$$H_A^B = \begin{bmatrix} 1 & 0 & 0 & 3 \\ 0 & 1 & 0 & 0 \\ 0 & 0 & 1 & 0 \\ 0 & 0 & 0 & 1 \end{bmatrix} \begin{bmatrix} 0 & -1 & 0 & 0 \\ 1 & 0 & 0 & 0 \\ 0 & 0 & 1 & 0 \\ 0 & 0 & 0 & 1 \end{bmatrix} = \begin{bmatrix} 0 & -1 & 0 & 3 \\ 1 & 0 & 0 & 0 \\ 0 & 0 & 1 & 0 \\ 0 & 0 & 0 & 1 \end{bmatrix}. \tag{E3.7.1}$$

(b) Frame C is obtained through a set of manipulations with respect to frame B's axes:

(i) Rotating frame B for an angle of −90° about its X-axis to get the I_1 frame

(ii) Rotating I_1 frame for 180° about frame B's Y-axis to get I_2 frame (I_2 is another intermediate frame)

(iii) Translating the I_2 frame for 4 units along frame B's X-axis to get frame C.

Hence

$H_B^C = \text{Trans}(4, 0, 0)\ \text{Rot}(Y, 180°)\ \text{Rot}(X, -90°),$

$$H_B{}^C = \begin{bmatrix} 1 & 0 & 0 & 4 \\ 0 & 1 & 0 & 0 \\ 0 & 0 & 1 & 0 \\ 0 & 0 & 0 & 1 \end{bmatrix} \begin{bmatrix} -1 & 0 & 0 & 0 \\ 0 & 1 & 0 & 0 \\ 0 & 0 & -1 & 0 \\ 0 & 0 & 0 & 1 \end{bmatrix} \begin{bmatrix} 1 & 0 & 0 & 0 \\ 0 & 0 & 1 & 0 \\ 0 & -1 & 0 & 0 \\ 0 & 0 & 0 & 1 \end{bmatrix}$$

$$= \begin{bmatrix} -1 & 0 & 0 & 4 \\ 0 & 0 & 1 & 0 \\ 0 & 1 & 0 & 0 \\ 0 & 0 & 0 & 1 \end{bmatrix}. \qquad \text{(E3.7.2)}$$

(c) $H_A{}^C = H_A{}^B \, H_B{}^C$

$$= \begin{bmatrix} 0 & -1 & 0 & 3 \\ 1 & 0 & 0 & 0 \\ 0 & 0 & 1 & 0 \\ 0 & 0 & 0 & 1 \end{bmatrix} \begin{bmatrix} -1 & 0 & 0 & 4 \\ 0 & 0 & 1 & 0 \\ 0 & 1 & 0 & 0 \\ 0 & 0 & 0 & 1 \end{bmatrix} = \overset{\begin{matrix} X_C & Y_C & Z_C \end{matrix}}{\begin{bmatrix} 0 & 0 & -1 & 3 \\ -1 & 0 & 0 & 4 \\ 0 & 1 & 0 & 0 \\ 0 & 0 & 0 & 1 \end{bmatrix}} \begin{matrix} X_A \\ Y_A \\ Z_A \\ \; \end{matrix}. \qquad \text{(E3.7.3)}$$

Observation: $H_A{}^C$ shows that

(i) Frame C's X-axis is opposite to frame A's Y-axis

(ii) Frame C's Y-axis is in the direction of frame A's Z-axis

(iii) Frame C's Z-axis is opposite to frame A's X-axis

(iv) The origin of frame C is away from origin of A frame by 3 units along frame A's X-axis, 4 units along frame A's Y-axis and 0 units along frame A's Z-axis.

Example 3.7 also illustrates a part of robot structural definition. A robot structure for which the coordinate frames of Figure E3.7.1 can be fit is presented in Figure 3.6. A, B and C are the coordinate frames representing each joint. Only Z-axis is shown in each joint and the other axes are as given in Figure E3.7.1.

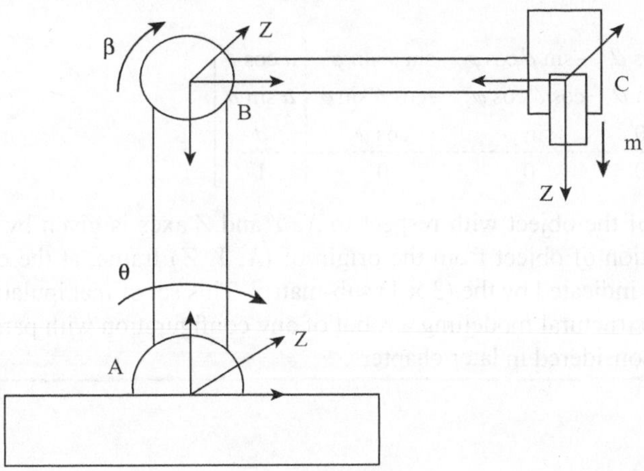

Figure 3.6 A Three-joint, Two DoF SCARA Robot

The students can try to enter the other two axes. The *Z*-axis represents the direction or axis of motion of a joint. Joint A has a rotation θ represented by the direction of *Z* of frame A, at joint B the rotation β is represented by the direction of *Z* of frame B and joint C has a translation *m* represented by the direction of *Z* of frame C. Figure 3.6 shows a three-joint, two DoF SCARA robot.

Example 3.7 has illustrated the procedure for determining the transformation matrix from a reference frame (*R*) to a new frame (*N*), that is, $H_R^{\ N}$.

If there is a chain of frames in the sequence, such as *M*, *P* and *Q*, appear in between frame *R* and frame *N*, then the matrix $H_R^{\ N}$ is obtained through consecutive transformation matrices as

$$H_R^{\ N} = H_R^{\ M} H_M^{\ P} H_P^{\ Q} H_Q^{\ N}. \tag{3.18}$$

EXAMPLE 3.8

Determine the resultant transformation matrix that represents a set of following manipulations on an object with respect to a reference frame (*X*, *Y*, *Z*):

(i) Rotation of $\varphi°$ about the *X*-axis,

(ii) Translation of *a* units along the *X*-axis

(iii) Translation of *d* units along the *Z*-axis

(iv) Rotation of $\theta°$ about the *Z*-axis.

What are the position and orientation of the object after the completion of all manipulations?

Solution:

The resultant transformation matrix is the product of transformation matrices of individual manipulations.

$$H = \text{Rot}(Z, \theta)\,\text{Trans}(0, 0, d)\,\text{Trans}(a, 0, 0)\,\text{Rot}(X, \varphi) = H_4 H_3 H_2 H_1, \tag{E3.8.1}$$

where H_i is the transformation matrix of *i*th manipulation.

$$H = \begin{bmatrix} \cos\theta & -\sin\theta & 0 & 0 \\ \sin\theta & \cos\theta & 0 & 0 \\ 0 & 0 & 1 & 0 \\ 0 & 0 & 0 & 1 \end{bmatrix} \begin{bmatrix} 1 & 0 & 0 & 0 \\ 0 & 1 & 0 & 0 \\ 0 & 0 & 1 & d \\ 0 & 0 & 0 & 1 \end{bmatrix} \begin{bmatrix} 1 & 0 & 0 & a \\ 0 & 1 & 0 & 0 \\ 0 & 0 & 1 & 0 \\ 0 & 0 & 0 & 1 \end{bmatrix} \begin{bmatrix} 1 & 0 & 0 & 0 \\ 0 & \cos\varphi & -\sin\varphi & 0 \\ 0 & \sin\varphi & \cos\varphi & 0 \\ 0 & 0 & 0 & 1 \end{bmatrix}$$

$$H = \left[\begin{array}{ccc:c} \cos\theta & -\sin\theta\cos\varphi & \sin\theta\sin\varphi & a\cos\theta \\ \sin\theta & \cos\theta\cos\varphi & -\cos\theta\sin\varphi & a\sin\theta \\ 0 & \sin\varphi & \cos\varphi & d \\ \hdashline 0 & 0 & 0 & 1 \end{array}\right]. \tag{E3.8.2}$$

The orientation of the object with respect to *X*, *Y* and *Z* axes is given by the (3 × 3) sub-matrix. The position of object from the origin of (*X*, *Y*, *Z*) frame, at the completion of all manipulations, is indicated by the (3 × 1) sub-matrix. This set of manipulations is important in the context of structural modelling a robot of any configuration with parameters $\{\theta, \varphi, a, d\}$. This will be considered in later chapters.

3.5 INVERSE TRANSFORMATION

H_A^B represents the forward transformation matrix that defines frame B with respect to frame A. Then H_B^A is known as inverse transformation matrix which defines frame A with respect to frame B. It is obvious that H_A^B and H_B^A are inverse to each other.

The transformation matrix

$$H_A^B = \begin{array}{c} \begin{matrix} X_B & Y_B & Z_B & \end{matrix} \\ \begin{bmatrix} 0 & -1 & 0 & 3 \\ 1 & 0 & 0 & 0 \\ 0 & 0 & 1 & 0 \\ 0 & 0 & 0 & 1 \end{bmatrix} \begin{matrix} X_A \\ Y_A \\ Z_A \\ \\ \end{matrix} \end{array}$$

is inverse of

$$H_B^A = [H_A^B]^{-1} = \begin{array}{c} \begin{matrix} X_A & Y_A & Z_A & \end{matrix} \\ \begin{bmatrix} 0 & 1 & 0 & 0 \\ -1 & 0 & 0 & 3 \\ 0 & 0 & 1 & 0 \\ 0 & 0 & 0 & 1 \end{bmatrix} \begin{matrix} X_B \\ Y_B \\ Z_B \\ \\ \end{matrix} \end{array}. \tag{3.19}$$

It can be verified that $H_A^B H_B^A = I$, the identity matrix.

In general, getting an inverse of a (4×4) matrix from first principles demands a good level of computations. However, we wish to have less computations in real time. The transformation of coordinates is performed by the robot computer and hence requires less computations to meet real-time control of robot arm. A simpler approach in getting the inverse is described in the following. This approach is suitable only when the coordinate axes are ortho-normal (i.e. 90°) to each other.

Given a transformation matrix H of the form

$$H = \begin{bmatrix} n_x & o_x & a_x & p_x \\ n_y & o_y & a_y & p_y \\ n_z & o_z & a_z & p_z \\ 0 & 0 & 0 & 1 \end{bmatrix}, \tag{3.20}$$

then the inverse transformation of H denoted by H^{-1} is defined as

$$H^{-1} = \begin{bmatrix} n_x & n_y & n_z & -(p \cdot n) \\ o_x & o_y & o_z & -(p \cdot o) \\ a_x & a_y & a_z & -(p \cdot a) \\ 0 & 0 & 0 & 1 \end{bmatrix}. \tag{3.21}$$

Here in the last column

vector $n = [n_x \, n_y \, n_z]^T$, normal vector

$o = [o_x \, o_y \, o_z]^T$, orientation vector

$a = [a_x \, a_y \, a_z]^T$, approach vector and

$p = p_x + p_y + p_z$, positional vector.

Then

$$(p{\cdot}n) = (p_x n_x + p_y n_y + p_z n_z),$$
$$(p{\cdot}o) = (p_x o_x + p_y o_y + p_z o_z),$$
$$(p{\cdot}a) = (p_x a_x + p_y a_y + p_z a_z).$$

Also from Equations (3.20) and (3.21), $H H^{-1} = I$, the identity matrix.

Since all three axes are ortho-normal to each other, the dot product $(n{\cdot}o) = (o{\cdot}a) = (a{\cdot}n) = 0$. The reader can verify this concept with any of the transformation matrix he/she develops.

It is easily accepted that the inverse transformation H^{-1} is to undo the operation accomplished by the transformation H.

3.5.1 Kinematic Chain

Kinematics is the study of motion of a robot without regard to forces or other factors that influence the motion. An industrial robot consists of a group of rigid bodies called as the links, connected together by joints. The joint can be revolute or prismatic. The revolute joints allow rotations of one link to another while the prismatic joints allow translation between links. The links are interconnected such that they are constrained to move relative to one another to position the end effecter. Due to the joint motions, each joint is considered as having a coordinate system or frame. With possibility of joint motions and the link lengths, each coordinate frame is related to one another. These relations in terms of

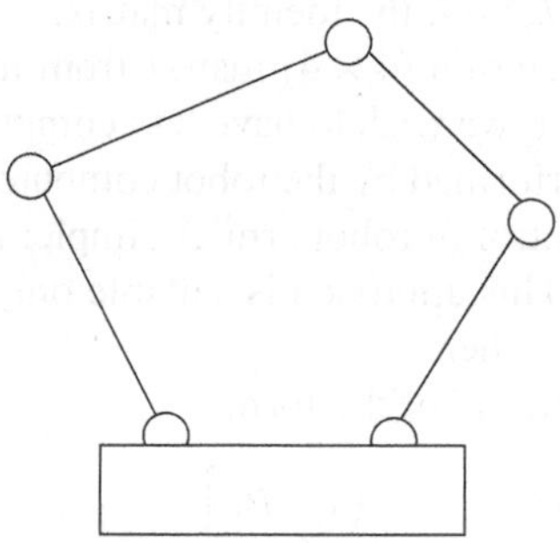

(a)

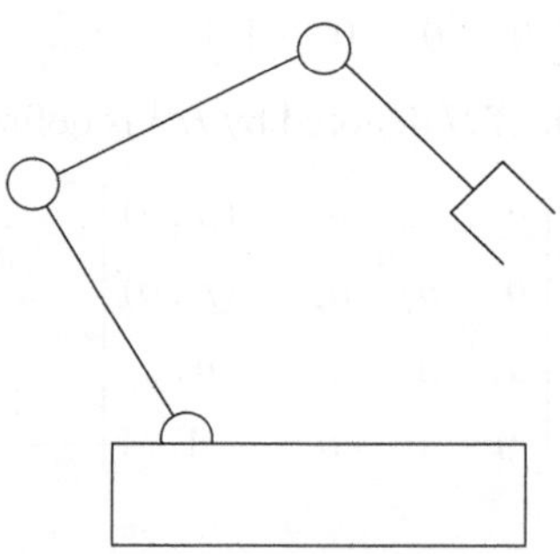

(b)

Figure 3.7 Kinematic Chains: (a) Closed Kinematic Chain and (b) Open Kinematic Chain

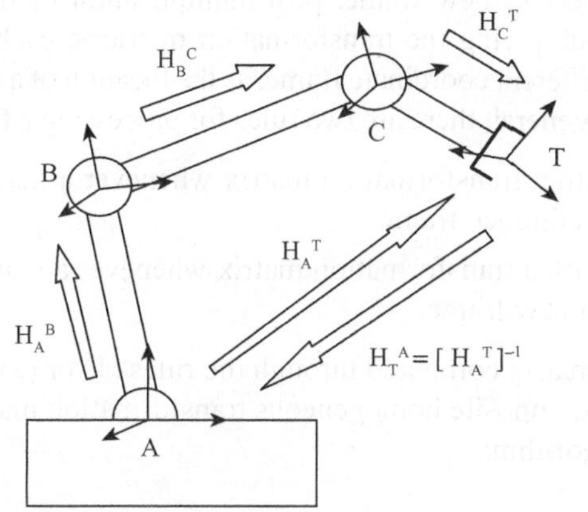

Figure 3.8 Robot and its Kinematic Chain

coordinate frames in a sequence provide the kinematic chain. The kinematic chains are of two types – closed kinematic chain and the open kinematic chain – as described in Figure 3.7.

In the closed kinematic chain, every link has two joints. All links are rigid and hence the joint movements are restricted. An end effector attached to an extended link will not produce a reasonably large work volume. On the other hand, in the open kinematic chain all links have two joints except the last link at which an end effecter is attached. This is a conventional robot having a larger work volume.

A conventional robot is depicted in Figure 3.8. This has three joints, A, B and C. Each joint is represented by its coordinate frame. The last link has an end effector. The point T is known as the tool centre point (TCP). The transformation matrix H connects the consecutive joints.

It is known that $H_A{}^T = H_A{}^B\, H_B{}^C\, H_C{}^T$. The kinematic chain closes via $H_T{}^A = [H_A{}^T]^{-1}$. That is, the product of all transformation matrices, $H_A{}^B\, H_B{}^C\, H_C{}^T\, H_T{}^A = H_A{}^A = I$, the identity matrix, thus establishing a closed kinematic chain.

3.6 COMPOSITE TRANSFORMATION MATRIX

In our discussions so far, we have considered translation and rotation with respect to one reference frame, the (X, Y, Z). This frame can be referred to as the absolute reference frame, R. Such a transformation with respect to R is known as absolute or forward transformation. We have shown that when the transformation matrices are pre-multiplied, the transformations are performed with respect to R. In contrast, post-multiplications may lead to wrong result.

In some manipulations, it is beneficial to keep another frame derived from R, as an intermediate reference frame. For example, while working with the end effector it will be advantageous to refer the manipulations of gripper with respect to wrist frame for which the transformation matrix has been already derived from the absolute reference frame. To cite another example, it is an usual practice to have the gripper frame as the reference while defining the location and orientation of an object. In such cases, when the transformation is considered with respect to new frame, post-multiplication of transformation matrix is essential. When post-multiplying the transformation matrices, each transformation is executed with respect to a different coordinate frame, as the location of a new frame changes with each transformation. In general, there are two rules for processing a transformation equation:

(1) Pre-multiply with a transformation matrix whenever a manipulation is performed with respect to reference frame.

(2) Post-multiply with a transformation matrix whenever a manipulation is performed with respect to a new frame.

Such a transformation matrix computed through the rules (1) or (2) or both for all manipulations is known as the composite homogeneous transformation matrix. This matrix can be obtained through an algorithm:

3.6.1 Algorithm for Composite Transformation

Let F be the fixed (original or absolute or reference) coordinate frame, and N be the new (transformed) coordinate frame.

Then

(a) Initialize the transformation matrix $H = I$, the identity matrix, which means F and N are coincident.

(b) Get a new transformation matrix, N, to represent a translation or a rotation using the fundamental transformation matrices.

(c) If N is to be rotated about or translated along the axis of F, pre-multiply H with F to get a modified H.

(d) If N is to be rotated about or translated along one of its own axes, post-multiply H with N to get a modified H.

(e) If there are more fundamental rotations or translations to be performed, go to step (c); else stop.

(f) The resulting composite homogeneous transformation matrix H maps N on to F.

Through the following two examples, we can establish the concept of composite transformation matrices.

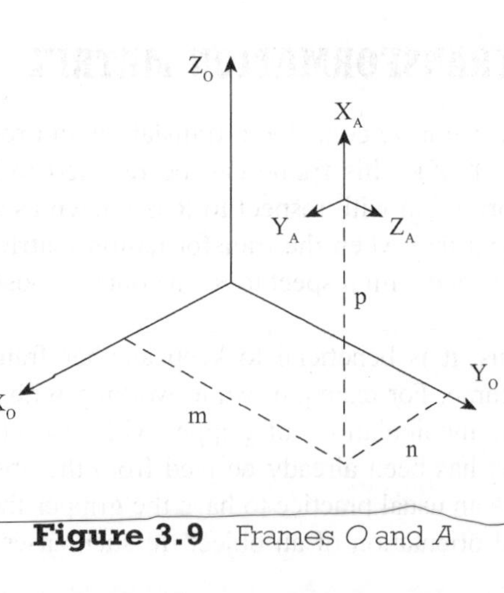

Figure 3.9 Frames O and A

There are two frames O and A as given in Figure 3.9. It is required to derive the transformation matrix that relates frame A with frame O. Two approaches are possible: (i) through set of translations and rotations pertaining to the axes of frame O only and (ii) through a set of translations and rotations pertaining to axes of new frames consecutively generated.

EXAMPLE 3.9

Manipulations with respect to frame O to get frame A (Figure E3.9.1):

Rule 1

(i) Rot(x, –90°) frame O to get frame 1, then
(ii) Rot(y, –90°) frame 1 to get frame 2, then
(iii) Trans(n, m, p) frame 2 with respect to frame O to get frame A.

Equation (E3.9.1) illustrates the steps of obtaining the resultant transformation matrix.

The resultant transformation matrix is obtained by pre-multiplications of individual transformation matrices.

$$H_O^{A} = \underset{\text{(iii)}}{\begin{bmatrix} 1 & 0 & 0 & n \\ 0 & 1 & 0 & m \\ 0 & 0 & 1 & p \\ 0 & 0 & 0 & 1 \end{bmatrix}} \underset{\text{(ii)}}{\begin{bmatrix} 0 & 0 & -1 & 0 \\ 0 & 1 & 0 & 0 \\ 1 & 0 & 0 & 0 \\ 0 & 0 & 0 & 1 \end{bmatrix}} \underset{\text{(i)}}{\begin{bmatrix} 1 & 0 & 0 & 0 \\ 0 & 0 & 1 & 0 \\ 0 & -1 & 0 & 0 \\ 0 & 0 & 0 & 1 \end{bmatrix}}$$

$$= \begin{bmatrix} 0 & 1 & 0 & n \\ 0 & 0 & 1 & m \\ 1 & 0 & 0 & p \\ 0 & 0 & 0 & 1 \end{bmatrix}. \tag{E3.9.1}$$

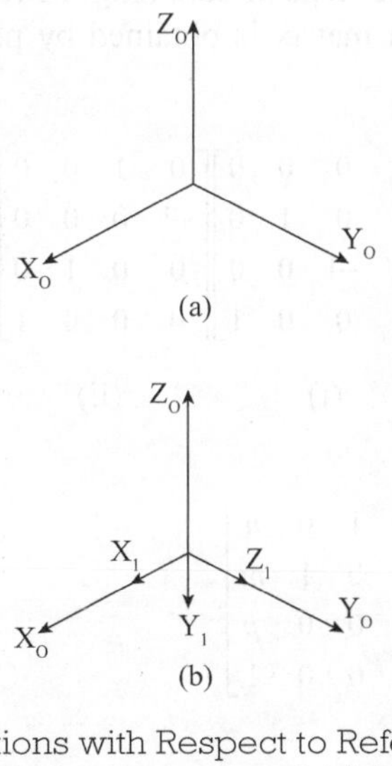

Figure E3.9.1 Manipulations with Respect to Reference Frame O: (a) Frame O, (b) Frame 1, (c) Frame 2 and (d) Frame A

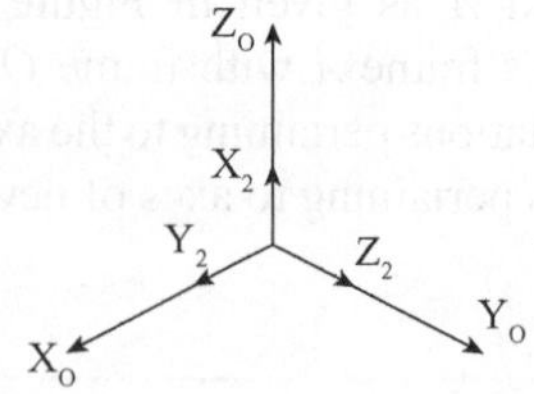

(c)

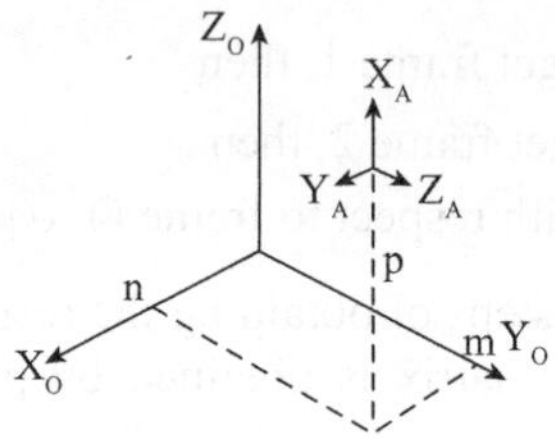

(d)

Figure E3.9.1 Continued

EXAMPLE 3.10

Manipulations with respect to the new frames (Figure E3.10.1).

Rule 2

(i) Rot(x_o, −90°) frame O to get frame P, then
(ii) Rot(z_o, −90°) to get frame Q, then
(iii) Trans(p, n, m) along (X_Q, Y_Q, Z_Q) to get frame A.

Equation (E3.10.1) illustrates the steps of obtaining the resultant transformation matrix.

The resultant transformation matrix is obtained by post-multiplications of individual transformation matrices.

$$H_O^{\ A} = \underset{\text{(i)}}{\begin{bmatrix} 1 & 0 & 0 & 0 \\ 0 & 0 & 1 & 0 \\ 0 & -1 & 0 & 0 \\ 0 & 0 & 0 & 1 \end{bmatrix}} \underset{\text{(ii)}}{\begin{bmatrix} 0 & 1 & 0 & 0 \\ -1 & 0 & 0 & 0 \\ 0 & 0 & 1 & 0 \\ 0 & 0 & 0 & 1 \end{bmatrix}} \underset{\text{(iii)}}{\begin{bmatrix} 1 & 0 & 0 & p \\ 0 & 1 & 0 & n \\ 0 & 0 & 1 & m \\ 0 & 0 & 0 & 1 \end{bmatrix}}$$

$$= \begin{bmatrix} 0 & 1 & 0 & n \\ 0 & 0 & 1 & m \\ 1 & 0 & 0 & p \\ 0 & 0 & 0 & 1 \end{bmatrix}. \tag{E3.10.1}$$

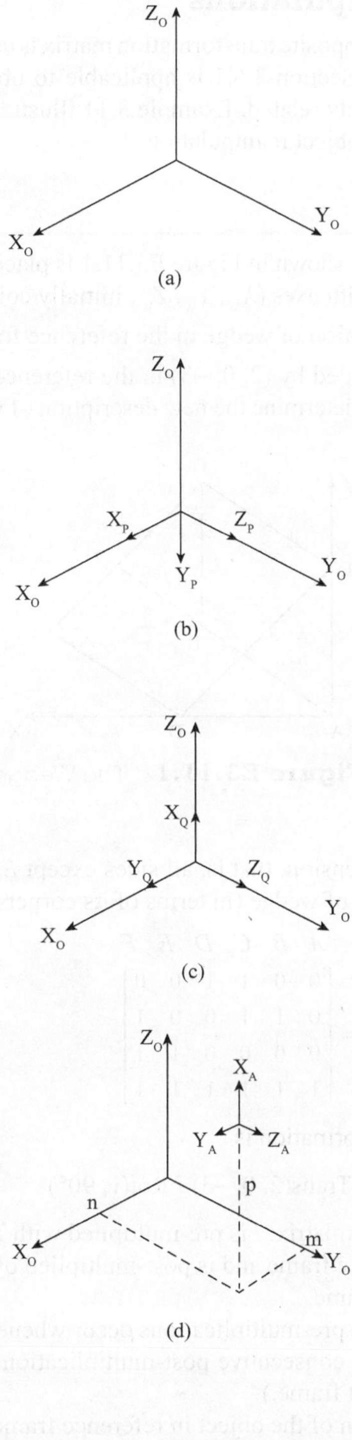

Figure E3.10.1 Manipulations with Respect to Consecutive Frames: (a) Frame O, (b) Frame P, (c) Frame Q and (d) Frame A

3.6.2 Object Manipulations

We shall examine how the composite transformation matrix is useful in object manipulations. The algorithm described in Section 3.6.1 is applicable to object manipulations in which Rules 1 and 2 are appropriately related. Example 3.11 illustrates how the composite transformation is developed into object manipulation.

EXAMPLE 3.11

The wedge of unit dimension shown in Figure E3.11.1 is placed in a reference frame O. Its (or object frame) frame W with axes (X_W, Y_W, Z_W) initially coincides with frame O.

(a) determine the description of wedge in the reference frame.

(b) If the wedge is translated by (2, 0, −3) in the reference frame and is rotated by 90° about its own Y-axis, determine the new description of wedge in the reference frame.

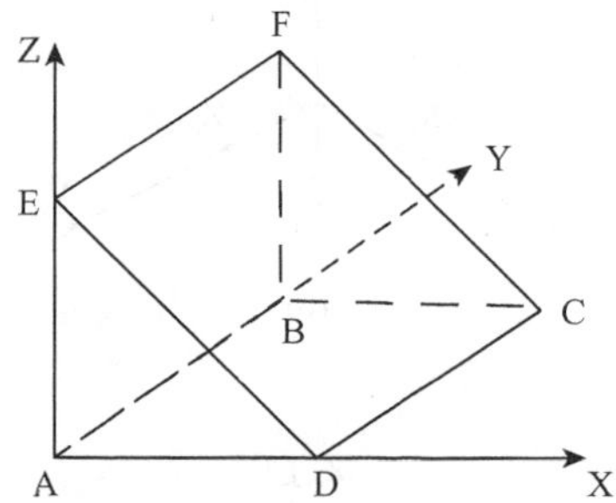

Figure E3.11.1 The Wedge

Solution:

(a) Wedge is of unit dimension, that is, all sides except ED and FC are of unit length; hence the description of wedge (in terms of its corners) in its original place is

$$W_O = \begin{array}{c} \begin{matrix} A & B & C & D & E & F \end{matrix} \\ \begin{bmatrix} 0 & 0 & 1 & 1 & 0 & 0 \\ 0 & 1 & 1 & 0 & 0 & 1 \\ 0 & 0 & 0 & 0 & 1 & 1 \\ 1 & 1 & 1 & 1 & 1 & 1 \end{bmatrix} \end{array}. \quad \text{(E3.11.1)}$$

(b) The composite transformation is

$$H = \text{Trans}(2, 0, -3)\ I\ \text{Rot}(y, 90°) \quad \text{(E3.11.2)}$$

where I is the identity matrix. I is pre-multiplied with Trans(2, 0, −3) due to manipulation in the reference frame and is post-multiplied by Rot(y, 90°) due to manipulation in the object frame.

(Note: Consecutive pre-multiplications occur whenever manipulations happen in reference frame; also consecutive post-multiplications occur whenever manipulations happen in object frame.)

The new description of the object in reference frame is

$$W_{new} = \text{Trans}(2, 0, -3)\ \text{Rot}(y, 90°)\ W_o$$

$$= \begin{bmatrix} 1 & 0 & 0 & 2 \\ 0 & 1 & 0 & 0 \\ 0 & 0 & 1 & -3 \\ 0 & 0 & 0 & 1 \end{bmatrix} \begin{bmatrix} 0 & 0 & 1 & 0 \\ 0 & 1 & 0 & 0 \\ -1 & 0 & 0 & 0 \\ 0 & 0 & 0 & 1 \end{bmatrix} \begin{bmatrix} 0 & 0 & 1 & 1 & 0 & 0 \\ 0 & 1 & 1 & 0 & 0 & 1 \\ 0 & 0 & 0 & 0 & 1 & 1 \\ 1 & 1 & 1 & 1 & 1 & 1 \end{bmatrix}$$

$$\begin{matrix} A & B & C & D & E & F \end{matrix}$$

$$= \begin{bmatrix} 2 & 2 & 2 & 2 & 3 & 3 \\ 0 & 1 & 1 & 0 & 0 & 1 \\ -3 & -3 & 4 & -4 & -3 & -3 \\ 1 & 1 & 1 & 1 & 1 & 1 \end{bmatrix}. \qquad \text{(E3.11.3)}$$

3.7 THE WRIST

Position and orientation are required in manipulating an object. Typically, the arm (with joints and links) of the robot controls the position of the end effector in space and the wrist controls the fingers to final orientation. The end effector has two names:

1. Gripper which grasps the object by any means
2. Tool which undertakes a processing operation.

With this definition, drilling a hole is a process of operation and grasping a cement bag by gripper is another process.

To get a six DoF, three DoF can be achieved by translation of the arms and another three DoF can be created by the rotation of the end effector. However, the functions of position and orientation cannot be fully decoupled to be separate functions of the arm and the wrist. Our palm and fingers can form the end effector. Prismatic joints that give translations are not normally found in human wrists.

It is complex to design a complete three-DoF wrist and such wrists are useful only in special applications. Figure 3.10 shows a typical design of a wrist having three DoF. However, several designs are available from different manufacturers.

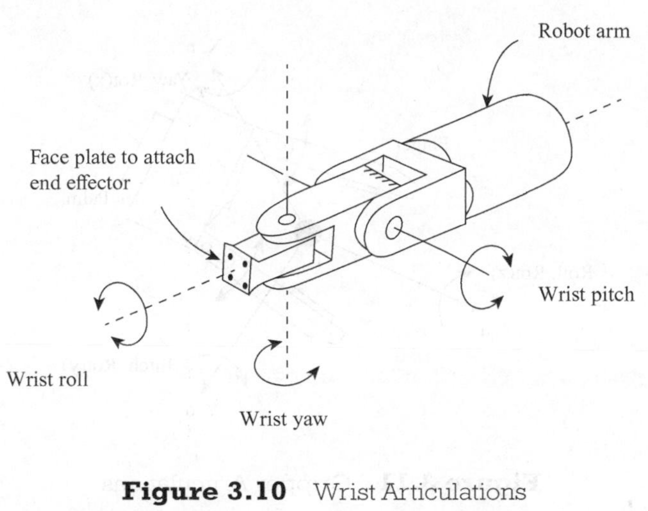

Figure 3.10 Wrist Articulations

The wrist has three rotations, the roll, the yaw and the pitch such that

(a) roll involves rotations of the wrist mechanism about the arm axis
(b) pitch involves the up and down rotations of the wrist
(c) yaw involves the right side or left side rotations

Figure 3.11 illustrates the palm and fingers (the gripper) attached at the face plate of wrist. The articulations of wrist allows the gripper to have three DoF–the roll, the yaw and the pitch. The gripper axes (*X*, *Y*, *Z*) are shown in the two finger form of wrist as in Figure 3.11. Its articulations are defined as follows:

(a) Roll is the rotation of the *Z*-axis that is directed from arm axis.
(b) The *Y*-axis is normally considered as an axis going through the fingers. A rotation about the *Y*-axis is known as pitch.
(c) The *X*-axis is perpendicular to both *Z* and *Y* axes. A rotation about the *X*-axis is called as yaw. In Figure 3.11, the *X*-axis is directed towards you.
(d) The origin of the gripper frame, T, is referred to as the TCP.

An interesting aspect is to be observed here. When the gripper is mounted to the end effector, it appears that the TCP (marked as T in Figure 3.11) will be changing its position whenever the wrist articulates. It is true unless compensated by extra articulations of arms. That is, when we command the robot to bring the TCP to a position in the 3D space, the robot joints adjust within themselves in such a way that the TCP reaches the defined position. Such adjustments in robot joints are possible in some of robot softwares so that we are guaranteed that the TCP reaches the point we define.

In some cases, the axis directions are also referred to as the following:

X-axis is known as '*n*', the normal axis (normal to *Y* and *X* axes).
Y-axis is known as '*o*', the orientation axis (axis goes through the fingers)
Z-axis is known as '*a*', the approach axis (axis approaching an object).

However, the fixing of TCP and the directions of vectors *n*, *o* and *a* depend on the geometry of the gripper.

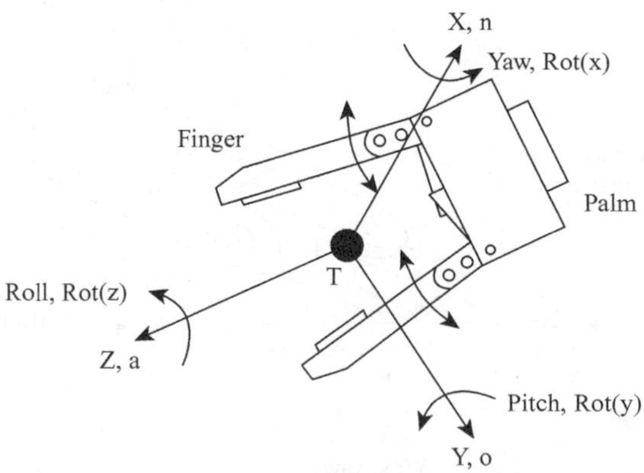

Figure 3.11 Gripper Articulations

EXAMPLE 3.12

Figure E3.12.1 describes a robot vision system. The robot is controlled on the basis of information received by the camera. It is required that robot has to align its gripper so that the object is properly gripped by the gripper. Two transformation matrices are given as follows:

$$T_C^O = \begin{bmatrix} 0 & 1 & 0 & 1 \\ 1 & 0 & 0 & 10 \\ 0 & 0 & -1 & 9 \\ 0 & 0 & 0 & 1 \end{bmatrix}; \; T_C^B = \begin{bmatrix} 1 & 0 & 0 & -10 \\ 0 & -1 & 0 & 20 \\ 0 & 0 & -1 & 10 \\ 0 & 0 & 0 & 1 \end{bmatrix}, \tag{E3.12.1}$$

where C, O and B stand for coordinate frames at the camera, object and base, respectively.

Then,

(a) What is the position of the centre of object with respect to the base coordinate frame?

(b) What is the orientation matrix if the robot has to pick-up the object from its top?

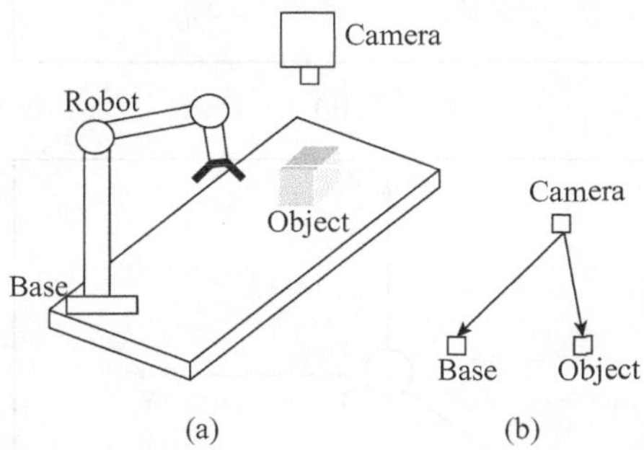

Figure E3.12.1 Robot Vision System: (a) Robot Workspace and (b) Transformations

Solution:

Let us first observe how the transformation matrices given in Equation (E3.12.1) relate to the respective coordinate axes:

$$T_C^O = \begin{array}{c} \begin{matrix} X_o & Y_o & Z_o & \end{matrix} \\ \begin{bmatrix} 0 & 1 & 0 & 1 \\ 1 & 0 & 0 & 10 \\ 0 & 0 & -1 & 9 \\ 0 & 0 & 0 & 1 \end{bmatrix} \begin{matrix} X_C \\ Y_C \\ Z_C \\ \\ \end{matrix} \end{array}. \tag{E3.12.2}$$

The origin of O is away by $[1, 10, 9]^T$ from the origin of C.

$$T_C^B = \begin{array}{c} \begin{matrix} X_B & Y_B & Z_B & \end{matrix} \\ \begin{bmatrix} 1 & 0 & 0 & -10 \\ 0 & -1 & 0 & 20 \\ 0 & 0 & -1 & 10 \\ 0 & 0 & 0 & 1 \end{bmatrix} \begin{matrix} X_C \\ Y_C \\ Z_C \\ \\ \end{matrix} \end{array}. \tag{E3.12.3}$$

The origin of B is away by $[-10, 20, 10]^T$ from the origin of C.

The above descriptions help us to illustrate the C, O and B frames as in Figure E3.12.2 as

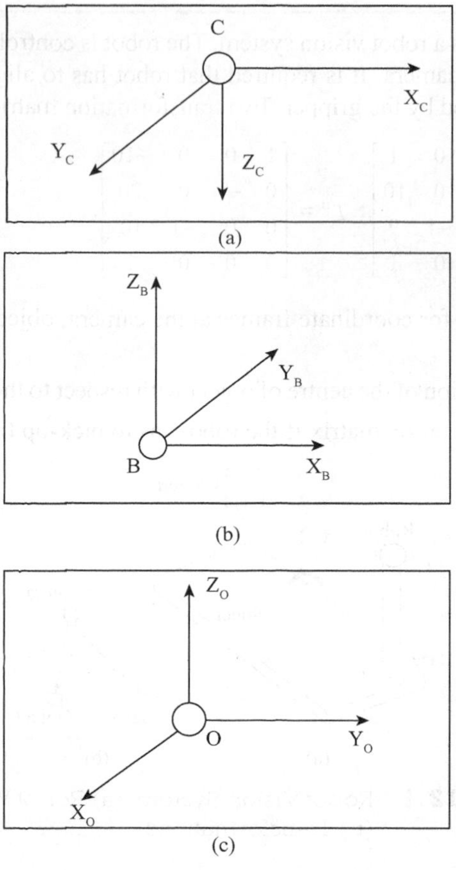

Figure E3.12.2 Base, Camera and Object Frames

(a) We have to find out the transformation of O from B, that is, T_{B}^{O} matrix. This can be derived as

$$T_{\text{B}}^{\text{O}} = T_{\text{B}}^{\text{C}} \; T_{\text{C}}^{\text{O}} = [\, T_{\text{C}}^{\text{B}} \,]^{-1} \; T_{\text{C}}^{\text{O}} = \text{inv}\begin{bmatrix} 1 & 0 & 0 & -10 \\ 0 & -1 & 0 & 20 \\ 0 & 0 & -1 & 10 \\ 0 & 0 & 0 & 1 \end{bmatrix} \begin{bmatrix} 0 & 1 & 0 & 1 \\ 1 & 0 & 0 & 10 \\ 0 & 0 & -1 & 9 \\ 0 & 0 & 0 & 1 \end{bmatrix},$$

where inv[·] is the inverse of matrix; hence

$$T_{\text{B}}^{\text{O}} = \begin{array}{c} \begin{matrix} X_{\text{o}} & Y_{\text{o}} & Z_{\text{o}} & \end{matrix} \\ \begin{bmatrix} 0 & 1 & 0 & 11 \\ -1 & 0 & 0 & 10 \\ 0 & 0 & 1 & 1 \\ 0 & 0 & 0 & 1 \end{bmatrix} \end{array} \begin{matrix} X_{\text{B}} \\ Y_{\text{B}} \\ Z_{\text{B}} \\ \\ \end{matrix}. \tag{E3.12.4}$$

The position of origin of frame O is at $[11, 10, 1]^{\text{T}}$ units from the origin of frame B.

(b) If the robot has to pick up the object from its top, there are two possibilities as indicated in Figure E3.12.3. In both cases, the object orientation is fixed at the frame (X_O, Y_O, Z_O).

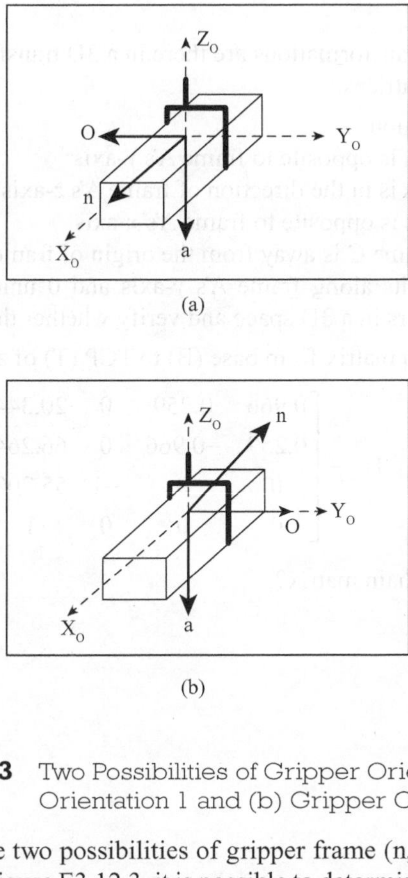

(b)

Figure E3.12.3 Two Possibilities of Gripper Orientations: (a) Gripper Orientation 1 and (b) Gripper Orientation 2

However, there are two possibilities of gripper frame (n, o, a) by which it can grip the object. From Figure E3.12.3, it is possible to determine the relationship between the object frame and the gripper frame.

<u>Possibility 1</u>

$$\begin{matrix} X_o & Y_o & Z_o \end{matrix}$$
$$\begin{bmatrix} 1 & 0 & 0 \\ 0 & -1 & 0 \\ 0 & 0 & -1 \end{bmatrix} \begin{matrix} n \\ o, \\ a \end{matrix}$$

<u>Possibility 2</u>

$$\begin{matrix} X_o & Y_o & Z_o \end{matrix}$$
$$\begin{bmatrix} -1 & 0 & 0 \\ 0 & 1 & 0 \\ 0 & 0 & -1 \end{bmatrix} \begin{matrix} n \\ o. \\ a \end{matrix}$$

REVIEW QUESTIONS

1. What is coordinate transformation? Why is it required?
2. What are the two parts when a robot is to handle an object?
3. What is 2D homogeneous transformation? Why or why not is this sufficient?

4. In 2D transformation space, how do you describe any object? Are there any restrictions?
5. How many DoF does the object undergoes in a 3D transformation space? Illustrate through a diagram.
6. How many basic transformations are there in a 3D transformation space? Illustrate through various matrices.
7. After a transformation
   (i) Frame C's x-axis is opposite to frame A's y-axis
   (ii) Frame C's y-axis is in the direction of frame A's z-axis
   (iii) Frame C's z-axis is opposite to frame A's x-axis
   (iv) The origin of frame C is away from the origin of frame A by 3 units along frame A's x-axis, 4 units along frame A's y-axis and 0 units along frame A's z-axis. Sketch the vectors in a 3D space and verify whether the above rules satisfy.
8. The transformation matrix from base (B) to TCP (T) of a robot is

$$H_B^T = \begin{bmatrix} 0.966 & 0.259 & 0 & 20.344 \\ 0.259 & -0.966 & 0 & 66.264 \\ 0 & 0 & -1 & 55.700 \\ 0 & 0 & 0 & 1 \end{bmatrix}.$$

What is the closing chain matrix?

KINEMATICS 4

The danger of the past was that men became slaves.
The danger of the future is that men may become robots.
Erich Fromm

This chapter covers the following key topics:

Introduction – Joint Coordinate Space – Kinematics and Inverse Kinematics – A Two-Joint Two-DoF Robot – Use of Homogeneous Transformations – Robot Vision System – Link Parameters – Joint Parameters – D-H Notation of Coordinate Frames – D-H Transformation Matrix – Symbolic Procedure – D-H Algorithm – Application Examples – Jacobian –Manipulator Jacobian – Jacobian Singularities

---

4.1 INTRODUCTION

In Chapter 3, we have described Cartesian coordinate systems in a three-dimensional space. Each robot joint is considered as having its own coordinate frame. A joint is indicated by the axis of rotation or axis of translation. These coordinate frames do not have same orientation. We have described the transformation matrices for connecting the joint coordinate frames. We have also derived the homogeneous transformation matrices. The transformation matrices give the orientation and the location of the next frame with respect to the frame we are working now. In this chapter, we will discuss kinematics, which is an interesting topic in robotics. We conclude this chapter with Jacobian, a matrix which represents the transform of infinitesimal changes of joint velocities to infinitesimal changes of end-effector position and orientation.

4.2 JOINT COORDINATE SPACE

A structure of a three degrees of freedom (DoF) robot is depicted in Figure 4.1. The current robot space is known as $\{\theta_1, \theta_2, \theta_3\}$ space. This is a typical structure of a robot having links

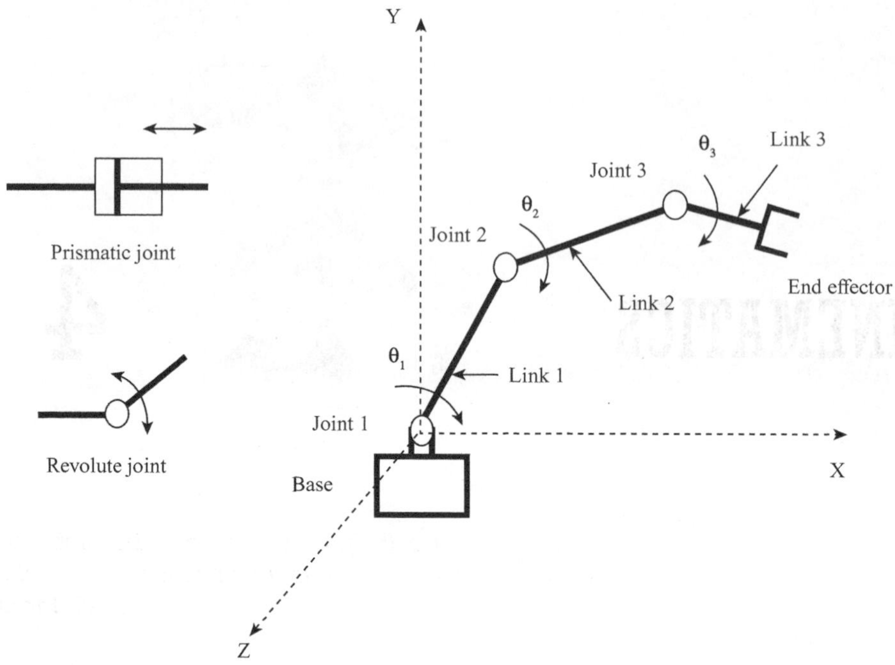

Figure 4.1 Joints and Links of a Robot

and joints. A way of identifying the links and joints is also shown. The structure has an end effector controlled by the motions of joints with respect to their individual axes. Figure 4.1 shows only rotational (revolute) joints; however, a robot may consist of translational (prismatic) joints also. The links connected to revolute joints are fixed in length and the links connected to prismatic joints are variable in length. The set $\{\theta_1, \theta_2, \theta_3\}$ represents the set of joint movements. Given the set $\{\theta_1, \theta_2, \theta_3\}$, it is possible to compute the position of gripper with the knowledge of link lengths. In other words, for a specific robot configuration, the set $\{\theta_1, \theta_2, \theta_3\}$ can be considered as representing a point in (*X*, *Y*, *Z*) space to which the gripper sets in. This is how the robot computer computes the position of gripper in real time. The set $\{\theta_1, \theta_2, \theta_3\}$ is referred to as joint coordinate frame. However, the students have a good concept of (*X*, *Y*, *Z*) coordinate frame. Let us call it world coordinate frame. Robot computer has a powerful software to convert joint coordinate frames to world coordinate frames and vice versa.

4.3 KINEMATICS AND INVERSE KINEMATICS

Let the set $\{\theta_1, \theta_2, \ldots..\theta_m\}$ represent a joint coordinate system. The value θ_i can be a rotation of a revolute joint or can be a translation of a prismatic joint. We use θ_i as a variable for both rotation (°) and translation (cm, m). Given the set of θ_i, $i = 1, 2, \ldots, m$, the computation of gripper location in the (*X*, *Y*, *Z*) space is known as direct (or forward) kinematics. On the other hand, given the gripper point as a location in the (*X*, *Y*, *Z*) space, computation of θ_i, $i = 1, 2, \ldots m$ is known as inverse kinematics. It will be shown that acquiring forward kinematics data is easier than acquiring inverse kinematics data. This can be illustrated by an example in the following section.

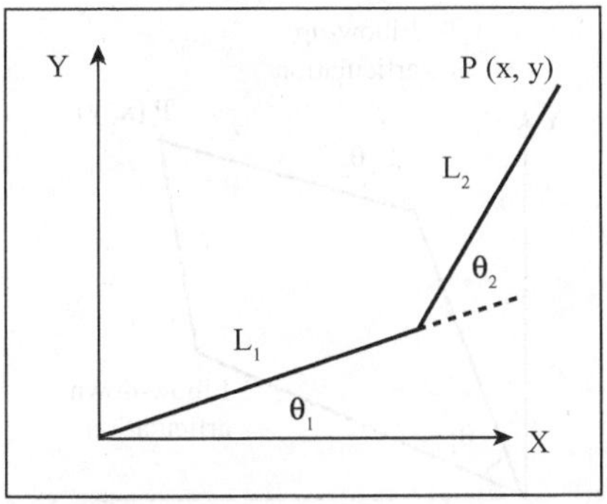

Figure 4.2 Forward Kinematics

4.3.1 A Two-Joint Two-DoF Robot

We will illustrate the forward kinematics and the inverse kinematics problems through a simple robot structure. Figure 4.2 shows a two-joint, two-dimensional (planar), two-DoF robot. The link lengths are L_1 and L_2 units. The angular positions of joint movements are θ_1 and θ_2. $P(x, y)$ is the gripper point. We use two-dimensional trigonometric equations to compute $P(x, y)$.

The forward kinematics problem is as follows:

Given the joint coordinates (θ_1, θ_2), compute the world coordinates of gripper point $P(x, y)$.

This can be solved directly by the following set of equations:

$$x = L_1 \cos \theta_1 + L_2 \cos (\theta_1 + \theta_2), \tag{4.1}$$

$$y = L_1 \sin \theta_1 + L_2 \sin (\theta_1 + \theta_2). \tag{4.2}$$

Thus, the point $P(x, y)$ can be easily computed. The solution to this forward kinematics problem is direct and unique.

The inverse kinematics problem is as follows:

Given the world coordinates of the gripper point $P(x, y)$, compute the joint coordinates (θ_1, θ_2).

The solution requires an involved process. We define two angles α and β as shown in Figure 4.3.

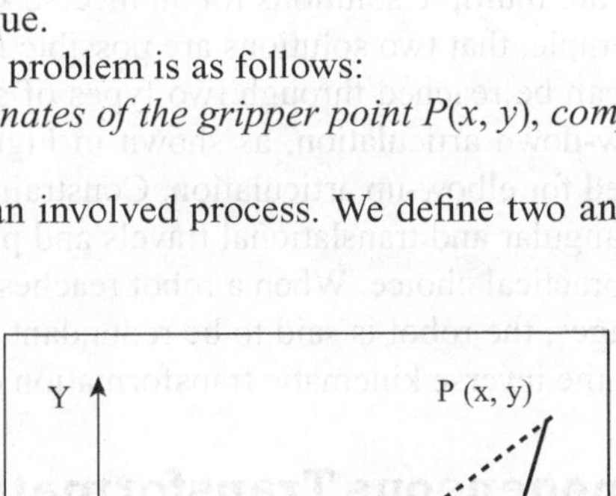

Figure 4.3 Inverse Kinematics

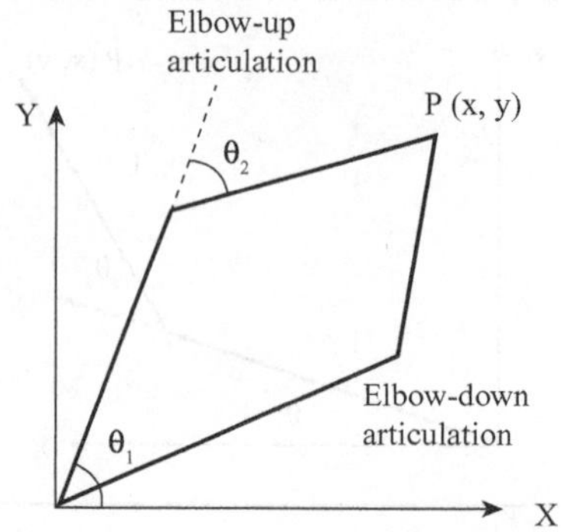

Figure 4.4 Two Ways of Reaching P(x, y)

Then, the angles θ_1 and θ_2 are computed through the following set of procedures :
Find θ_2 from

$$x^2 + y^2 = L_1^{\,2} + L_2^{\,2} + 2\,L_1\,L_2 \cos\,\theta_2. \tag{4.3}$$

Substitute θ_2 and find α from

$$\tan\,\alpha = (L_2\,\sin\,\theta_2)/(L_2\,\cos\,\theta_2 + L_2). \tag{4.4}$$

Find β from

$$\tan\,\beta = y/x \tag{4.5}$$

and then find θ_1 from

$$\theta_1 = (\beta - \alpha). \tag{4.6}$$

It is observed that several trigonometric functions are to be solved while solving inverse kinematic problems. Determining angles from trigonometric functions leads to multiple solutions and hence there are multiple solutions for an inverse kinematic problem. We can easily observe, in this example, that two solutions are possible from Equation (4.3) for θ_2. Hence, the point $P(x, y)$ can be reached through two types of articulations, the elbow-up articulation and the elbow-down articulation, as shown in Figure 4.4. In Figure 4.4, the angles (θ_1, θ_2) are indicated for elbow-up articulation. Constraints such as mechanical end stops, limitations of arm angular and translational travels and presence of obstacles in the work space can lead to a practical choice. When a robot reaches a position with more than one configuration of linkages, the robot is said to be redundant. Redundancy occurs when more than one solution to the inverse kinematic transformation exists.

4.3.2 Use of Homogeneous Transformation

We will solve the above problem using the three-dimensional homogeneous transformation matrices discussed in Chapter 3. Let a point object be placed at the origin, O, of a reference frame (X, Y, Z) (Figure 4.5). The object frame is (x, y, z). Initially the object frame (x, y, z) and the reference frame (X, Y, Z) are coincident. The object undergoes a set of following articulations to reach the gripper point, P:

(i) A rotation of θ_1° about the z-axis
(ii) A translation of L_1 units along the x-axis
(iii) A rotation of θ_2° about the z-axis
(iv) A translation of L_2 units along the x-axis

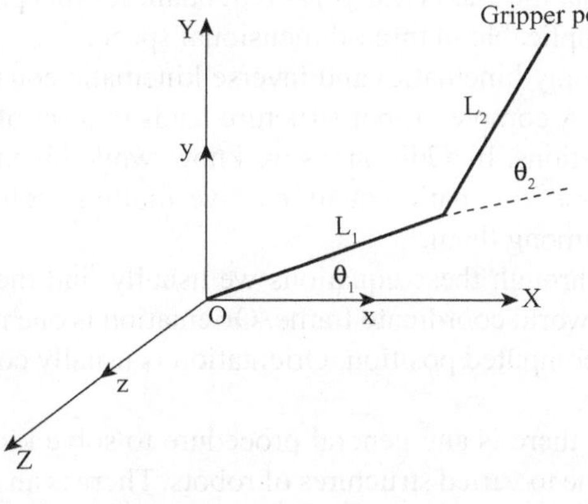

Figure 4.5 Transformations of P

Note that all transformations are performed with respect to the object frame (*x*, *y*, *z*). Hence, the fundamental transformation matrices are post-multiplied for every articulation (Rule 2, Section 3.6, page 52). The homogeneous transformation matrix from *O* to *P* is as follows:

$$H_0^{\mathrm{P}} = \mathrm{Rot}(z, \theta_1)\,\mathrm{Trans}(L_1, 0, 0)\,\mathrm{Rot}(z, \theta_2)\,\mathrm{Trans}(L_2, 0, 0) \tag{4.7}$$

$$= \begin{bmatrix} \cos\theta_1 & -\sin\theta_1 & 0 & 0 \\ \sin\theta_1 & \cos\theta_1 & 0 & 0 \\ 0 & 0 & 1 & 0 \\ 0 & 0 & 0 & 1 \end{bmatrix} \begin{bmatrix} 1 & 0 & 0 & L_1 \\ 0 & 1 & 0 & 0 \\ 0 & 0 & 1 & 0 \\ 0 & 0 & 0 & 1 \end{bmatrix} \begin{bmatrix} \cos\theta_2 & -\sin\theta_2 & 0 & 0 \\ \sin\theta_2 & \cos\theta_2 & 0 & 0 \\ 0 & 0 & 1 & 0 \\ 0 & 0 & 0 & 1 \end{bmatrix} \begin{bmatrix} 1 & 0 & 0 & L_2 \\ 0 & 1 & 0 & 0 \\ 0 & 0 & 1 & 0 \\ 0 & 0 & 0 & 1 \end{bmatrix}. \tag{4.8}$$

By simplifying we get

$$H_0^{\mathrm{P}} = \begin{bmatrix} \cos(\theta_1+\theta_2) & -\sin(\theta_1+\theta_2) & 0 & L_2\cos(\theta_1+\theta_2)+L_1\cos(\theta_1) \\ \sin(\theta_1+\theta_2) & \cos(\theta_1+\theta_2) & 0 & L_2\sin(\theta_1+\theta_2)+L_1\sin(\theta_1) \\ 0 & 0 & 1 & 0 \\ 0 & 0 & 0 & 1 \end{bmatrix}. \tag{4.9}$$

Let the gripper point $P(x, y, z)$ be given in the homogeneous vector form equal to $[p_x, p_y, p_z, 1]^{\mathrm{T}}$. Then

$$P(x, y, z) = [p_x, p_y, p_z, 1]^{\mathrm{T}} = \begin{bmatrix} L_2\cos(\theta_1+\theta_2)+L_1\cos\theta_1 \\ L_2\sin(\theta_1+\theta_2)+L_1\sin\theta_1 \\ 0 \\ 1 \end{bmatrix} = \begin{bmatrix} p_x \\ p_y \\ p_z \\ 1 \end{bmatrix}. \tag{4.10}$$

The two equations to solve kinematics and inverse kinematics problems are as follows:

$$p_x = L_1\cos\theta_1 + L_2\cos(\theta_1+\theta_2), \tag{4.11}$$

$$p_y = L_1\sin\theta_1 + L_2\sin(\theta_1+\theta_2). \tag{4.12}$$

It can be easily observed that the z-axis (i.e. p_z) is redundant for this problem; however, the procedure is general and applicable in three-dimensional space.

Direct kinematic (or simply kinematic) and inverse kinematic equations depend on the specific structure of robot. A complex robot structure leads to a set of complex kinematic and inverse kinematic equations. In addition, as we know, while kinematic equations offer unique solutions, the inverse kinematic equations give multiple solutions and the robot system has to choose one among them.

Another aspect is that through these equations we usually find the position of gripper with respect to a reference world coordinate frame. Orientation is one more aspect to detect while the gripper is at the computed position. Orientation is usually computed by the same or by some other equations.

One can check whether there is any general procedure to solve kinematic equations to circumvent complexities due to varied structures of robots. There is an approach in defining a unique set of parameters that will be suitable in defining any complex structure of robots. This offers a standard procedure in solving kinematic equations irrespective of robot structure. This will be discussed later in this chapter.

The robot continuously uses both direct and inverse kinematic equations. We mainly work in the familiar world coordinate system (X, Y, Z) and our data, such as positions of objects in work space are in world coordinate system. However, the robot has to work in joint coordinate system $\{\theta_1, \theta_2, \theta_3, \ldots, \theta_m\}$ because for every position of object, the robot computer has to compute the joint angles and instruct the joint motors to reach these joint angles. Hence, the robot computer has to be powerful and faster in computing joint angles through several complex inverse kinematic equations and in taking decisions for an acceptable solution. Another important aspect is that these computations and decision makings are to be performed in real time. The vision-controlled robot, a typical industrial robot system, is an example requiring real-time coordinate transformations.

4.3.3 Robot Vision System

Figure 4.6 describes a vision-controlled robot system. The system consists of a robot, a camera, a conveyor and a robot computer. The robot and the camera are interfaced with robot computer. The camera 'looks at' the object coming on the conveyor, acquires the

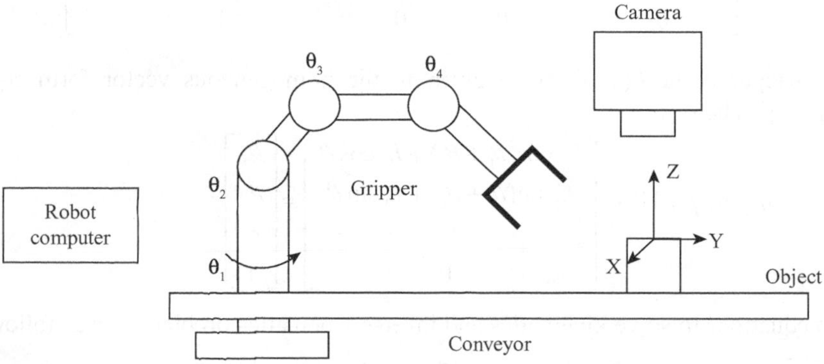

Figure 4.6 Application of Inverse Kinematics

image, processes the image and informs object position in (X, Y, Z) to the robot computer. The computer uses this world coordinate data, transforms it in to joint coordinate data $(\theta_1, \theta_2, \theta_3, \theta_4)$. It then controls the positions of the motors at the base and at the joints to their respective angles for picking-up the object. The robot picks up the object and places it in a different location (X_1, Y_1, Z_1). This also needs the solution of inverse kinematic equations to get a new set of $(\theta_1, \theta_2, \theta_3, \theta_4)$. Since the object's position is time dependent, all these efforts are to be performed in real time so that the whole process is fast enough. The robot will miss the object for picking if there is a delay in these efforts. Delays, either in image processing, world to joint coordinate transformation and also in mechanical system responses, are natural. A successful pick and place of objects is possible if the conveyor speed is adjusted to compensate such delays.

4.4 LINK PARAMETERS

The robot can be modelled as a kinematic chain of rigid links interconnected by revolute and prismatic joints. Figure 4.7 shows the links (L) and joints (J) and also their respective coordinate frames. Position and orientation of each joint is described by its coordinate frames. In a kinematic chain, the coordinate frame of a joint is described relative to the coordinate frame of the previous joint. Hence, starting from the robot base, the coordinate frame for each joint can be developed. The process of obtaining the coordinate frame of a joint requires a set of four parameters. Two of them are associated with the links and the other two with the joints. They are named as link parameters and joint parameters, respectively.

The process of deciding the link parameters is described in Figure 4.8. Note the convention of representing axis numbers, joint numbers and link numbers. Link n connects Axis $(n - 1)$ and Axis (n). Axis $(n - 1)$ is the same as Joint (n).

The link is arbitrary and can have any shape, length and twist. Owing to this, Joint n and Joint $(n + 1)$ are not necessarily in one plane. There exists a unique mutually perpendicular

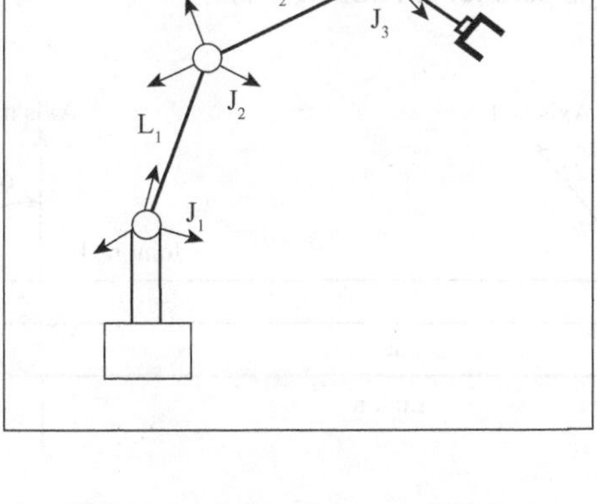

Figure 4.7 Joint and Links

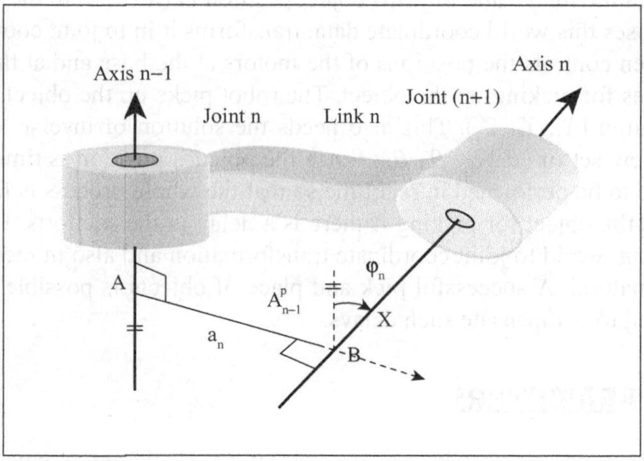

Figure 4.8 Description of Link Parameters $\{a_n, \varphi_n\}$

line *AB* to both axes. *AB* is known as the common normal between two axes. In addition, *AB* indicates the minimum distance between the axes and is known as link length (a_n).

Line $A^{P}_{(n-1)}$ is parallel to Axis (n – 1). The angle between this parallel and Axis (n) on a plane perpendicular to the common normal *AB* is known as link twist (φ_n). The effective distances *AB* of Joint n and its twist is measured by link parameters $\{a_n, \varphi_n\}$. The link parameters are constants for a given link.

The link length, a_n, measured from *A* to *B* is considered as positive; link twist φ_n is positive when measured from $A^{P}_{(n-1)}$ to Axis n. Owing to the arbitrary shape of the link either a_n or φ_n or both can be negative.

Sometimes, Axis (n – 1) and Axis n can intersect. In such case, a_n is zero. When the axes are parallel, $\varphi_n = 0$. If the joint n is prismatic, then the link length a_n is variable.

For an industrial robot, the axes are usually parallel with a_n equal to physical length of the link. However, there can be twists in the link. Figure 4.9 shows a link having a 90° twist. The directions of measurement of a_n and φ_n are also indicated. Here, a_n is the same as physical length of the link; link twist $\varphi_n = 90^\circ$ is measured from $A^{P}_{(n-1)}$ to Axis n about a_n (extended), where $A^{P}_{(n-1)}$ is parallel to Axis (n – 1).

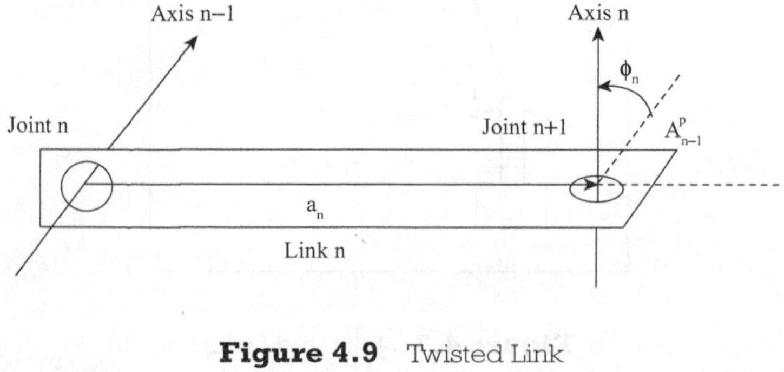

Figure 4.9 Twisted Link

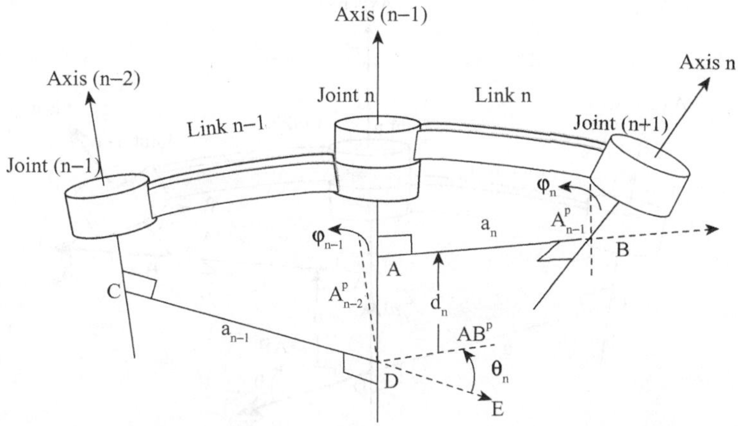

Figure 4.10 Description of Joint and Link Parameters $\{\theta_n, d_n, a_n, \varphi_n\}$

4.4.1 Joint Parameters

Figure 4.10 describes the process of obtaining rest of the parameters of the *n*th joint, that is, joint angle (θ_n) and joint distance (d_n). Figure 4.10 shows three joints, ($n - 1$), n and ($n + 1$) with their respective axes, Axis ($n - 2$), Axis ($n - 1$) and Axis (n). Joint n has two levels, upper and lower. *CD* is a common normal from Axis ($n - 2$) to Axis ($n - 1$); *AB* is a common normal from Axis ($n - 1$) to Axis n. *DE* is the extension of *CD*. The lines $A^P_{(n-2)}$ and $A^P_{(n-1)}$ are parallel to Axis ($n - 2$) and Axis ($n - 1$) respectively. In addition, AB^P is parallel to *AB*.

4.5 D-H NOTATION OF COORDINATE FRAMES

In this section, we use the set of four kinematic parameters to define the coordinate frame (X, Y, Z) of a joint. The definition is based on Denavit–Hartenberg (D-H) notation. D-H notation is followed universally for defining the coordinate frame of the joints irrespective of any complex-shaped robot links. We also follow this notation extensively herein after in this book.

All four kinematic parameters $\{\theta_n, d_n, a_n, \varphi_n\}$ for joint n and link n are described in Figure 4.10. This process is followed from the robot base, through each joint, to the gripper to get the entire set of robot parameters for all joints and for all links. This set of all parameters is used for developing a kinematic model for the robot.

We have to follow Figure 4.10 closely. We use the word 'axis' with two different meanings, one to represent the axis at a joint and the other to represent axis of a coordinate frame. To avoid misinterpretation, we use the terms such as 'Axis ( )' to represent the axis number at a joint and '( )-axis' or '( )' to represent coordinate frame axis.

In D-H notation, we define the coordinate frame (X_{n-1}, Y_{n-1}, Z_{n-1}) corresponding to Axis ($n - 1$) as follows:

(i) Point *D* is the origin of frame (X_{n-1}, Y_{n-1}, Z_{n-1})
(ii) Axis ($n - 1$) fixes Z_{n-1}

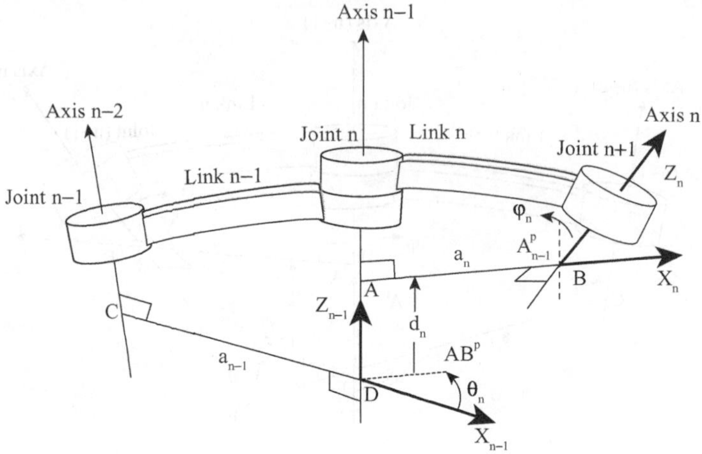

Figure 4.11 Relation between Consecutive Frames

(iii) The direction of common normal from C to D defines X_{n-1}.
(iv) A suitable line perpendicular to both X_{n-1} and Z_{n-1} in the right-hand sense defines Y_{n-1} (not shown in Figure 4.10).

In a similar way, we define the coordinate frame (X_n, Y_n, Z_n) corresponding to Axis n as follows:

(i) Point B is the origin of frame (X_n, Y_n, Z_n)
(ii) Axis n fixes Z_n
(iii) The direction of common normal from A to B defines X_n
(iv) A suitable line perpendicular to both X_n and Z_n in the right-hand sense defines Y_n (not shown in Figure 4.10).

Now, we have to establish the relationship between the two consecutive coordinate frames.

Figure 4.11 shows only the relevant details corresponding to these frames. The frames are shown in bold lines. Y_{n-1} and Y_n are not shown but are existing.

4.6 D-H TRANSFORMATION MATRIX

In order to establish the relationship between the two coordinate frames (X_{n-1}, Y_{n-1}, Z_{n-1}) and (X_n, Y_n, Z_n), we assume a point object placed at origin D (Figure 4.11). We will manipulate the point object so that it undergoes a set of selected motions to move the point object from location D to B and to match its frame exactly to the frame (X_n, Y_n, Z_n). The manipulations of object and the effects are listed in Table 4.1.

Note that all rotations and translations are with respect to current coordinate frame (X_{n-1}, Y_{n-1}, Z_{n-1}) and not with any absolute reference frame. Hence, Rule 2 (Chapter 3, Section 3.6, page 52) is followed to obtain the homogeneous coordinate transformation matrix which defines the relationship between frames (X_{n-1}, Y_{n-1}, Z_{n-1}) and (X_n, Y_n, Z_n).

The final result of coordinate transformation matrix is to move origin D and its frame to origin B. Its frame is

Table 4.1 Manipulations and their Effects

No.	Manipulation of point object	Effect
1	Rotate by θ_n° about Z_{n-1}	X_{n-1} has its new direction along AB^P
2	Translate by d_n units along Z_{n-1}	The object is moved to A; its X_{n-1} is directing along AB
3	Translate by a_n units along X_{n-1}	The object has reached B; its X_{n-1} is along X_n; its Z_{n-1} is along A^P_{n-1}
4	Rotate by φ_n° about X_{n-1}	Z_{n-1} is now directed along Z_n. The frame (X_{n-1}, Y_{n-1}, Z_{n-1}) is rotated by θ_n°, translated by d_n, translated by a_n and rotated by φ_n°; the frame (X_n, Y_n, Z_n) matches exactly

$$H_D^{\,B} = \begin{bmatrix} \cos\theta & -\sin\theta & 0 & 0 \\ \sin\theta & \cos\theta & 0 & 0 \\ 0 & 0 & 1 & 0 \\ 0 & 0 & 0 & 1 \end{bmatrix} \begin{bmatrix} 1 & 0 & 0 & 0 \\ 0 & 1 & 0 & 0 \\ 0 & 0 & 1 & d \\ 0 & 0 & 0 & 1 \end{bmatrix} \begin{bmatrix} 1 & 0 & 0 & a \\ 0 & 1 & 0 & 0 \\ 0 & 0 & 1 & 0 \\ 0 & 0 & 0 & 1 \end{bmatrix} \begin{bmatrix} 1 & 0 & 0 & 0 \\ 0 & \cos\phi & -\sin\phi & 0 \\ 0 & \sin\phi & \cos\phi & 0 \\ 0 & 0 & 0 & 1 \end{bmatrix}$$

$$= \begin{bmatrix} \cos\theta & -\sin\theta\cos\varphi & \sin\theta\sin\varphi & a\cos\theta \\ \sin\theta & \cos\theta\cos\varphi & -\cos\theta\sin\varphi & a\sin\theta \\ 0 & \sin\varphi & \cos\varphi & d \\ 0 & 0 & 0 & 1 \end{bmatrix}, \tag{4.13}$$

where $H_D^{\,B}$ is the D-H transformation matrix at $\theta = \theta_n$°, $\varphi = \varphi_n$°, $a = a_n$ and $d = d_n$, for all $n = 0, 1, 2, \ldots, m$.

4.7 SYMBOLIC PROCEDURE

We follow a symbolic procedure as given in Table 4.2 to determine the kinematic parameters $\{\theta, d, a, \varphi\}$ once the structure of joints and links are given. This procedure is proved to be easily remembered.

O_{n-1} is the origin of the frame (X_{n-1}, Y_{n-1}, Z_{n-1}) indicated as D in Figure 4.11. Table 4.1 can be compared with the definitions of the kinematic parameters given in Figure 4.11.

O_n is the origin of frame (X_n, Y_n, Z_n) indicated as B in Figure 4.11. The symbol ($Z_{n-1} \odot X_n$) means the point of intersection of two lines Z_{n-1} and X_n. In some cases, Z_{n-1} and X_n may not intersect, then the point ($Z_{n-1} \odot X_n$) cannot be located, which is a special situation.

4.8 D-H ALGORITHM

The D-H method is useful in obtaining a kinematic model of a robot. Application of the D-H method requires the robot to be initially set at its home position or home state. At the home state, all joints positions (revolute as well as prismatic) are usually held at zero values. Owing to mechanical restrictions, some joints may reach minimum position but not the exact zero position; in this case, a new zero position is to be defined by adding a constant displacement. This normally happens when a revolute joint and a prismatic joint have the same origin.

Figure 4.12 describes how the joints (J), joint motions (θ), coordinate frames (X, Y, Z), origins (O) and the links (L) are labelled from the base to the end effector.

The D-H algorithm has the following phases:

Phase 1: Labelling the links and joints and assigning of joint frames
Phase 2: Determining kinematic parameters, $\{\theta_i, d_i, a_i, \varphi_i\}$
Phase 3: Computing the joint frame transformations (kinematic model)
Phase 4: Computing the position and orientation of end effector for a given set of joint positions, $\{\theta_i, d_i, a_i, \varphi_i\}$

We will discuss this in the following ways with respect to Figure 4.12:

We assume that there are J_i joints where $i = 1, 2, 3, 4, 5$. J_1 is at the base and J_m ($m = 5$) is at the end effector (Figure 4.12).

PHASE 1: Labelling the links and joints and assignment of joint frames

Step 1 Assign $Z_0, Z_1, \ldots, Z_{m-1}$ axes along the direction of motion of $J_1, J_2, \ldots J_m$ respectively. Therefore, Z_0-axis coincides with direction of motion of Joint J_1 and so on.

Step 2 Choose X_0, Y_0, and Z_0 axes at Joint J_1 in any convenient way to form a right-hand frame (X_0, Y_0, Z_0). The origin is O_0.

Let $i = 1$.

Step 3 If Z_{i-1} and Z_i intersect, choose O_i, the origin at the intersection point. If Z_{i-1} and Z_i do not intersect, choose O_i as the intersection of Z_i at the common normal between Z_{i-1} and Z_i.

Step 4 If Z_{i-1} and Z_i intersect, choose X_i-axis to be perpendicular to both Z_{i-1} and Z_i axes. The direction is arbitrary.

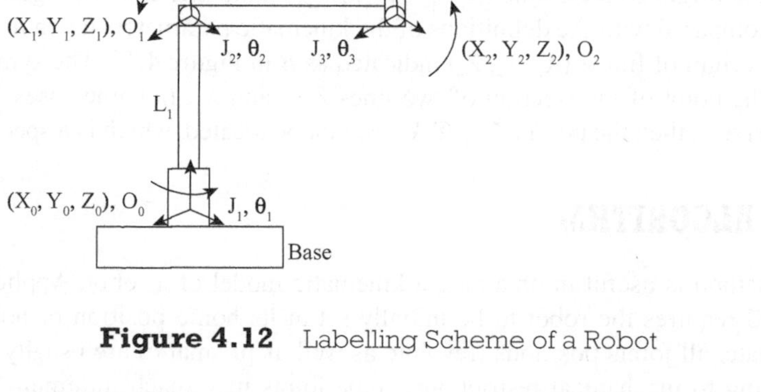

Figure 4.12 Labelling Scheme of a Robot

If Z_{i-1} and Z_i do not intersect, choose X_i-axis along the common normal with X_i directed away from Z_{i-1}.

Step 5 Choose Y_i-axis perpendicular to both Z_i and X_i to form a right-hand frame.

Repeat steps 3 to 5 for $i = 2, 3, 4, 5$.

Step 6 Determine the end-effector frame (X_m, Y_m, Z_m) in the following way:

- Choose the origin of end-effector frame at the tool centre point (TCP)
- Set Z_m-axis parallel to Z_{m-1}-axis, then set X_m-axis to be orthogonal to both Z_{m-1} and Z_m axes and set the direction of X_m to be away from Z_{m-1}
- If Z_{m-1} and Z_m are aligned, then set X_m to be perpendicular to both Z_{m-1} and Z_m with any chosen direction.
- Finally, set Y_m-axis to form the right-hand frame (X_m, Y_m, Z_m). Y_m is not shown in Figure 4.12.

PHASE 2: Determining kinematic parameters $\{\theta_n, d_n, a_n, \varphi_n\}$
Apply the symbolic procedure of Table 4.2 and determine the kinematic parameters $\{\theta_i, d_i, a_i, \varphi_i\}$ for $i = 1, 2, 3, 4, 5$.

PHASE 3: Computing the joint frame transformations (kinematic model)
Equation (4.14) determines the homogeneous transformation matrices of consecutive frames for every set of kinematics parameters. These parameters are inherent to the robot when it is in its home position. At home position, the robot is yet to be controlled to do the required task. The matrices are

$$H^{i}_{i-1} = \begin{bmatrix} \cos\theta_i & -\sin\theta_i\cos\varphi_i & \sin\theta_i\sin\varphi_i & a_i\cos\theta_i \\ \sin\theta_i & \cos\theta_i\cos\varphi_i & -\cos\theta_i\sin\varphi_i & a_i\sin\theta_i \\ 0 & \sin\varphi_i & \cos\varphi_i & d_i \\ 0 & 0 & 0 & 1 \end{bmatrix}, \tag{4.14}$$

where $\{\theta_i, d_i, a_i, \varphi_i\}$, $i = 1, 2, 3, 4, 5$.

PHASE 4: The homogeneous transformation matrix is obtained from the base to the centre of the end effector (TCP) as

$$H_1^{5} = H_1^{2}\, H_2^{3}\, H_3^{4}\, H_4^{5}, \tag{4.15}$$

where H_1^5 is known as arm matrix or solution to forward kinematics of a 5-link robot manipulator at its home position.

All the above phases (PHASES 1–4) are required for determining parameters of each axis and finally the homogeneous transformation matrix.

4.9 APPLICATION EXAMPLES

We will demonstrate the application of the D-H algorithm and the symbolic procedure through examples.

Table 4.2 A Symbolic Procedure (n = 1, 2, ... m (the TCP))

Procedure	Explanation
Z_{n-1} θ_n X_n X_{n-1}	Angle of rotation from X_{n-1} to X_n measured about Z_{n-1} **(Joint angle, θ_n)**
Z_{n-1} d_n $Z_{n-1} \odot X_n$ O_{n-1}	Distance from origin O_{n-1} to point ($Z_{n-1} \odot X_n$) measured along Z_{n-1} **(Joint distance, d_n)**
X_n a_n O_n $Z_{n-1} \odot X_n$	Distance from point ($Z_{n-1} \odot X_n$) to the origin O_n measured along X_n **(Link length, a_n)**
X_n φ_n Z_n Z_{n-1}	Angle of rotation from Z_{n-1} to Z_n measured about X_n **(Link twist, φ_n)**

Table 4.2 guides us to determine the orientation of link, the kinematic parameter table and the transformation matrix from home position. Carefully go through each entry and you may use them in several other problems.

EXAMPLE 4.1

Figure E4.1.1 shows a two-link (i.e. L_1 and L_2) manipulator. A task requires the rotations of joints such as joint 1 by 30° and joint 2 further by 15°.

Determine the rotation (or orientation) matrix and the position vector for P (TCP).

Figure E4.1.1 Two-Link Robot

Solution:

Phase 1: Assignment of coordinate frames

Figure E4.1.2 indicates the origins of coordinate frames O_0, O_1 and O_2 (= P). The development of the coordinate frames (or simply frames) is as follows:

1. Fix Z_0 and Z_1 to indicate the directions of motion.
2. Assign X_0 in a suitable way so that the frame O_0 is formed.
3. Fix the frame O_1 at the intersection of Z_1 with common normal between Z_0 and Z_1.
4. Select X_1 to be along the common normal and direct it away from Z_1.
5. There is no motion at frame O_2 (= P). There is a choice of selecting Z_2. Select Z_2-axis parallel to Z_1-axis. Choose X_2 in line with X_1 and away from X_1.
6. Y_0, Y_1 and Y_2 are to be in right-hand sense at every frame to reach O_2.(= P).

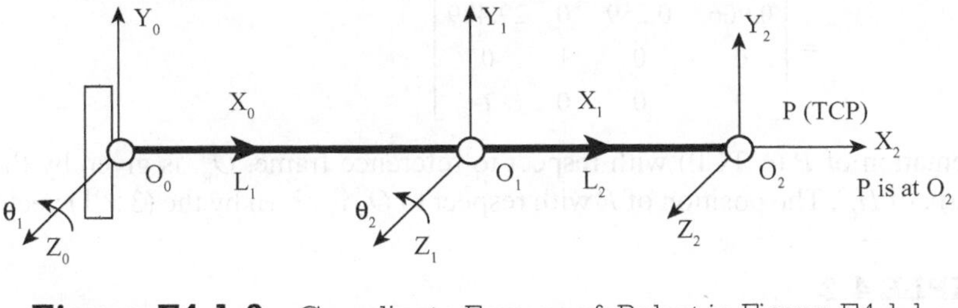

Figure E4.1.2 Coordinate Frames of Robot in Figure E4.1.1

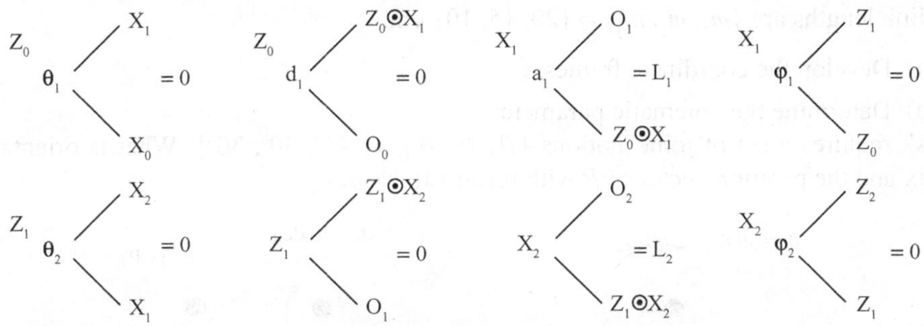

Figure E4.1.3 Determination of Kinematic Parameters of Figure E4.1.2 (Z_2, the TCP)

Phase 2: Kinematic parameters

We will use the symbolic procedure to derive the kinematic parameters. The kinematic parameters are listed in Table E4.1.1.

Table E4.1.1 Kinematic Parameter Table

Joint i	θ_i (degrees)	d_i (cm)	a_i (cm)	φ_i (degrees)
1	θ_1	0	$L_1 = 30$	0
2	θ_2	0	$L_2 = 45$	0

Phases 3 and 4: Transformation matrix for the desired task position
(both phases are required in determining the following equations)

$$H_0^2 = H_0^1 H_1^2. \tag{E4.1.1}$$

with $\theta_1 = 30°$, $\theta_2 = 45°$, $L_1 = 30$ cm and $L_2 = 15$ cm, the transformation matrix is derived as

$$H_0^2 = H_0^1 H_1^2 = \begin{bmatrix} 0.866 & -0.5 & 0 & 25.981 \\ 0.5 & 0.866 & 0 & 15.0 \\ 0 & 0 & 1 & 0 \\ 0 & 0 & 0 & 1 \end{bmatrix} \begin{bmatrix} 0.707 & -0.707 & 0 & 10.607 \\ 0.707 & 0.707 & 0 & 10.607 \\ 0 & 0 & 1 & 0 \\ 0 & 0 & 0 & 1 \end{bmatrix}$$

$$= \begin{bmatrix} 0.259 & -0.966 & 0 & 29.863 \\ 0.966 & 0.259 & 0 & 29.489 \\ 0 & 0 & 1 & 0 \\ 0 & 0 & 0 & 1 \end{bmatrix}. \tag{E4.1.2}$$

The orientation of P (= TCP) with respect to reference frame, O_R, is given by the (3 × 3) sub-matrix of H_0^2. The position of P with respect to O_R is given by the (3 × 1) vector of H_0^2.

EXAMPLE 4.2

Figure E4.2.1 indicates a TRR (Twist-Rotation-Rotation) robot in its home position. The link lengths are $\{m_1, m_2, m_3\} = \{20, 15, 10\}$ cm.

(a) Develop the coordinate frames.

(b) Determine the kinematic parameters.

A task requires a set of joint motions $\{\theta_1, \theta_2, \theta_3\} = \{45°, 30°, 30°\}$. What is orientation matrix and the position vector of P with respect to base?

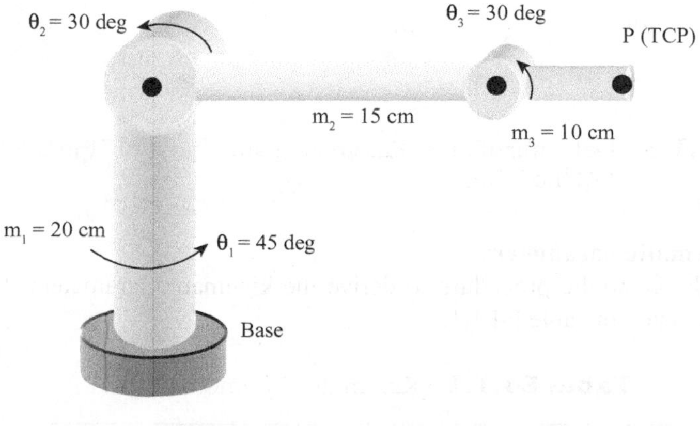

Figure E4.2.1 TRR Robot

Solution:
The coordinate frames are developed as in Figure E4.2.2. P is the TCP.

1. Z_0, Z_1 and Z_2 indicate the directions of axis movements; but Z_3 does not have movement. Z_3 is to be chosen. A choice is made to set Z_3 parallel to Z_2.

2. X_0, X_1 and X_2 are made parallel to each other. X_3 does not have movement. X_3 is to be chosen. A choice is made to be along the common normal between Z_2 and Z_3 and is away from Z_2.
3. Y_i, $i = 0, 1, 2, 3$ are selected to form right-angled frames. $O_3 = P$ which is a TCP

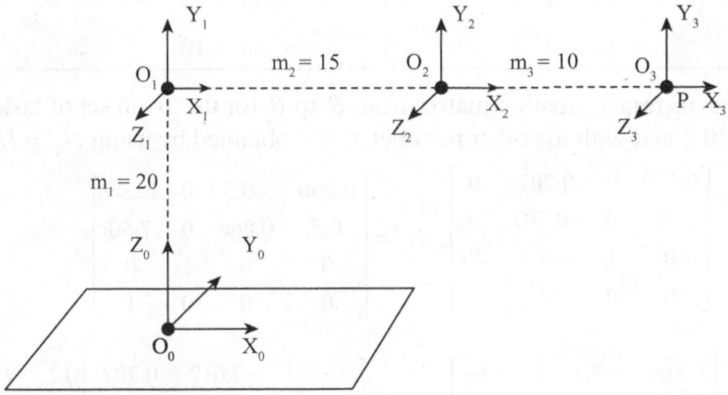

Figure E4.2.2 Coordinate Frames of the TRR Robot. $O_3 = P$ which is a tool center point.

Determination of kinematic parameters:

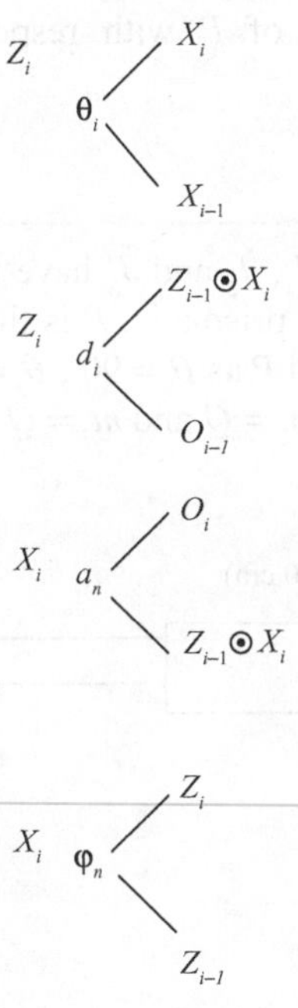

Joint angle: θ_i is an angle between x_i−1 and x_i it is measured about the line of z_i.

Joint distance: d_i is found by using the distance measurement of O_{i-1} and the intersection of (z_{i-1}, x_i) using z_i

Link length: a_i is measured by the angle of intersection of (z_{i-1}, x_i) and the point O_i using x_i

Twist angle: φ_i can be measured from z_{i-1} to z_i using x_i

Dummy variables $i = 1, 2, 3$ are listed in Table E4.2.1.

Figure E4.2.3 Determination of Kinematic Parameters of TRR Robot

Figure E4.2.3 shows the procedure for deriving kinematic parameters. Derived parameters are listed in Table E4.2.1.

Table E4.2.1 Kinematic Parameter Table for TRR Robot

Joint i	θ_i (degrees)	d_i (cm)	a_i (cm)	φ_i (degrees)
1	θ_1	$m_1 = 20$	0	90
2	θ_2	0	$m_2 = 15$	0
3	θ_3	0	$m_3 = 10$	0

The homogeneous transformation matrix from θ_1 to θ_3 for the given set of task movements of {45°, 30°, 30°} and with the other parameters are obtained by using $H_0^{\,3} = H_0^{\,1}\, H_1^{\,2}\, H_2^{\,3}$.

$$H_0^{\,1} = \begin{bmatrix} 0.707 & 0 & 0.707 & 0 \\ 0.707 & 0 & -0.707 & 0 \\ 0 & 1 & 0 & 20 \\ 0 & 0 & 0 & 1 \end{bmatrix}, \quad H_1^{\,2} = \begin{bmatrix} 0.866 & -0.5 & 0 & 12.99 \\ 0.5 & 0.866 & 0 & 7.50 \\ 0 & 0 & 1 & 0 \\ 0 & 0 & 0 & 1 \end{bmatrix},$$

$$H_2^{\,3} = \begin{bmatrix} 0.866 & -0.5 & 0 & 8.66 \\ 0.5 & 0.866 & 0 & 5.00 \\ 0 & 0 & 1 & 0 \\ 0 & 0 & 0 & 1 \end{bmatrix}, \quad H_0^{\,3} = \left[\begin{array}{ccc|c} 0.3535 & -0.612 & 0.707 & 12.719 \\ 0.3535 & -0.612 & -0.707 & 12.719 \\ 0 & 0 & 1 & 20 \\ \hline 0 & 0 & 0 & 1 \end{array}\right]. \qquad \text{(E4.2.1)}$$

The orientation matrix of P with respect to base is given by the (3 × 3) sub-matrix of $H_0^{\,3}$ shown in Equation (E4.2.1). The position of P with respect to base frame is $[12.719 \;\; 12.719 \;\; 20]^T$ cm.

EXAMPLE 4.3

This robot is presented in Figure E4.3.1. Joints J_1, J_2 and J_3 have the same origin at Q. The first two of them are revolute and the third is prismatic. P is the TCP. Determine the transformation matrix connecting the frames Q and P as $\theta_1 = 90°$, $\theta_2 = 90°$ and $m_3 = 10$ cm. m_0 is fixed, no motion. The origins are at $m_1 = Q$, $m_2 = Q$ and $m_3 = Q$ but variable by 10 cm

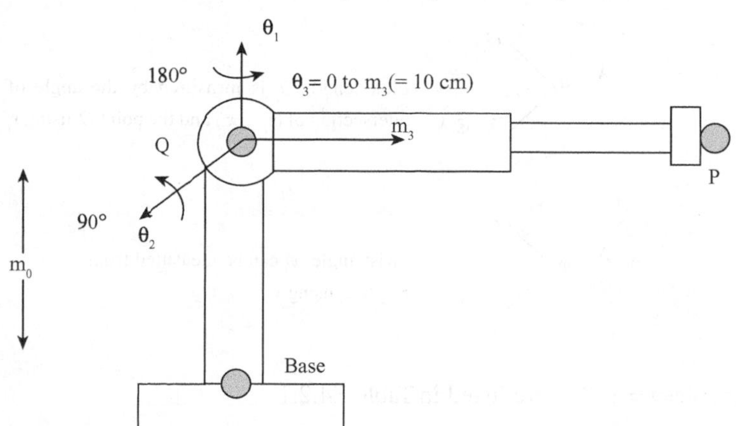

Figure E4.3.1 A Robot with a Prismatic Joint

Solution:

1. m_0 is a fixed member with no movement.
2. We initially assume that the joints are separated by m_1, m_2 and m_3.
3. Later we combine them so that all movements are from same point.
4. The coordinate frames of the robot are given as follows.

Detailed frame descriptions are given in Figure E4.3.2.

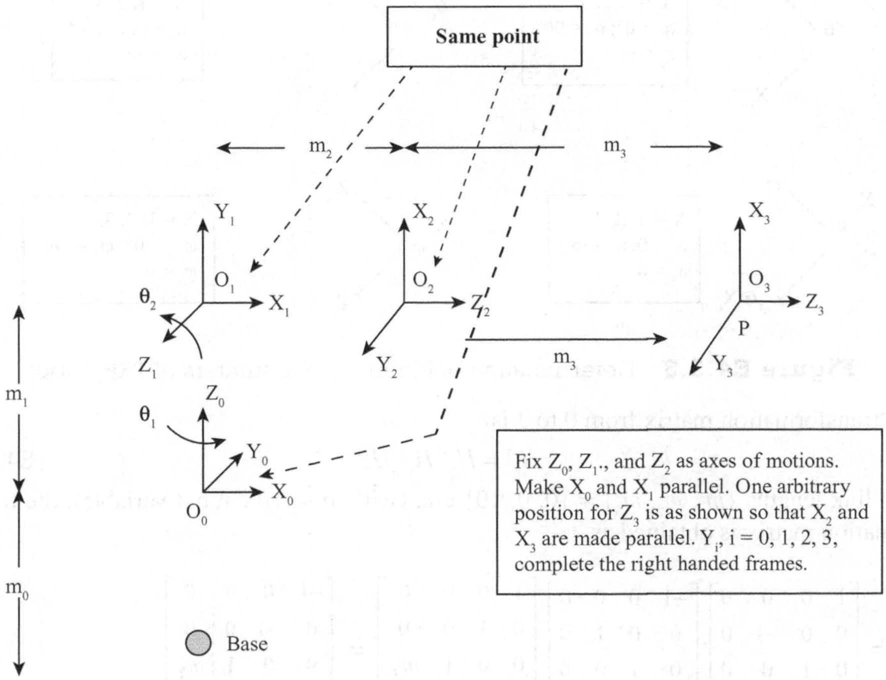

Figure E4.3.2 Coordinate Frames of Robot of Figure E4.3.1

The kinematic parameters of the robot are listed in Table 4.3.1.

Table E4.3.1 Kinematic Parameters of Robot in Figure E4.3.1

Joint i	θ_i (degrees)	d_i (cm)	a_i (cm)	φ_i (degrees)
1	θ_1	m_1	0	90
2	θ_2	0	m_2	90
3	θ_3	m_3, [linear variable] = 10 cm	0	0

1. Obtaining $\{\theta_i, \varphi_i\}$, $i = 1, 2, 3$, is easier.
2. It is also easier to get the parameter sets $\{a_1, d_1\}$ and $\{a_3, d_3\}$ for joints 1 and 3.
3. $m_1 = m_2 = 0$ and m_3 is variable and, consequently, the revolute motion $\theta_3 = 0$.
4. The inset of Figure E4.3.2 gives a brief explanation on how the frames are formed.
5. Z_2 has to be along the translated movement of joint 3 and hence we cannot make X_2 parallel to X_1.

PHASE 1: Labelling the links and joints; assignment of joint frames

Step 1 Assign Z_0, Z_1, Z_3 axes along the direction of motion of J_1, J_2, J_3 respectively. Therefore, Z_0-axis coincides with direction of motion of joint J_1 and so on.

To get the parameters $\{a_2, d_2\}$, we have to determine the point of intersection of common normal between Z_1 and X_2.

Determination of kinematic parameters is shown in Figure E4.3.3.

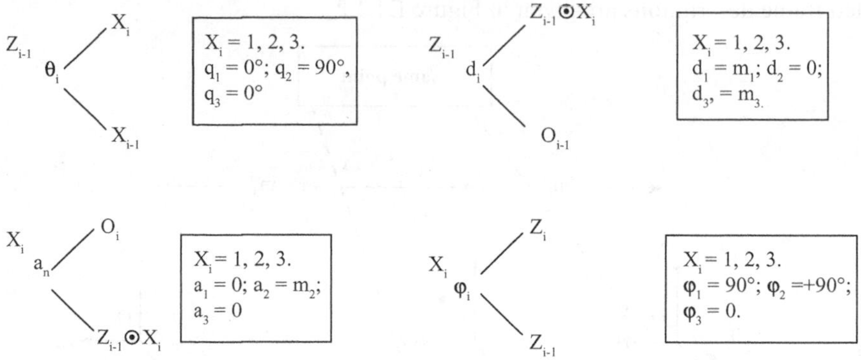

Figure E4.3.3 Determination of Kinematic Parameters of TRR Robot

The transformation matrix from 0 to 3 is

$$H_0^3 = H_0^1 H_1^2 H_2^3. \tag{E4.3.1}$$

with link lengths $\{m_1, m_2, m_3\} = \{0, 0, 10\}$ cm, (with $m_3 = 10$ cm but variable), the transformation matrix is obtained as

$$H_0^3 = \begin{bmatrix} 1 & 0 & 0 & 0 \\ 0 & 0 & -1 & 0 \\ 0 & 1 & 0 & 0 \\ 0 & 0 & 0 & 1 \end{bmatrix} \begin{bmatrix} -1 & 0 & 0 & 0 \\ 0 & 0 & 1 & 0 \\ 0 & 1 & 0 & 0 \\ 0 & 0 & 0 & 1 \end{bmatrix} \begin{bmatrix} 1 & 0 & 0 & 0 \\ 0 & 1 & 0 & 0 \\ 0 & 0 & 1 & m_3 \\ 0 & 0 & 0 & 1 \end{bmatrix} = \left[\begin{array}{ccc|c} -1 & 0 & 0 & 0 \\ 0 & -1 & 0 & 0 \\ 0 & 0 & 1 & m_3 \\ \hline 0 & 0 & 0 & 1 \end{array}\right]. \tag{E4.3.2}$$

1. Equation (E4.3.2) indicates that the position of P is $[0\ 0\ m_3]^T$ as m_3 is units along reference z axis.
2. Its orientation with respect to reference frame O_R is the 3×3 sub-matrix indicated in Equation (E4.3.2).
3. The reader can easily sketch and verify the final position and the orientations of each co-ordinate frame with respect to O_R.

EXAMPLE 4.4

An important type of robot manipulator is the vertical axis jointed arm robot, the SCARA (Selective Compliance Assembly Robot Arm) manufactured by Adept Technologies Inc. A four-axis direct drive Adept One robot is shown in Figure E4.4.1. The robot drive mechanisms have no gears or other mechanical power transmission systems, thus eliminating gear friction and backlash. Its drives are from low speed, high torque brushless DC motors. Adept One SCARA robots allow for clean, precise, high-speed operation.

Figure E4.4.1 also shows the set of joint movements and joint distances. The joint movements are as follows:

(i) Base rotation, θ_1° (45° about Z_1)
(ii) Elbow rotation, θ_2° (-60° about Z_2)
(iii) Vertical extension, d_3 cm (12 cm along Z_3)
(iv) Tool roll, θ_4° (90° about Z_4)

The joint distances are as follows:
$m_1 = 87.7$ cm, $m_2 = 42.5$ cm, $m_3 = 37.5$ cm,
$m_4 = $ (10 to 20 cm, variable) and $m_5 = 20.0$ cm.

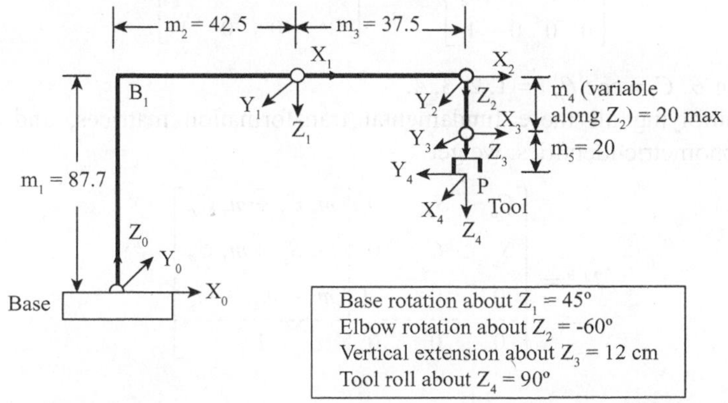

Figure E4.4.1 Coordinate Frames of SCARA

(a) Determine the set of kinematic parameters of Adept One SCARA.
(b) Find the position of the tool from its base when the variables are indicated as above.

Solution:

1. On careful observation of the selected coordinate frames of Figure E4.4.1, we can conclude that the intersection points ($Z_{i-1} \odot X_i$), $i = 1, 2, 3, 4$ are clearly specified.
2. With this information, we can complete the kinematic parameter set as in Table E 4.4.1.

Table E4.4.1 Kinematic Parameters of SCARA

Joint J	θ_i (degrees)	d_j (cm)	a_j (cm)	φ_i (degrees)
1	45	m_1	m_2	180
2	-60	0	m_3	0
3	12 cm	m_4=12	0	0
4	90	m_5	0	0

The transformation matrix from base (O_0) to TCP (P) is obtained from:

$$H_0^P = H_0^1 H_1^2 H_2^3 H_3^P,$$

where

$$H_0^1 = \begin{bmatrix} C_1 & S_1 & 0 & m_2 C_1 \\ S_1 & -C_1 & 0 & m_2 S_1 \\ 0 & 0 & -1 & m_1 \\ 0 & 0 & 0 & 1 \end{bmatrix} \quad H_1^2 = \begin{bmatrix} C_2 & -S_2 & 0 & m_2 C_2 \\ S_2 & C_2 & 0 & m_3 S_2 \\ 0 & 0 & 1 & 0 \\ 0 & 0 & 0 & 1 \end{bmatrix}$$

$$H_2^3 = \begin{bmatrix} 1 & 0 & 0 & 0 \\ 0 & 1 & 0 & 0 \\ 0 & 0 & 1 & m_4 \\ 0 & 0 & 0 & 1 \end{bmatrix} \quad H_3^{\mathrm{P}} = \begin{bmatrix} C_4 & -S_4 & 0 & 0 \\ S_4 & C_4 & 0 & 0 \\ 0 & 0 & 1 & m_5 \\ 0 & 0 & 0 & 1 \end{bmatrix} \qquad \text{(E4.4.1)}$$

where $S_i = \sin\theta_i$, $C_i = \cos\theta_i$, $i = 1, 2, 3, 4$.

After multiplying all these fundamental transformation matrices, and simplifying through trigonometric identities, we get

$$H_0^{\mathrm{P}} = \left[\begin{array}{ccc:c} C_\alpha & S_\alpha & 0 & m_2 C_1 + m_3 C_\beta \\ S_\alpha & -C_\alpha & 0 & m_2 S_1 + m_3 S_\beta \\ 0 & 0 & -1 & m_1 - m_4 - m_5 \\ \hdashline 0 & 0 & 0 & 1 \end{array}\right], \qquad \text{(E4.4.2)}$$

where $\alpha = (\theta_1 - \theta_2 - \theta_4)$, $\beta = (\theta_1 - \theta_2)$.

Substituting $\{m_1, m_2, m_3, m_4, m_5\} = \{87.7, 42.5, 37.5, 12.0, 20.0\}$ cm and $\{\theta_1, \theta_2, \theta_4\} = \{45°, -60°, 90°\}$ into Equation (E4.4.2), we get the transformation matrix from base to tool as

$$H_0^{\mathrm{P}} = H_{\mathrm{Base}}^{\mathrm{Tool}} \left[\begin{array}{ccc:c} 0.966 & 0.259 & 0 & 20.344 \\ 0.259 & -0.966 & 0 & 66.264 \\ 0 & 0 & -1 & 55.700 \\ \hdashline 0 & 0 & 0 & 1 \end{array}\right]. \qquad \text{(E4.4.3)}$$

4.10 JACOBIAN

Direct kinematic models are discussed in this section. This establishes the relationship between the robot's joint positions such as displacements and orientations. This relationship permits the static control of robot to place the end effector at a specified location, thus making a possibility of its end effector to traverse a specific path in space. This makes the end effector to reach a final location with proper joint velocities. This requires the coordination of instantaneous end-effector velocity and joint velocities. One way to achieve this is to take time derivatives of kinematic equation of the robot. We do the transformations from joint velocities to the end-effector velocity described by a matrix named as the *Jacobian [J]*.

In other words, the Jacobian, *J*, is one of the important tools for characterization of differential motions of the robot. Inverse of Jacobian, J^{-1}, is another important aspect. The

Jacobian may at times lose rank and may not be possible to compute the inverse. Such a position of gripper is known as singular position or simply singular. In this section, some singularities are also discussed. In addition, the Jacobian is also useful in describing forces applied to joints and their responses we arrived at.

4.10.1 Manipulator Jacobian

The Jacobian is a multi-dimensional form of derivatives. Suppose we have six functions each of which is a function of six independent variables such that

$$y_1 = f_1(x_1, x_2, \ldots, x_6),$$
$$y_2 = f_2(x_1, x_2, \ldots, x_6),$$
$$\ldots$$
$$\ldots$$
$$y_6 = f_6(x_1, x_2, \ldots, x_6), \tag{4.16}$$

where $[x_1, x_2, \ldots x_6]^T$ is a vector of six independent variables, $[f_1, f_2, \ldots f_6]^T$ is a vector of six dependent variables and $[y_1, y_2, \ldots y_6]^T$ is a vector of dependent variables on $[f_1, f_2, \ldots f_6]^T$. In a vector matrix form

$$y = f^T x. \tag{4.17}$$

From Figure 4.13, it can be observed that the end-effector velocities (linear and angular) are linearly related to the joint velocities. The end-effector velocity is described in Cartesian space. An infinitesimal change in joint position is created by infinitesimal joint motions. It will be shown that the relationship between joint motions produced by end-effector position and orientation is a matrix. As these changes take place in infinitesimal time, we have to define a matrix Jacobian between instantaneous end-effector velocity in Cartesian space and instantaneous joint velocities in joint space.

This Jacobian between differential changes in Cartesian space and joint space is linear and can be expressed as

$$T_e = J\,\theta', \tag{4.18}$$

where T_e is the torque of the end effector; θ' is the joint velocity and J is the Jacobian. They are defined as

$$T_e = [v_e^T\ w_e^T]^T = J\,\theta', \tag{4.19}$$

where

$$J = [Jv^T, Jw^T]^T. \tag{4.20}$$

The matrix J is defined as

$$J^T = [J_{vx}\ J_{vy}\ J_{vz}\ J_{wx}\ J_{wy}\ J_{wz}], \tag{4.21}$$

where $\{vx, vy, vz\}$ and $\{wx, wy, wz\}$ are the components of directions in X, Y and Z axes for linear joint velocity and angular joint velocity.

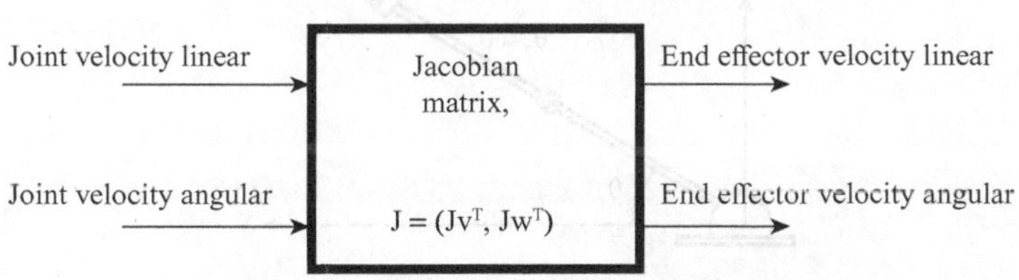

Figure 4.13 Differential Model

4.10.2 Jacobian Singularities

The manipulator Jacobian, J, may become rank-singular (determinant vanishing) at some configuration in Cartesian space. In such cases, the inverse Jacobian does not exist. These singularities are said to be 'J non-invertible' and the configuration is itself called singular. In such cases, J is not of full rank and its column vectors are linearly dependent. That is, there exists at least one direction when the end effector does not move.

The study of manipulator singularities is of great significance for the following reasons:

(a) Not possible to give an arbitrary motion: It is not possible to give an arbitrary motion to the end effector. That is, singularities represent structural configuration in which at least one structural mobility vanishes.

(b) No solution will exist: At one singularity, no solution exists for the inverse Jacobian problem.

(c) Requires a high velocity at the joint space: Small velocities in the neighbourhood of a singularity, especially in the Cartesian space, require very high velocities in the joint space. This will pose a problem when the manipulator is required to track a trajectory that passes closer to the singularity.

The manipulator loses at least one DoF at a singular configuration.

The singular configurations are classified into following two categories based on locations of end effector.

(i) The boundary singularities: These singularities will occur when the end effector is on the boundary of work space. That is when the manipulator is either fully stretched out or completely contracted. Let us consider the case of two-link, two-DoF planar manipulator shown in Figure 4.14.

In this configuration, the two links are in a straight line. The end effector can only be moved in a direction, not perpendicular, but in line with the links. This boundary singularity can occur if the design of manipulator is not accurate. This singularity can produce angle error, possibly, to a minimum extent. Such an occurrence of singularity depends on the size of limbs.

(ii) Interior singularities: These singularities can occur when the end effector is within the work space of the manipulator. They are visible when the joints are collinear. They can also be seen when the joints are in a specific configuration when the manipulator cannot solve these problems of singularities. They cause serious problems for designing path planning.

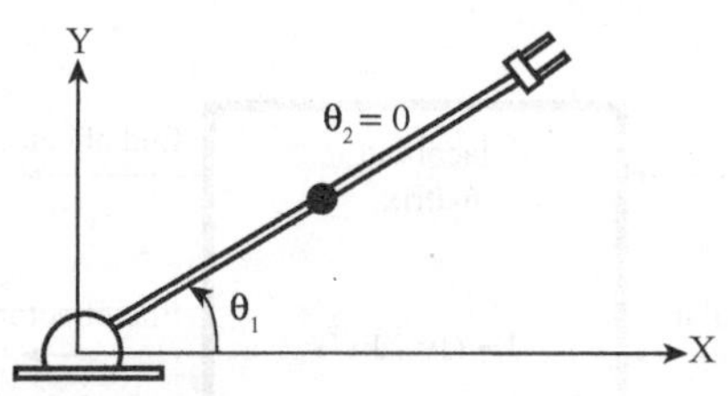

Figure 4.14 Two-Link and Two-DoF Robot

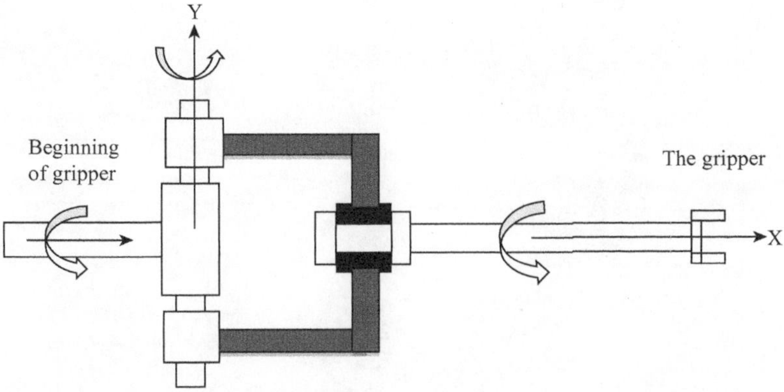

Figure 4.15 Spherical Wrist as a Singularity Configuration

Figure 4.15 describes the interior singularity with a gripper at a singularity configuration. This avoids the planned movements in a variety of path planning at particular locations of gripping force.

REVIEW QUESTIONS

1. What is the difference between kinetics and kinematics? Explain in a few words.
2. Illustrate prismatic joint, revolute joint and joint coordinate frame with diagrams.
3. Discuss the difference between (i) forward kinematics and (ii) inverse kinematics.
4. What is homogeneous transformation? Explain the uses of homogeneous transformation.
5. What is robot vision system? Explain with a diagram.
6. What are link and joint parameters a_n, θ_n, φ_n, d_n? Illustrate with a diagram.
7. Explain symbolic procedure with diagrams.
8. How do you attach various labels on a robot? Explain using a five-jointed robot.
9. Explain the following with at least five diagrams: (i) a single link manipulator coordinate frame, (ii) parameter table and (iii) transformation matrix for home position.
10. What is home position of a robot?
11. Explain the importance of kinematic parameter table.
12. What are the Jacobian matrix and the Jacobian robot singularities?
13. Explain singularities in Jacobian matrix with a diagram.

ROBOT SENSORS 5

What this power is I cannot say; all I know is that it exists
Alexander Graham Bell

This chapter covers the following key topics:

Introduction – Internal and External Sensors – Applications of Robot Sensors – Desirable Features of Robot Sensors – Proximity and Tactile Sensors – Proximity Sensors – Touch Sensors – Slip Sensors – Range Sensors – Opto Range Sensors – Ultrasonic Range Sensors – Force Sensors – Vision System for Inspection.

5.1 INTRODUCTION

Sensors are important devices that go with any automation system. Without sensors, feedback control is not possible. Robot is an automation device and the control is inherent. Without a sensor, the robot is just a mechanical structure. Sensors make the robot intelligent for performing certain decision-making activities. Usually, the terms 'transducers' and 'sensors' are interchangeably used by the robot engineers, but they do have some differences.

A transducer is a device that converts one type of physical variable into another form of physical variable. For example, the concept of temperature change leading to metal expansion is a temperature transducer. The cause and effect, in this case, are both non-electrical. The concept of vibrations of a metal conductor in a magnetic field results in induced voltage in the metal conductor. The cause is of non-electrical and the effect is of electrical, thus this is a vibration transducer.

On the other hand, the sensor is a transducer that converts a physical variable into electric variable, specifically a measurable voltage. This conversion can be straight (one-to-one level); or, in some cases, the conversion can take two levels of processing to get the final output voltage. Thermocouple (temperature to voltage) is a thermal sensor, whereas tachometer (velocity to voltage) is a speed sensor. These are one-to-one level sensors. The strain gauge converts a change in force into change in resistance which is then converted to change in voltage (through a bridge circuit); this is a force sensor of two stages.

Sensors applied in robotics are similar to industrial sensors. However, robot sensors differ from specifications and other physical requirements such as dimension and weight. We study sensors and robot sensors in this chapter.

5.2 INTERNAL AND EXTERNAL SENSORS

Following are two principle kinds of robot sensors:

(i) Internal sensors: These are fitted internal to the robot structure. Sensors for position and velocity of a servo motor are the internal sensors. They are inherent to robot's basic control system. Their main function is to assist in creating feedback for achieving an accurate servo control system for robot joint manipulation. Potentiometers, resolvers and tacho-generators are internal sensors for a robot to measure position and velocity.

(ii) External sensors: Sensors other than internal sensors are termed as external sensors. Sensors to detect the presence of an object in a conveyor (proximity sensors) to indicate the distance between object and gripper (range sensors) and to exert a suitable grip to objects (force sensors) are some examples of external sensors. External sensors help a robot to interact with environmental changes. In some applications, a camera (vision sensor) system can be used to monitor certain areas within the robot work space in order to locate where exactly an object is present or not present.

5.3 APPLICATIONS OF ROBOT SENSORS

Robot sensors, especially the external sensors, have the following applications:

(i) Safety monitoring: Sensors are employed to ensure safety to human being, to equipments and also to other robots working in a robot work cell[1]. Workers' safety is the prime concern. The sensors are also used for workers safety monitoring.

(ii) Obtaining position-related information: Sensors are used to coordinate the sequence of activities of different pieces of equipment in the work cell. This is a major application of the robot sensors. These sensors are required to interface the external environment to perform required tasks depending on environmental changes. Even the presence of a new object or absence of an object in the work space of a robot is considered as an environmental change.

One simple example is the identification of objects coming on a conveyor. Sensors are required to detect the presence and orientation of the object at a specific location along the conveyor. Sensors pass information to robot controller. The robot is controlled to orient its gripper, pick up the object and to place the object on to a different location, may be on another conveyor; another set of sensors detect the presence of object on the other conveyor and the sensor signals are sent out to other equipments to indicate that the object has been placed. This initiates further tasks. Thus, a network of sensors can provide these position-related information so that robot and other equipments take up the tasks in several stages of work cycle[2].

---

1 A robot work cell is a place in an industry where a group of robots and other equipments/machines work together in a specified sequence to produce a finished or a partly finished product.

2 A robot work cycle is defined as the set of sequential operations in a robot work cell that produce a finished/unfinished product.

(iii) Quality control: Sensor permits automatic inspection of finished objects for quality control. Sensors are required for outer damage detection (e.g. vision sensors), inside flaw detection (e.g. a sonar device), dimension measurement (e.g. a set of probes), etc.
(iv) Software interlock: Interlocks are used to coordinate the sequence of activities of different equipments within the robot work cell. Completion of one work cycle in the work cell has to start the next work cycle. A robot work cell has several components of tasks to be completed by the robot and by other equipments. Sensors assist in maintaining the sequence of tasks in a work cell.

5.4 DESIRABLE FEATURES OF ROBOT SENSORS

The required features of robot sensors depend on specific application of robot in industrial tasks. The robot used in component mounting in an electronics industry needs sensors for its several tasks. For example, inserting electronic components in their respective bases on a printed circuit board has to be very fast for agile manufacturing.

Such a robot needs sensors for identifying a component, picking the component with certain gripping pressure, identifying the location where the component is to be inserted and inserting the component in its base with a certain force. Sensors for this robot application should have features of fast response, accuracy, reliability and precision. Table 5.1 lists some of the desirable features of commonly used sensors.

5.5 PROXIMITY AND TACTILE SENSORS

In this section, we describe several types of sensors, especially proximity and tactile sensors, commonly employed in industrial robotics. The proximity sensors are used for detecting the presence of objects without touching; the tactile sensors detect the presence

Table 5.1 Some Desirable Features of Robot Sensors

Features	Details
Accuracy	Accuracy in a measurement is defined as the difference between expected value and measured value. This can be the statistical mean value of several measurements. Smaller the mean value, higher the accuracy. Accuracy should be as high as possible
Precision	Precision means that there is little or no random variability in measured values even when the measurement is repeated several times. Precision has to be high
Speed of response	The sensor should be capable of responding to fast changes in the measured variable. Ideally, the measurement is to be instantaneous
Operating range	The sensor is to be accurate and precise even when the measured variable varies to a larger range unpredictably. Sensor should have large operating range with consistent accuracy and precision
Calibration	Drift is gradual loss in accuracy over a time. All sensors are to be calibrated every now and then to minimize the drift. The calibration procedure should consume minimum time and should not be complex and require specialized people

(*Continued*)

Table 5.1 Continued

Features	Details
Reliability	Sensor performance is to be highly reliable in its life span. Should not fail during its operation
Cost	The cost has several components: cost of purchase, install, operate, repair and calibrate. Each of these cost is to be as low as possible.
Weight, and volume	Robot sensors may be mounted externally at robot joints, limbs and at gripper. In such a case, the weight and space occupied are to be as minimum as possible; mounting a sensor on a robot should not modify the characteristics of robot

of objects by touch. These sensors can range from a simple switch to complex elements which create environmental changes.

5.5.1 Proximity Sensors

The general operating parameters for a proximity sensor are listed in Table 5.2. There are two ways by which the parameters can be measured—(i) when the object is moving along the reference axis and (ii) when the object is moving perpendicular to the reference axis. Figure 5.1 illustrates these measurements.

Working nature of proximity sensors is of two types: (i) sensing due to modification of emitted signal and (ii) sensing due to disturbance in the environment. Reflection of light, a modification of emitted signal, is an example of type (i) and induced voltage in a coil due to disturbance in magnetic field is an example of type (ii). These features are described in the following:

Optical proximity sensor: This consists of two components: a light emitter and a light receiver. This proximity sensor is of type (i). Figure 5.2 shows how the optical proximity sensor works. In a through-beam optical sensor (Figure 5.2(a)), light passes across the conveyor belt and reaches the receiver; the receiver has an in-built light sensitive circuit. The circuit gives a voltage of a specific value (Logic 1). When the object passing the conveyor

Table 5.2 General Operating Parameters of a Proximity Sensor

Reference plane	Plane of reference on the proximity sensor from which all measurements are made. Usually a vertical plane to sensor axis passing through some point in the sensor
Reference axis	An axis through the sensor along which the measurements are made
Sensing objects	Objects to be sensed in terms of specified shape, size and material
Sensing distance	The distance from reference plane along the reference axis to the sensing object
Resetting distance	The sensing distance at which the sensor changes its output from ON to OFF as the object is withdrawn from the sensor
Setting distance	The sensing distance at which the sensor changes its output from OFF to ON as object is brought nearer to sensor

(Continued)

Table 5.2 Continued

Differential distance	The difference between resetting distance and the setting distance; this is a measure of hysteresis
Response time	The time required for the senor to change its output when the object is suddenly withdrawn from or brought nearer to the sensor
Frequency response	The maximum frequency of sensor output ON-OFF waveform that equals the rate at which the object is made to move across the differential distance
Temperature and voltage specifications	The range of variations in temperature and voltage within which the specified setting and resetting distances are the same as when such variations are absent

interrupts the emitted light, the receiver produces a zero voltage (Logic 0) indicating that there was an object just passed across the optical sensor. Thus, the optical sensor works as an optical proximity switch. Such a switching feature is very useful in counting the objects that are passing through the sensor.

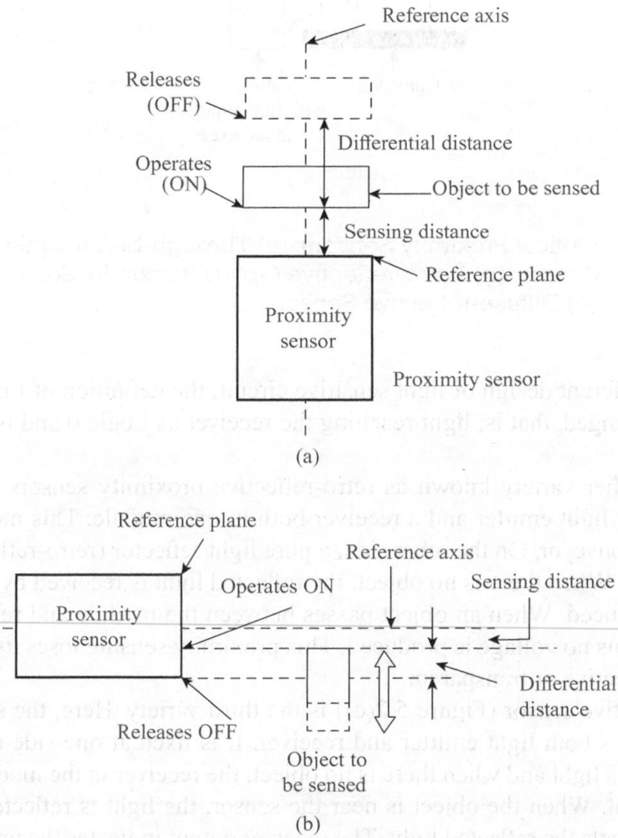

Figure 5.1 Proximity Sensor Measurements: (a) Object Moving Along Reference Axis and (b) Object Moving Perpendicular to Reference Axis

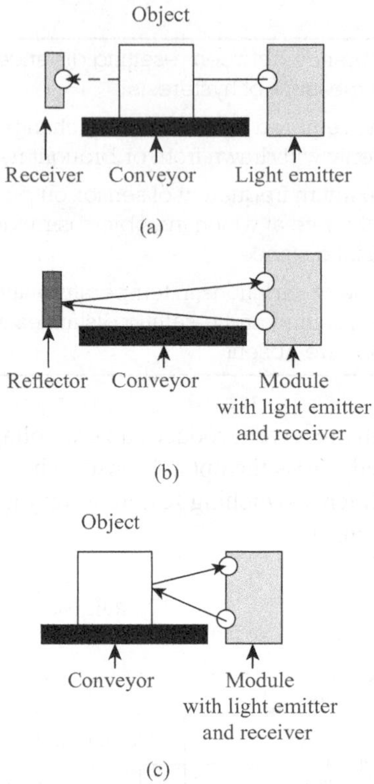

Figure 5.2 Optical Proximity Sensors: (a) Through-beam Optical Proximity Sensor, (b) Retro-reflective Optical Proximity Sensor and (c) Diffuse-reflective Sensor

Through a different design of light sensitive circuit, the definition of Logic 1 and Logic 0 can be interchanged, that is, light reaching the receiver as Logic 0 and light interruption as Logic 1.

There is another variety known as retro-reflective proximity sensors (Figure 5.2(b)). The sensor has a light emitter and a receiver both in one module. This module is fixed at one side of the conveyor. On the other side, a pure light reflector (retro-reflector) is aligned with the module. When there is no object, the reflected light is received by the module and a voltage is produced. When an object passes between the module and reflector, the light is interrupted; thus no voltage is produced. This proximity sensing loses its accuracy when the object is reflective or transparent.

Diffuse-reflective sensor (Figure 5.2(c)) is the third variety. Here, the senor is a single module which has both light emitter and receiver. It is fixed at one side of the conveyor. The module emits light and when there is no object, the receiver in the module does not get the reflected light. When the object is near the sensor, the light is reflected by the object and the receiver gets the reflected light. The receiver output indicates the presence of object near the sensor. One requirement is that the object should have a light reflective surface. In addition, the accuracy of sensor diminishes if the object is too near or too far from the sensor.

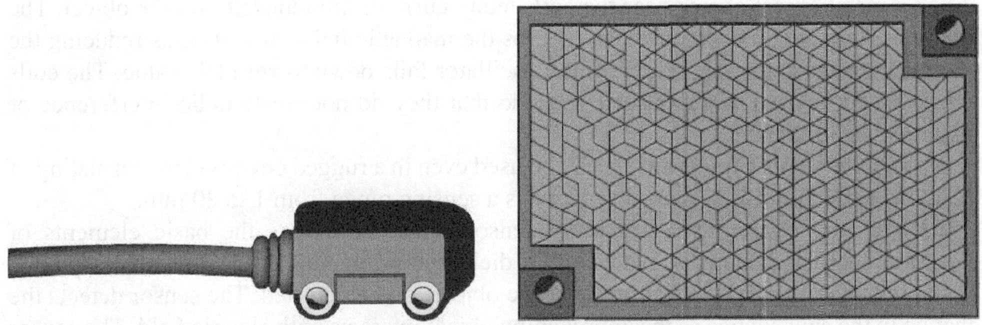

Figure 5.3 Retro-reflective Proximity Switch and Reflector

A commercial version of the retro-reflective proximity switch is shown in Figure 5.3.

Electromagnetic proximity sensor: This sensor is useful when the object is of electrically conductive material, typically iron and steel. Figure 5.4 shows a commercially available inductive proximity sensor. Sensing is due to induced eddy currents in the object. The schematic diagram of an inductive sensor is shown in Figure 5.5. The entire circuit and the coils are within the module of the sensor. When the object is absent, the gain of the circuit is set by adjusting the potentiometer P till the circuit oscillates. The oscillator has two coils (L_1 and L_2) of equal magnetic strengths and the coupling between two coils is minimum. The output of oscillator is rectified to a constant DC by the diode D and the capacitor C.

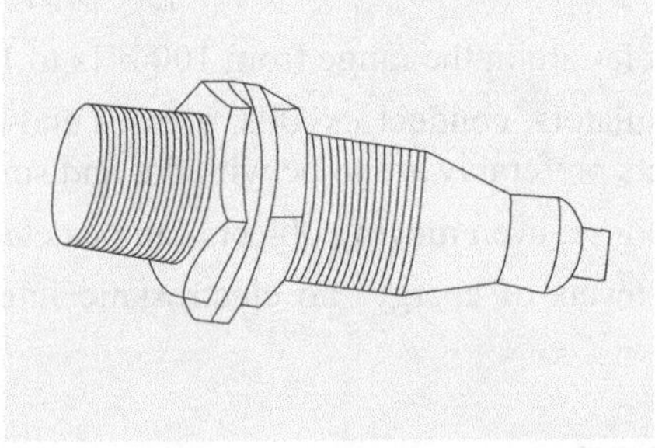

Figure 5.4 Inductive Proximity Sensor

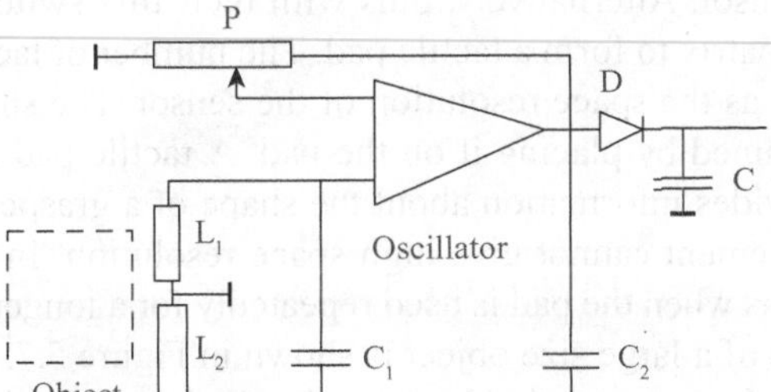

Figure 5.5 Schematic Diagram of Inductive Proximity Sensor

When a metal object comes near the coils, eddy currents are induced into the object. The magnetic field due to eddy currents disturbs the magnetic fields of coils thus reducing the energy from the coils. The output of the oscillator falls down to zero DC value. The coils are designed to have low level of energy so that they do not create radio interference or generate heat in the object.

The inductive proximity sensors can be used even in a rugged environment consisting of dust, dirt and humidity. A typical sensor has a sensing range from 1 to 20 mm.

Electrostatic (Capacitive) proximity sensor: This sensor has the basic elements of capacitor such as conductive plates and a dielectric producing an electric field arranged in some manner so that a metal or dielectric object can be detected. The sensor detects the change in the capacitance as the approaching object interacts with electric field. The sensor has an operational amplifier circuit (Figure 5.6(a)) that oscillates when an object modifies the electric field.

Two types of capacitance sensors are available—dielectric and conductive. The dielectric type (Figure 5.6(b)) has two capacitor plates integrated into the oscillating circuit. When an object is in the vicinity of the sensor, the capacitance is raised and the resulting oscillations indicate the presence of the object. This type is used to detect all types of materials as long as the dielectric constant is more than 1.2. The second type, conductive (Figure 5.6(c)), has only one capacitor plate integrated into the oscillator in the sensor. The effect of second plate is realized when an object of conductive material enters the field and raises the capacitance to start oscillations. Conductive sensors are excellent for *looking through* an insulation material such as rubber, glass or paper, to detect the presence of any conductive material hidden in the object.

The resultant oscillations are rectified and filtered to a binary DC value for proximity switching purposes. The capacitive sensor has the following typical characteristics:

(a) Oscillator frequencies are in the range from 100 kHz to 1 MHz.
(b) Objects can be insulators, conductors, oils, greases and materials with high moisture content; objects preferably are to be with flat and smooth surface.
(c) Fast detection response; even moving objects can be detected.
(d) Operation in low levels of energy; no electrostatic interference to other objects nearby.

5.5.2 Touch Sensor

A touch (or tactile) sensor generally is a form of a switch. It is expected that the touch on the sensor is made with a minimum force. A simple microswitch with a projecting pin can be used as a tactile sensor. Alternatively, pins with their tiny switches at one end can be arranged in a regular matrix to form a tactile pad. The number of tactile sensors on the pad per unit area is known as the space resolution of the sensor. The shape of a manufactured product can be determined by placing it on the pad. A tactile pad attached to the fingers of a robot gripper provides information about the shape of a grasped object. The pad with mechanical pin arrangement cannot give high space resolution. In addition, there can be switch contact problems when the pad is used repeatedly for a longer time. A typical tactile pad to check the shape of a large size object is shown in Figure 5.7.

Tactile sensors of today use resistive-based technologies where the sensor acts as a variable resistor in an electrical circuit. There are several types of pressure sensitive resistors which can be integrated into a pad. A small deflection on the sensor pad causes implanted

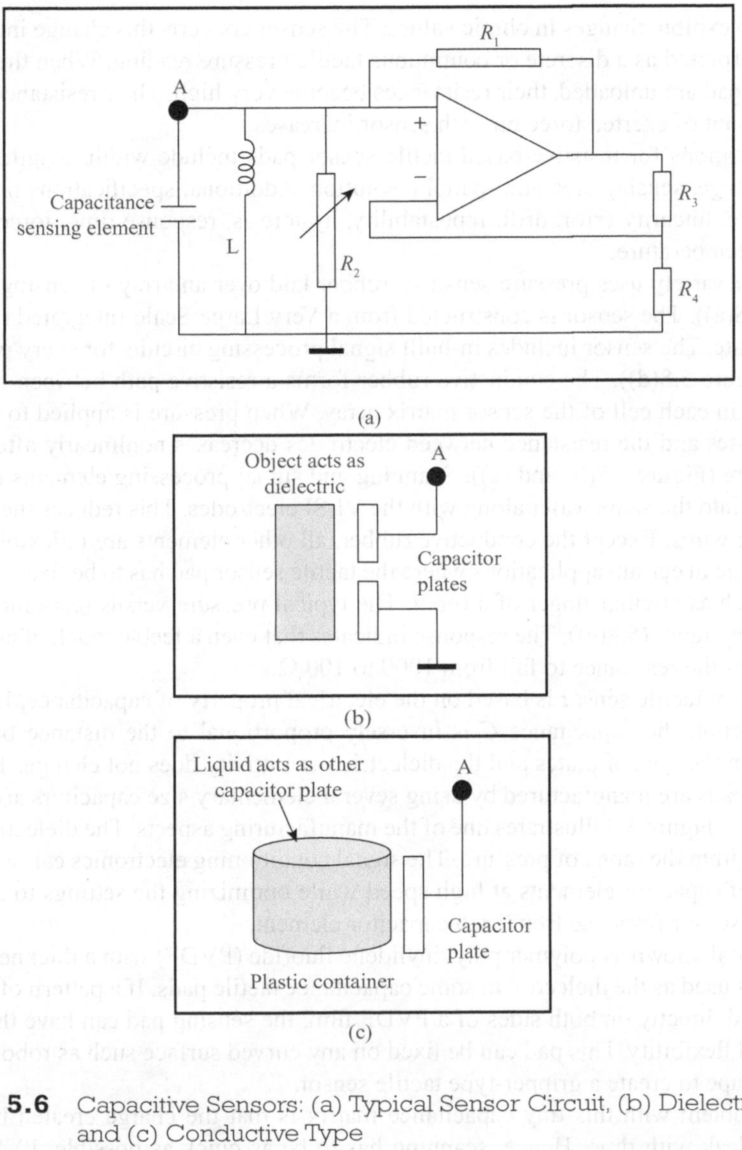

Figure 5.6 Capacitive Sensors: (a) Typical Sensor Circuit, (b) Dielectric Type and (c) Conductive Type

Figure 5.7 Microswitch Matrix Tactile Pad

resistors to exhibit changes in ohmic values. The sensor converts this change into a voltage that is interpreted as a discrete or continuous tactile pressure reading. When the sensors in the tactile pad are unloaded, their resistances become very high. Their resistances decrease as the amount of exerted force on each sensor increases.

Specifications for resistive-based tactile sensor pads include width, length, thickness, pressure range, sensing area and spatial resolution. Additional specifications include saturation force, linearity error, drift, repeatability, hysteresis, response time, force range and operating temperature.

Another variety uses pressure sensitive rubber laid over an array of sensing electrodes (Figure 5.8(a)). The sensor is constructed from a Very Large Scale Integrated (VLSI) silicon substrate. The sensor includes in-built signal processing circuits for every pair of electrodes (Figure 5.8(d)). The conductive rubber forms a resistive path between consecutive electrodes in each cell of the sensor matrix array. When pressure is applied to the rubber, it compresses and the resistance between electrodes decreases nonlinearly after a threshold pressure (Figure 5.8(b) and (c)). Scanning and signal processing elements can also be integrated into the same wafer along with the VLSI electrodes. This reduces the number of connecting wires. Except the conductive rubber, all other elements are inflexible. This is a disadvantage in certain applications where the tactile sensor pad has to be fixed on a curved surface such as circular finger of a robot. The typical pressure versus resistance response in shown in Figure (5.8(e)). The response indicates that even a feeble touch of about 10 gm force makes the resistance to fall from 1000 to 100 Ω.

Capacitive tactile sensor is based on the electrical property of capacitance. In a parallel plate capacitor, the capacitance C is inversely proportional to the distance between the plates when the area of plates and the dielectric permittivity does not change. The capacitor tactile pads are manufactured by using several elementary size capacitors arranged in a matrix form. Figure 5.9 illustrates one of the manufacturing aspects. The dielectric material is elastic within the range of pressure. The signal conditioning electronics can scan through the array of capacitor elements at high speed while optimizing the settings to achieve the maximum sensor response from each capacitor element.

A material known as polymer polyvinylidene fluoride (PVDF) with a thickness as small as 28 μm is used as the dielectric in some capacitance tactile pads. If a pattern of electrodes is deposited directly on both sides of a PVDF film, the sensing pad can have the property of physical flexibility. This pad can be fixed on any curved surface such as robot fingers of circular shape to create a gripper-type tactile sensor.

One problem with this tiny capacitance matrix is that the charge created by a tactile force can leak with time. Hence, scanning has to be as quick as possible. PVDF is temperature sensitive and creates additional charge due to changes in temperature. Since the charges and consequent voltage are feeble, the signal processing circuits have to be very near to the sensor, preferably in-built.

5.5.3 Slip Sensors

An important capability of human hand and fingers is to detect the slipping of object when it is being handled. The biological receptors at our fingers help continuous grasping of the object with adequate force so that the object does not slip; and when slipping occurs, the force exerted by the fingers instantaneously increases to avoid slipping. This ability has to be incorporated in the robot fingers when the robot has to handle varieties of objects.

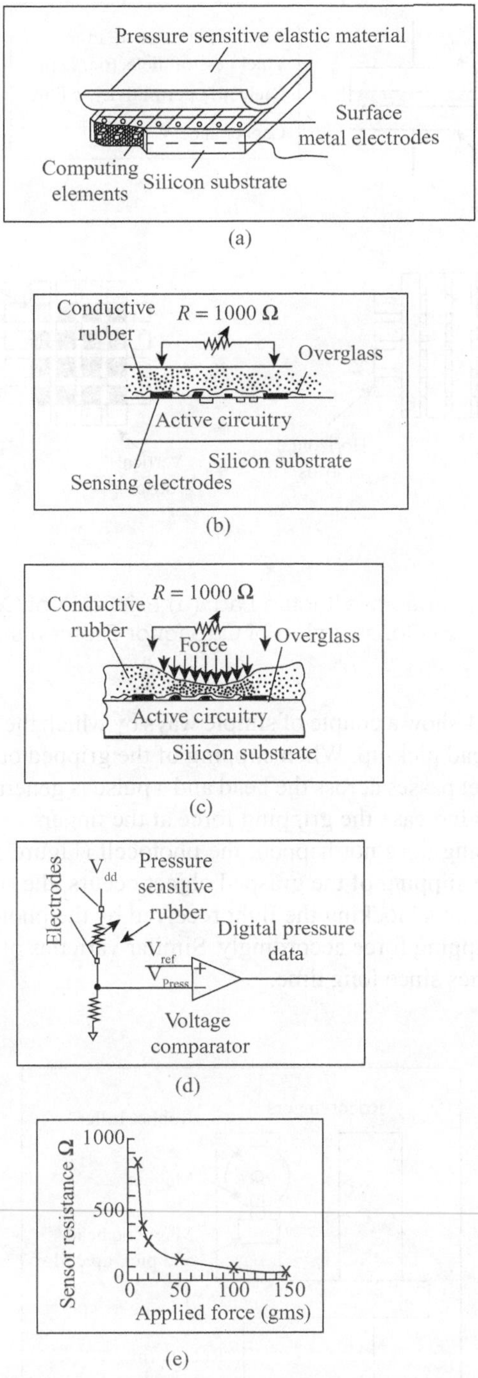

Figure 5.8 Resistance-based Tactile Sensing Array: (a) Sensor Structure, (b) Sensor with no Force, (c), Sensor under Pressure, (d) Electronic Circuit and (e) Resistance versus Applied Force (From Website of OMRON)

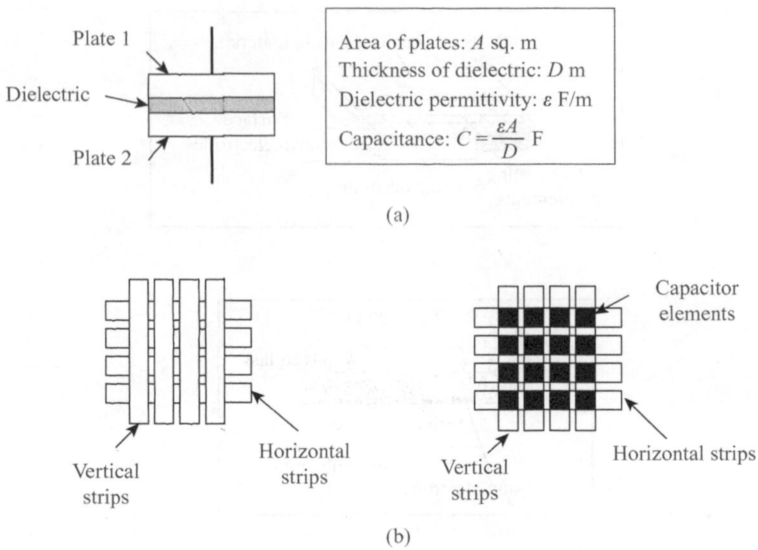

Figure 5.9 Capacitive Tactile Pad: (a) Principle of Capacitance and (b) Construction of Capacitor Elements

Figures 5.10 and 5.11 show a couple of simple ways by which the slipping can be sensed. One uses a magnetic head pick-up. When slipping of the gripped object occurs, the rubber roller rotates, the magnet passes across the head and a pulse is generated in the pick-up coil. This pulse is utilized to increase the gripping force at the fingers.

As long as the slipping does not happen, the photocell (Figure 5.11) receives the light continuously. When the slipping of the grasped object occurs, the rubber roller rotates, the position of slit changes thus blocking the light received by the photocell. This pulse is utilized to increase the gripping force accordingly. Similar varieties of slip sensors have been used in industrial robotics since long time.

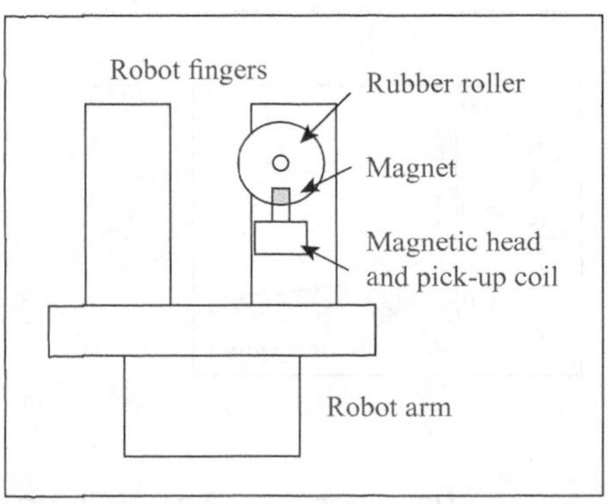

Figure 5.10 Simple Slip Sensor (Magnetic Type)

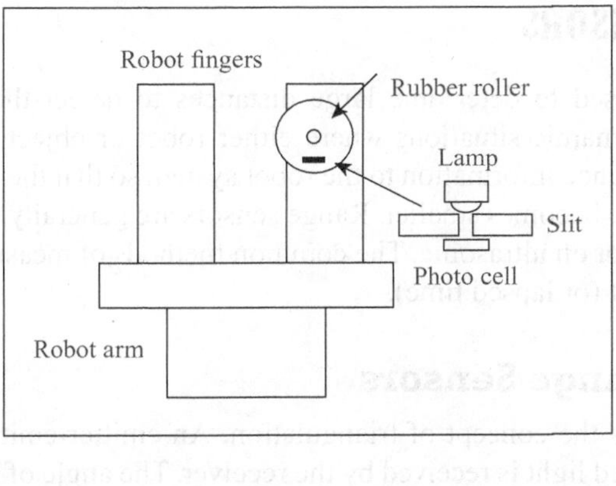

Figure 5.11 Simple Slip Sensor (Opto-interrupter Type)

These sensors, though effective, have the disadvantage of not detecting the second and further (multiple) slipping of object. In addition, on continuous usage, the sensor produces wear and tear of rotating parts and a possible misalignment of sensor elements even when there is no grasped object.

Robots of today are required to handle objects of different shapes and weights. A requirement has been felt in detecting multiple slipping of the grasped object and in increasing the gripping force accordingly. Further, by gripping, the robot is to exhibit some details of the shape of object among objects of varied shapes. This requirement is necessary in sorting out the objects. These facilities can be incorporated in the robot gripper by using matrix sensor elements such as microswitch arrays and resistive, capacitive and piezoelectric pads. For displaying the shape of object, in industries, a high resolution in the array of sensors, as discussed earlier, is not required. Thus, the array sensors can be made economical and can serve the purpose of robot understanding the shape.

Figure 5.12 shows the shape of the gripped object in terms of tactile force sensor elements. The sensor elements are active when they touch the object. The active sensor elements can be identified in terms of a binary image[3] matrix. This matrix image is repeatedly tested for its consistency at specified time intervals. Any slipping of object results in changes in binary image matrix and the computer increases the gripping force accordingly.

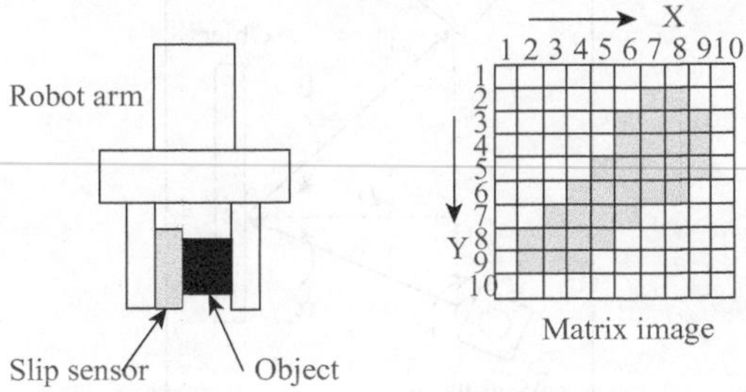

Figure 5.12 Slip Sensing

[3] In binary image, the pixel colours are either in black or in white, considered as (0 or 1) or as (1 or 0).

5.6 RANGE SENSORS

Range sensors are used to determine large distances to detect the obstacles. They are usually applied in dynamic situations where either robot or object is moving. These are meant to provide advance information to the robot system so that the robot can decide what to be done if the range becomes smaller. Range sensors are generally based on light (visible light, infrared, laser) or on ultrasonic. The common methods of measurement are triangulation and time of flight (or lapsed time).

5.6.1 Opto Range Sensors

Figure 5.13 describes the concept of triangulation. An emitter emits a ray of light on an object and the reflected light is received by the receiver. The angle of incidence is α and the angle of reflection is β. If the reflection is perfect, the reflected light reaches the receiver. If the reflection is not perfect, the reflected light does not reach the receiver. In this case, the emitter has to be rotated to some angle till the light reaches the receiver.

The distance between the receiver and the emitter is $m = m_1 + m_2$. D is the range to be computed. Then

$$\tan\beta = \frac{D}{m_1} \quad \text{and} \quad \tan\alpha = \frac{D}{m_2}. \tag{5.1}$$

Then, the range is

$$D = \frac{m \tan\alpha \tan\beta}{\tan\alpha + \tan\beta}. \tag{5.2}$$

The angle of reflection (β) and the distance between the receiver and the emitter (m) are fixed in a system. There is a provision for measuring the angle α; thus, D can be computed. The computation of angle is repeated at specific intervals to verify the changes in D, if any. The requirement of a computing element such as a microprocessor is evident in this application.

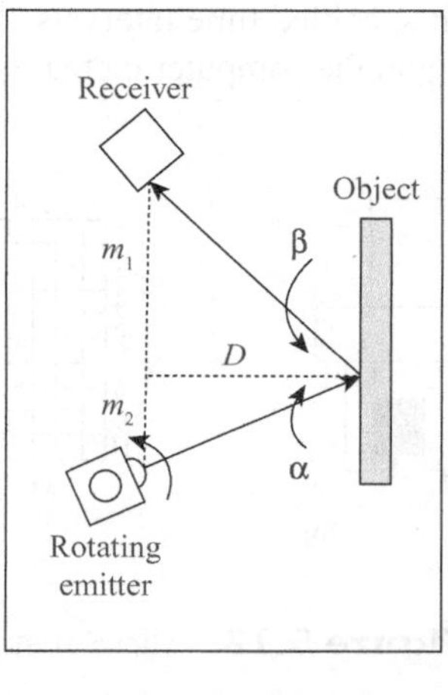

Figure 5.13 Triangulation Method of Range Sensing

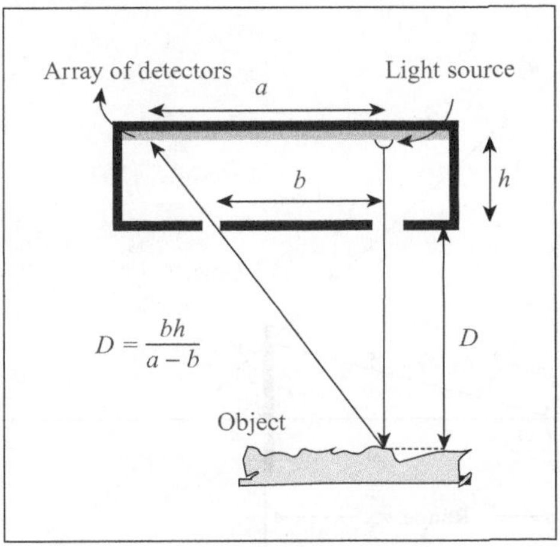

Figure 5.14 Optical Short-range Sensor Detecting Seam

Figure 5.14 illustrates a more practical way of measuring the range, *D*. This also follows the triangulation method. The array of photodetectors is known as line scan array. The position at which the reflected beam strikes the array is proportional to the distance to the object. This sensor is effective for short-range measurements; long-range measurements may not be possible since the reflected ray of light may go out of the line scan array. This sensor is useful in testing the quality of seam welding especially to find any possible gaps in the welding line.

Nguyen and Blackburn designed a range finder simpler in implementation since measurement of angles is not required as in some other ranging devices. Figure 5.15(a) and (b) show the schematic diagram and the triangulation configuration, respectively, for the range finder. The device consists of a laser transmitter and a laser receiver (camera). If an object is present at point P at a distance z from camera, the laser beam is reflected and the laser camera receives a point on its screen. The distance nz is to be measured as the range. An onboard microprocessor repeatedly computes z as the object moves within a specified range.

Figure 5.15(b) shows the Z and Y coordinates, the origin being O; the point o is a reference point on the camera screen. u_1, u_2 and u are points on the screen corresponding to P_1, P_2 and P, respectively. A close observation of Figure 5.15(b) shows that z can be derived as

$$z = \frac{yf}{u}, \tag{5.3}$$

where f is the known focal length of camera and u is a measured distance from the reference point o on the camera screen.

P is a point at which object appears and can vary between P_1 and P_2; the line P_1P_2 in the Y–Z plane is described by

$$y = mz + c, \tag{5.4}$$

where c is the point of intersection of laser beam with the Y-axis, and

$$m = \frac{y_1 - y_2}{z_1 - z_2}. \tag{5.5}$$

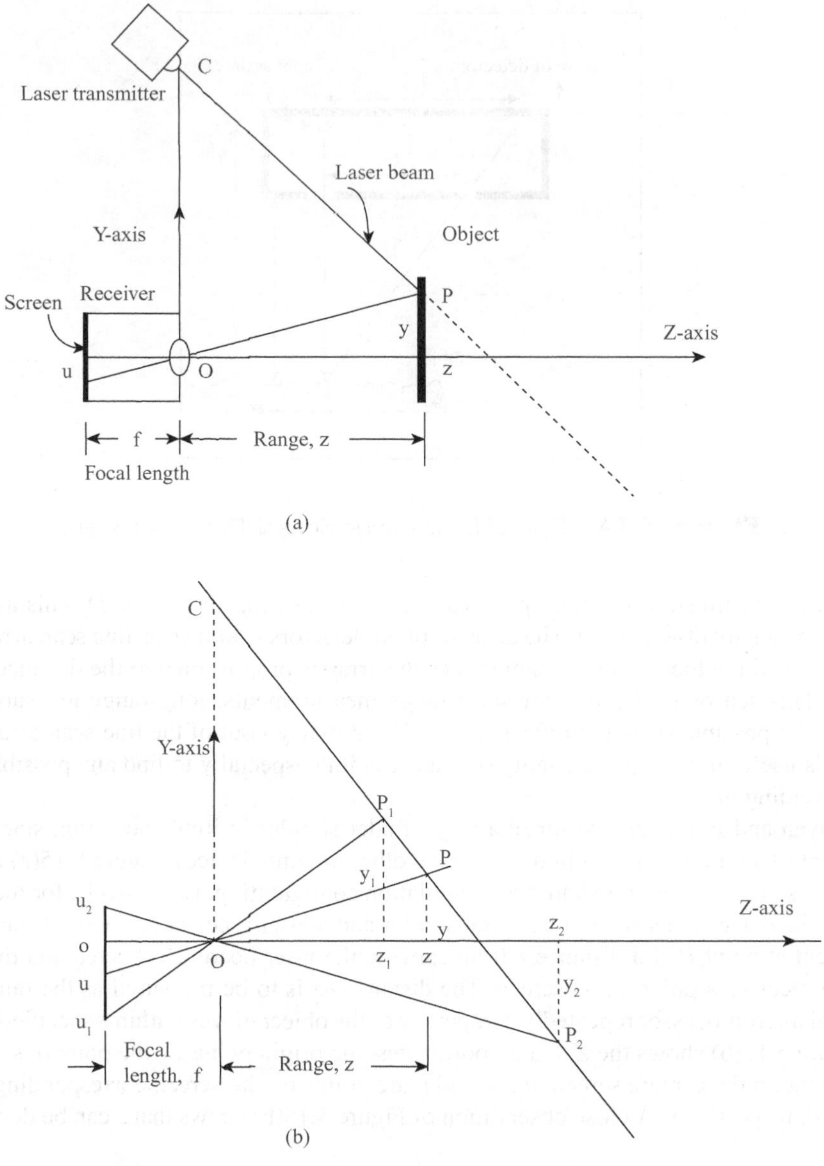

Figure 5.15 Laser Range Finder: (a) Schematic Diagram of Range Sensor and (b) Triangulation Configuration

A simple manipulation of Equations (5.3)–(5.5) gives the range

$$z = \frac{fc}{u - mf}, \tag{5.6}$$

where the variables and constants on the left-hand side are known and measurable. The variable *u* can be measured in pixel off-set.

5.6.2 Ultrasonic Range Sensors

Ultrasonic sensors are rugged, simple in implementation, inexpensive and low powered. They are used in many sensing applications such as motion detection, intruder alarms, navigational aid for mobile robots, internal flaw detection and mapping of environment. The sensor has an emitter of ultrasonic signal and a receiver, both usually integrated in one module. The ultrasonic signals used in varieties of current devices have a frequency range of 20 kHz to about 2 MHz which is well above the audible sound frequency range of 20 Hz–20 kHz. The manufacturers use materials such as lead zirconate titanate piezoelectric crystals for producing range sensors.

Most ultrasonic sensors measure the range (distance) using the time-of-flight method. The sensor emits a short duration of high-frequency ultrasound, called burst, which travels certain distance and is reflected back after encountering an object. The reflected signal is received by the receiver. The time-of-flight is the duration between the instant at which the ultrasonic signal is emitted and the instant at which it is received by the receiver The distance between the sensor and the object is half of the total distance travelled by the burst which is equal to the time-of-flight multiplied by the speed of ultrasonic wave. Such a measurement has a limited resolution since the speed of ultrasound depends on the density of medium and its temperature and both may vary during the time-of-flight. The velocity of ultrasound in air is computed from

$$v = (331.4 + 0.6\ T)\ \text{m/s}, \tag{5.7}$$

where T is the temperature of air medium (in °C). Since the wave length $\lambda = (v/f)$, a 20 kHz of ultrasound has a wavelength of 16.5 mm frequency and at 2 MHz, the wavelength is 1.65 mm. Hence, the resolution in measuring the distance is better at high frequencies. However, high-frequency signal attenuates faster than low-frequency signals; faster attenuation severely limits the measurement of range.

Figure 5.16(a) shows an ultrasonic distance sensor (28015, Parallax Inc., PING) consisting of ultrasonic transmitter, receiver and the measuring circuit, suitable for applications with Basic Stamp microcontroller. The frequency of operation is 40 kHz with burst at every 200 μs and a measurement range of 2 cm–3.3 m. A set of ultrasonic sensors (Polaroid Corp.) is presented in Figure 5.16(a) and (b) and for larger power applications, the sensor element is larger in size with an increased weight. Typical beam pattern (polar curve) of a sensor is shown in Figure 5.16(c). The pattern shows that the sensor is very effective when the object is straight in front. There are limitations in range measurement if the object is slanting with respect to axis of sensor.

Some sensor has reasonable accuracy for a range of 1 m as long as the object's slanting angle is within ±30°.

5.7 FORCE SENSORS

Force is a vector which has a direction and magnitude. Force has a translational effect that deforms an object by creating a directional strain. Strain is defined as the deformation per unit length and is measured in millimetre per millimetre. Strain is related to

$$\varepsilon = \frac{\Delta l}{l}, \tag{5.8}$$

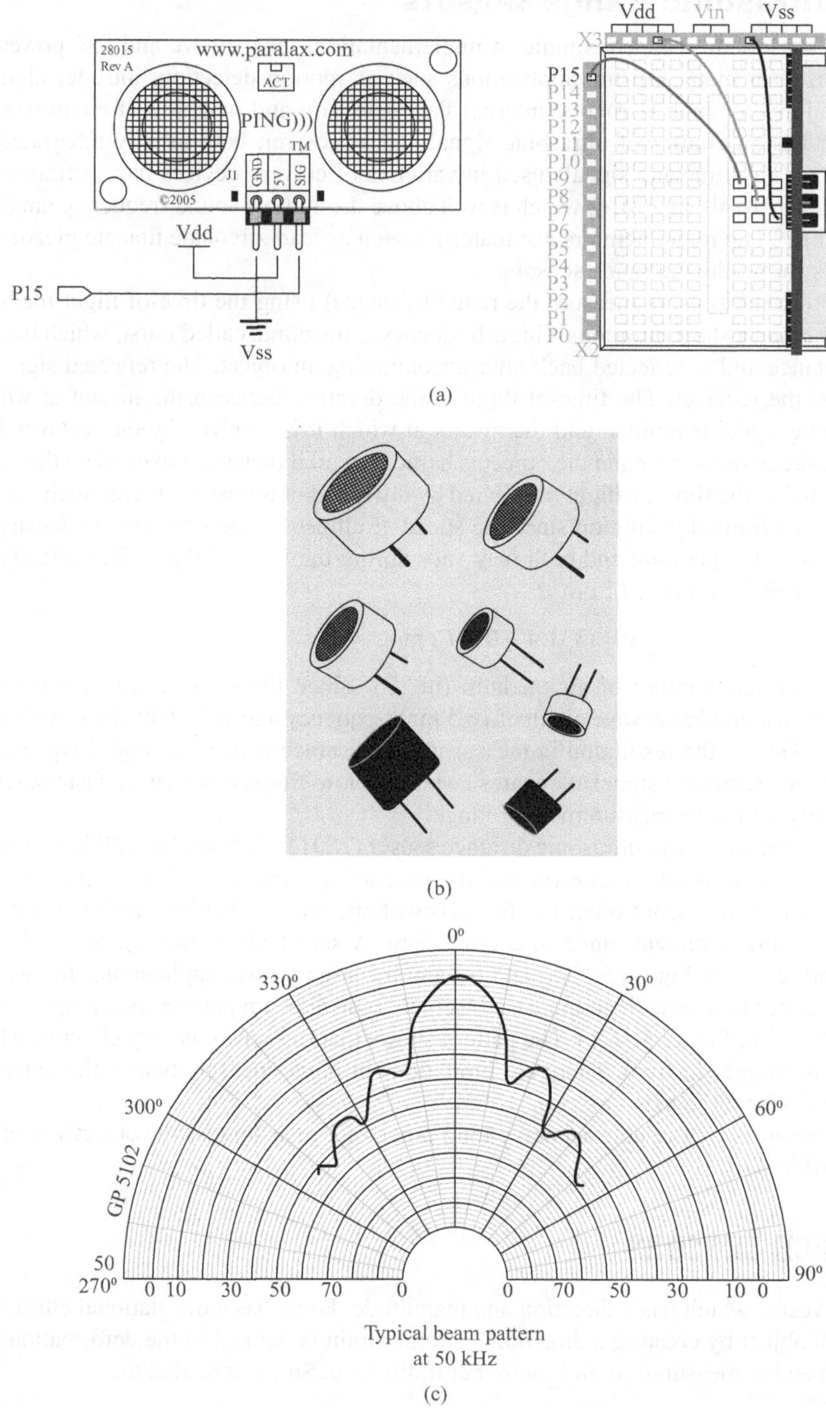

Figure 5.16 Ultrasonic Range Sensor Details: (a) Ultrasonic Distance Sensor Circuit, (b) Sensor Elements and (c) Typical Ultrasonic Beam Pattern (db Versus Angle)

where Δl is the change in length (deformation) and l is the length. If Δl is positive due to tensile force, then ε is positive; ε can be negative due to compressive force and Δl is negative. In most metals, strain is measured to be 0.005 mm/mm.

Strain gauge is a very useful device to measure strain and force. Strain gauge is basically a long wire having a specific resistance R (in ohm) expressed as

$$R = \frac{\rho l}{A}, \tag{5.9}$$

where ρ is the resistivity of the metal (in ohm/m), l is the length (in m) and A is the area of cross section (in m^2).

When the metallic wire with a minimum thickness (a foil) is subjected to a tensile stress (a pulling force from each end) acting along the wire (active axis), the length l increases and the area of cross section A decreases. Change in the cross section area of the foil is negligible. Then, the change in resistance, ΔR, is directly related to strain. For a specific material of the strain gauge, this relation can be expressed as

$$G = \frac{\Delta R/R}{\varepsilon} = \frac{\Delta R/R}{\Delta l/l}, \tag{5.10}$$

where G is known as the gauge factor of the strain gauge. A general purpose strain gauge has a value of G around 2.0.

For easier handling, the long foil is folded several times and encapsulated in a plastic packing. A metal foil strain gauge consists of a grid of thin metallic film bonded to a thin insulating backing called the carrier matrix. The unforced resistance of most common strain gauge is designed to be 120 Ω.

Various manufacturing parameters of a strain gauge are illustrated in Figure 5.17. The change in resistance due to force is maximum only when the force is along the active axis. For a strain gauge as in Figure 5.17, the active axis is horizontal. Its vertical axis is a passive axis which is not used in force measurements. To measure the force in different axes at the same time, different patterns of strain gauges are used. Figure 5.18 shows a variety of strain gauge patterns. Figure 5.18(a) is a metallic strain gauge with its active axis vertical; Figure 5.18(b) shows its active axes in the directions of X, Y along 45°; Figure 5.18(c) shows axes along

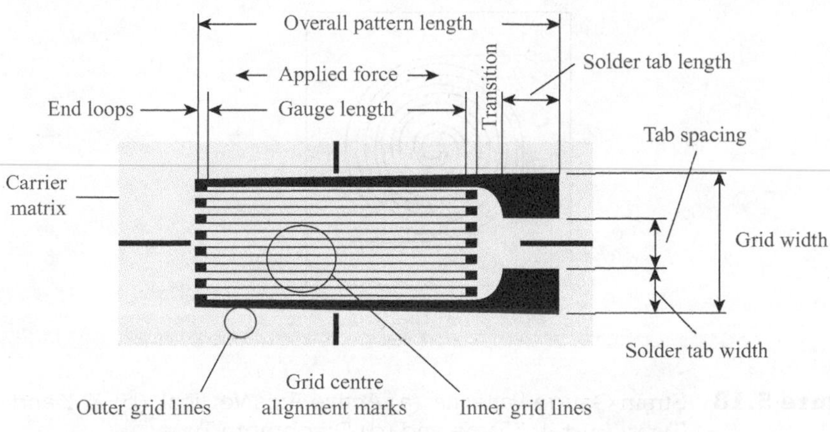

Figure 5.17 Strain Gauge Manufacturing Details

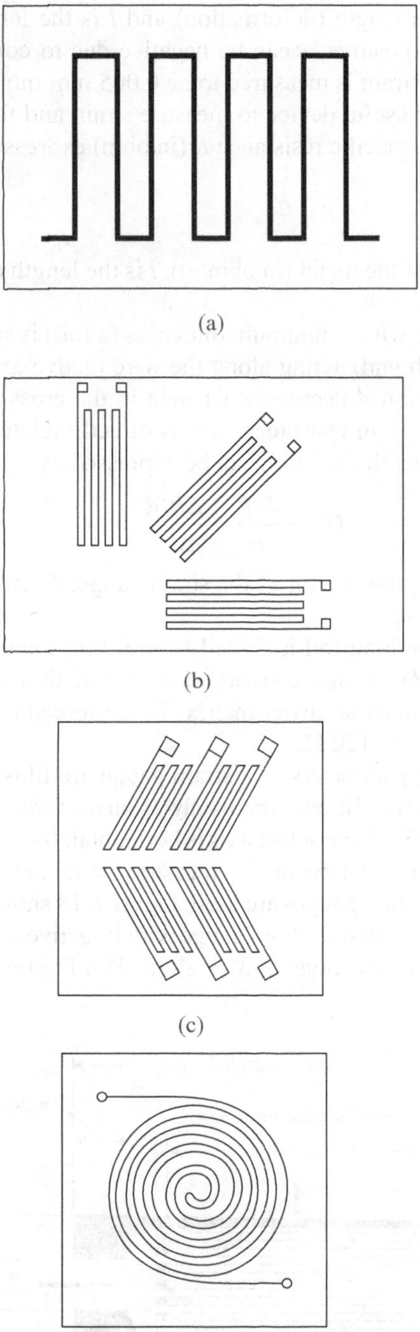

Figure 5.18 Strain Gauge Patterns: (a) Active Axis Vertical, (b) X, Y and 45° Type, (c) ± 45° Type and (d) Diaphragm Type

± 45° which is normally applied in shaft torque measurement and Figure 5.18(d) shows the diaphragm type, wherein strain gauge can measure the total force applied in all directions. This strain gauge can be used on the surface of a sphere to measure force due to varied internal pressure within the sphere.

The strain gauge is applied to measure strain on a surface subjected to force. For measuring the strain at a selected point on a surface, the strain gauge is permanently cemented to the surface at the selected point by using special glue. The active axis of the strain gauge has to be along the direction of strain to be measured. The strain changes the resistance of the strain gauge. A positive (tensile) strain increases the resistance of strain gauge and a negative (compressive) strain reduces the resistance. Figure 5.19 shows a cantilever beam subjected to a force. The strain at the top surface of the beam is tensile and is compressive at the bottom surface. The change in strain is very small and the resulting change in resistance of the strain gauge is also very small. This resistance change can be converted to a change in voltage using a Wheatstone bridge. The bridge voltage can then be amplified to any required level.

The bridge shown in Figure 5.19 is known as quarter bridge since the strain gauge is fixed on one arm of the bridge. The bridge is balanced initially (i.e. the bridge output, V_o is made to zero) with proper resistances and with no force at the beam. With a known force applied to the beam, the resistance of strain gauge increases resulting in a change in V_o. Then, the bridge output voltage V_o can then be calibrated to obtain a relation $f = kV_o$, where f is the applied force and k is a constant that depends on the material of beam, its dimensions and the location at which the force is applied.

Sensitivity of Quarter Bridge

To compute the sensitivity of the quarter bridge, let us assume that the bridge is balanced with resistances $R_1 = R_2 = R_3 = R$, where R is the unstrained resistance value of the strain gauge; then, $V_o = 0$. When a force is applied, the resistance of strain gauge is increased to ΔR and the change in bridge voltage is ΔV such that

$$\Delta V = V_i \left\{ \frac{R + \Delta R}{2R + \Delta R} - \frac{1}{2} \right\} = V_i \left\{ \frac{\Delta R/R}{4 + 2\Delta R/R} \right\}. \tag{5.11}$$

Since $(2\Delta R/R) << 4$, and from Equation (5.11)

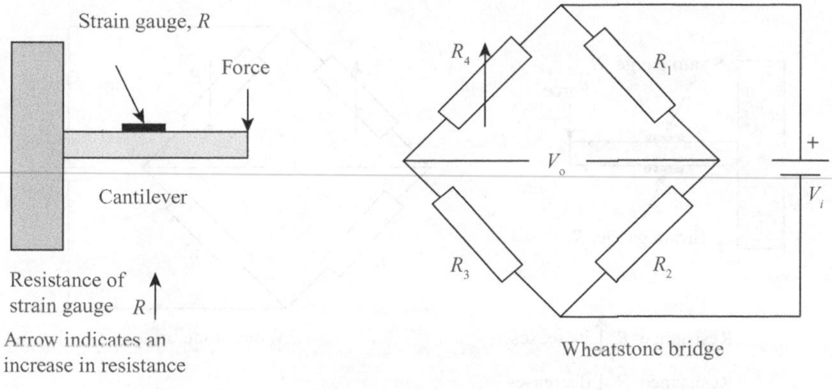

Figure 5.19 Quarter-bridge Force Measurement

$$\Delta V = \frac{G_f \varepsilon}{4} V_i, \tag{5.12}$$

where G_f is the gauge factor, ε is the strain and V_i is the bridge input. Then, we can define the strain gauge bridge sensitivity in volt per unit strain as

$$S = \frac{\Delta V}{\varepsilon} = \frac{G_f}{4} V_i. \tag{5.13}$$

Sensitivity of Half Bridge

The sensitivity of the bridge can be improved by adding another strain gauge, this time at the bottom of the cantilever. The bottom strain gauge increases its resistance when force is applied. Increasing and decreasing strain gauge resistances, R_4 and R_1, respectively, are in the arms of each side of the bridge as shown in Figure 5.20. The strain gauges R_4 and R_1 are identical with same resistance value R when the bridge is unloaded so that $R_1 = R_2 = R_3 = R_4 = R$. Hence, the variations in resistances due to applied force are $R_4 + \Delta R$ and $R_1 - \Delta R$. The identical strain gauges have the same variations in resistances.

A similar approach as above gives the sensitivity of the bridge as

$$S = \frac{G_f}{2} V_i. \tag{5.14}$$

Sensitivity of Full Bridge

The full bridge has strain gauges in both sides of its arm. Sensitivity of the strain gauge bridge can be further increased by using full-bridge configuration of force sensing. All four arms of the bridge have identical strain gauges which are fixed to the cantilever as shown in Figure 5.21. All strain gauges are identical having the same value of unstrained resistance and having the same change in resistance per unit strain. It should be noted that strain gauges of increasing resistance are to be connected to the opposite arms of the bridge and strain gauges of decreasing resistance to the other opposite sides. A similar approach discussed above offers the value of bridge sensitivity as

$$S = G_f V_i. \tag{5.15}$$

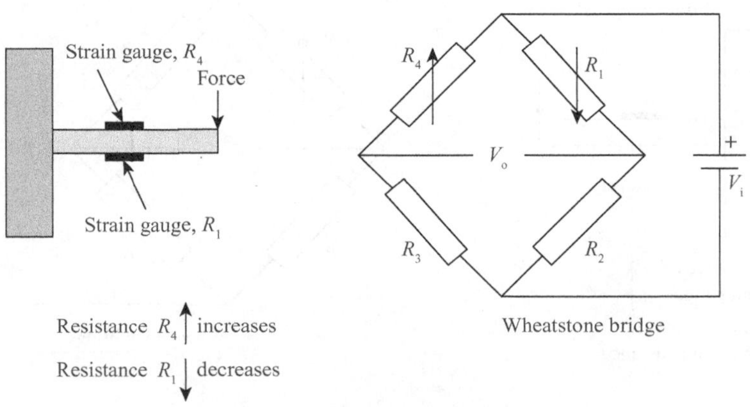

Figure 5.20 Half-bridge Force Measurement

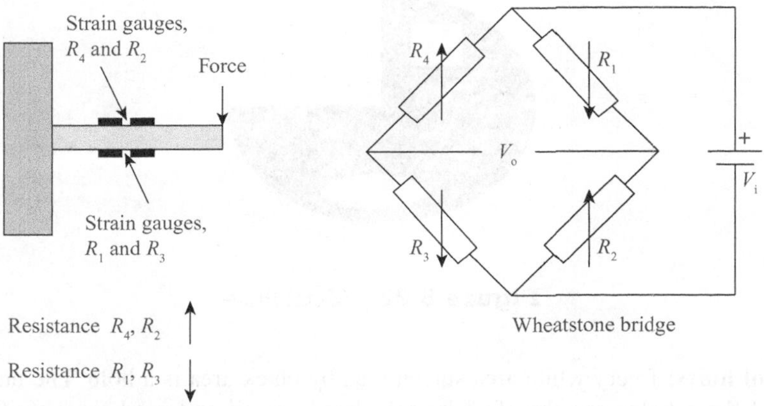

Figure 5.21 Full-bridge Force Measurement

5.8 VISION SYSTEM FOR INSPECTION

Robot computer has data of vision in its memory. Once the image has been stored in the computer memory, analysis of these data begins. The most common approach is to examine the object and compare this image with models held on file. These reference models are to be taught to vision system by showing samples to the camera and to program the computer to remember by measuring their various attributes.

Suppose a vision system has now in its memory a record of scene in terms of array of pixels. For easiness of expansion, binary data are used. We can assume that the pixel is presented either by a '0' or by a '1' in each memory location. The system can then assess the attributes of the scene or components in view. The following parameters are examples of some of the attributes and how the vision computer recognizes them.

Distance: Distance between the camera and the object; an important parameter that recognizes the final size of the object image. The camera should not be moved closer/farther during recognizing the attributes of object.

Area: When it is considered that the binary image is made up of white pixels representing the object, then the sum of the white pixels is one of the attributes of the object. If a number of countable holes are present, then a number of black pixels occur on the surface of the object. The same is applied to the white pixels and black pixels reversed.

Perimeter: This operation can again be carried out as the image scan continues. Each scan line changes from black to white (or white to black), which is the indication of the component edge. By totalling the number of changes, an indication of perimeter length is achieved. The changes that occur due to holes within the object can be eliminated, if possible by a software.

(Perimeter squared ÷ Area): This is an indication of compactness of object; this will not change as the magnitude of the image changes.

Centroid: The centroid is the position of centre area of the object. This is the position given when the first moment of area about both the *X* and *Y* axes is divided by the object area giving *x* and *y* coordinates of the centre area.

Figure 5.22 Metal Plate

Number of holes: Every white area surrounded by black area is a hole. The number and position of these holes can therefore be calculated, stored and used for identification or inspection purposes.

Maximum and minimum radii: These are maximum and minimum distances from the centroid to the farthest and nearest points on the perimeter.

Minimum enclosing rectangle: This is obtained by observing the maximum and minimum points on the X and Y axes of the perimeter of the object.

Lighting: The most important aspect of industrial vision system is the lighting that gives the necessary contrast. This is to enable differentiation from background. The background can be a conveyor or a place which gives a contrast to differentiate. One method that does not use natural light above the component is that of structured lighting. In this technique, a thin line of light is projected on the object to distinguish the component from moving conveyor.

The above features are minimum to distinguish the component from background. This list may not be sufficient as there are more features to look upon and the experience of the industrial engineer is the one to decide. For example, from our experience let us decide the pattern of Figure 5.22. Assume the black disc as a metal plate.

REVIEW QUESTIONS

1. What are the differences between transducers and sensors?
2. What are the main differences between industrial sensors and robot sensors?
3. List and explain main applications of robot sensors.
4. List and explain some desirable features of robot sensors.
5. List and explain some of the general operating features of proximity sensors.
6. With proper diagrams, explain the differences of (i) optical, (ii) electromagnetic and (iii) electrostatic proximity sensors.
7. Explain the differences between (i) a tactile pad and (ii) slip sensors with diagrams.
8. List and explain two different approaches of range sensing with diagrams.
9. Explain the advantages of ultrasonic range sensing compared with other types.
10. Write a note on force sensing.
11. What are the features of vision system for inspection? List at least five features.

ROBOT CONTROL

6

If anyone can control his tongue, it proves that he has a perfect control over himself in every other way

James 3.2 (Living Bible)

This chapter covers the following key topics:

Introduction – Euler–Lagrange Equation – Joint Motion – Linear Control Systems – Second-order Systems – State-space Equations – Lyapunov Stability – Lyapunov First Method – Lyapunov Second Method – Control Unit – Electric, Hydraulic and Pneumatic Drives – Industrial Vision System – Inspection Using Industrial Vision – Camera

6.1 INTRODUCTION

A robot commanded to carry out motions in finishing the required task has been discussed in the previous chapters. In this chapter, we learn how the joint interactions are arrived at to reach the task point. There are two varieties of reaching the task points— (i) non-contact type and (ii) contact type.

Non-contact type requires an environment such as air around the robot. This makes the robot articulation meaningful in the beginning and in completing the task finally. Contact type requires certain materials to physically process or handle. Molten metal is one of the materials which can be considered as an example for contact type. This can be picked up and poured out by a spoon-type tool. We know how robot utilizes its ability in stacking cement bag in a lorry. This application needs a larger sized tool or gripper. Both of these applications require accuracy and repeatability but are required to a reasonable level. In this chapter, we study such and further more robot applications.

6.2 EULER–LAGRANGE EQUATION

Figure 6.1 shows many blocks around the robot (manipulator). Each block works for a specific purpose. The trajectory planner block creates point-to-point tasks. The robot has to reach these points that can be constants such as for pick and place robot or can be variable

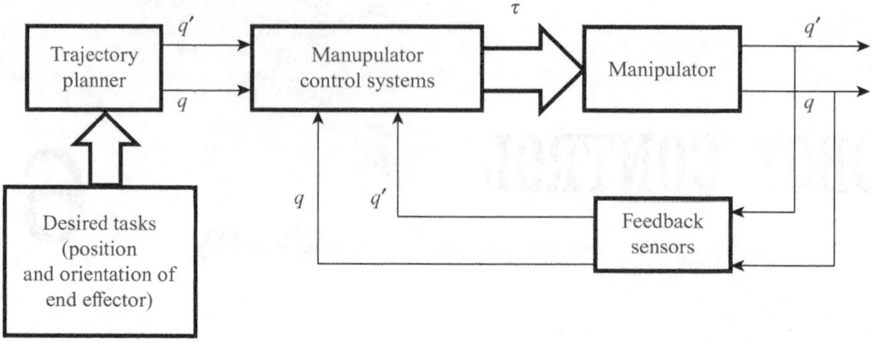

Figure 6.1 Block Diagram of Robot Control System

time functions such as tracking some curved surfaces. The manipulator control systems block creates the required torque τ for the robot. The manipulator goes according to this torque and produces position, q, and orientation, q', vectors that are required by robot for a particular task. These two vectors enter the feedback sensors block. Desired position and orientation of end-effector block gives data for required positions and orientations. The output signals from each block are vectors. This representation is the fittest example for a linear and a nonlinear system.

Euler–Lagrangian formulation is one kind of solving a nonlinear differential equation. Let us discuss on this formulation. This formulation can give errors due to following fixed and variable set points:

(a) Errors (q_{ij}) at every joint
(b) Nonlinearities (M_i) which are the sources of these errors
(c) G_i is the effect of gravitational pull, can be time varying.

The closed-loop dynamical equation connecting torque τ_i and all the above variables is given by

$$\tau_i = \sum_j M_{ij} q_{ij}'' + \sum_j \sum_k h_{ijk} q_j' q_k' + G_i, \qquad i = 1, 2, \ldots, m, \qquad (6.1)$$

where m is the total number of joints; i is the main variable; M_i, h_i and G_i are system nonlinearities; τ_i is the torque; q' and q'' are velocity and acceleration, respectively and j and k are the dummy variables.

The first term on the right side of Equation (6.1), which has single Σ indicates the potential energy. The potential energy depends on coefficient M_{ij}. The term with single Σ is the summation of all variables which work on q_{ij}''. The second term is the kinetic energy produced by q'_i and q'_k having double $\Sigma\Sigma$ and its coefficient is h_{ijk}. The final terms are G_i that indicate gravitational forces on joints.

Equation (6.1) does not include the effects of friction, backlash and also the effect of un-modelled and un-measurable vibrations (noise). These effects are highly nonlinear and difficult to model. An assumption is made here, that is, the effects of all these un-measurable and un-modelled nonlinearities contribute only a little to the torque, τ_i. Hence, Equation (6.1) has a little value. We are, hence, forced to use closed-loop control system to model the joints.

6.3 JOINT MOTION

It is to be noted that Equation (6.1) has to be computed in real time. In the system of Equation (6.1), the variables to be manipulated are, in a way, independent and do not produce a control within themselves. The main error is the steady-state error. Steady-state error, we mean, is the final error which can be corrected by some other means, and not by itself. The way of correction is that of using PID control or adaptive algorithm or by other ways. PID control is the concept of proportional (P) or integral (I) or derivative (D) control or a combinational form of controls. The adaptive algorithm is based on feedback error and its derivatives. This is a recursive way of reaching the output and takes time even in this era of fast computation. This is the method which a human uses when he (/she) is exposed to various control possibilities.

6.3.1 Linear Control Systems

In any closed-loop system, there will be a set of feedback error signals. They are to be corrected to zeros. There are number of ways that the final error signals can be corrected. Some of the simple control strategies are as follows:

(i) The on/off step control
(ii) Multiple step control
(iii) PID controls
(iv) Combinations such as P or (P and D) or (P and I) or (P, I and D)
(v) The (D and I) control never exists

In manipulator control, there are many other situations where better controls can be derived. They are partitioned control, adaptive control, computed torque control and so on. These controls are still in research stages and hence not suitable to discuss here.

The linear control schemes are particularly suited to systems whose behaviour can be mathematically modelled by a linear differential equation. Equation (6.1) is highly nonlinear and mostly time varying. Equation (6.1) can be linearized so that the signals can be approximated as linear functions. Therefore, any manipulator modelled as a second-order linear control system has to be within these restrictions.

The approximate linear model of an individual joint is achieved by ignoring the configuration-dependent nature of inertia and gravity forces. These simplified models are considered as satisfactory for pick and place and trajectory tracking purposes. Due to its high gear ratios, a vast majority of drives are considered as simpler and fitting. Moreover, there are many approximations in their trajectory computation which will compensate at gear ratios within themselves.

In addition, the natural and voluntary configurations of any industrial robotic systems are multi-input and multi-output (MIMO) systems. They can be, always, simplified to be a number of coupled single-input and single-output (SISO) systems. When there are n-joints, there have to be n-joint motions and all of them are, one way, dependent on each other. However, there are some methods which uncouple this coupled system. By the term 'uncoupling' we mean that the state variables can be made independent of each other. State variables can be uncoupled by selected ways such as state variable transformation. One restriction is that this is simplified as n-independent joints and the technique for simplified n-loop control can be applied. Readers who are beginners on second-order plants are recommended to refer books on dynamical control systems.

6.4 SECOND-ORDER SYSTEMS

This section helps to brush up introductory concepts of control theory and that of specifically the second-order systems. Here the effect of input–output characteristics of second-order plants is studied. If $x(t)$ is the input and $y(t)$ is the output of a second-order plant, then the relationship between them is

$$(\mathrm{d}^2y(t)/\mathrm{d}t^2) + a_1(\mathrm{d}y(t)/\mathrm{d}t) + a_0 y(t) = a_0 x(t), \tag{6.2}$$

where a_1 and a_0 are constants. The right-hand side of Equation (6.2) is intentionally set as a_0 in order to have unity gain.

The Laplace transform of this equation, with zero initial conditions, is

$$s^2 Y(s) + a_1 s\, Y(s) + a_0\, Y(s) = a_0\, X(s), \tag{6.3}$$

where $X(s)$ and $Y(s)$ are, respectively, the Laplace transforms of $x(t)$ and $y(t)$. The open-loop transfer function of the system is

$$\frac{Y(s)}{X(s)} = \frac{a_0}{\left(s^2 + a_1 s + a_0\right)}. \tag{6.4}$$

A more familiar way of writing Equation (6.4) is

$$G(s) = \frac{Y(s)}{X(s)} = \frac{\omega_n^2}{s^2 + 2\zeta\omega_n s + \omega_n^2}, \tag{6.5}$$

where ζ is the damping ratio (dimensionless) and ω_n is the natural frequency (rad./s) of the plant.

The damping ratio ζ is the ratio between normal damping (i.e. 1) and the damping given in the problem. Equation (6.5) indicates a mathematical model describing the plant dynamics. Many physical plants such as mass–spring–damper systems, LCR electrical circuits, fluid systems and thermal systems describe such characteristics.

Let $X(s) = R(s) = (1/s)$. Consider the closed-loop system with unity negative feedback

$$\frac{Y(s)}{R(s)} = \frac{\omega_n^2}{s(s^2 + 2\zeta\omega_n s + \omega_n^2)}. \tag{6.6}$$

The characteristic equation of second order is the one in the denominator, but without the term s.

The second-order characteristic equation of this linear closed-loop plant is written in a simplified form as

$$s^2 + 2\zeta\omega_n s + \omega_n^2 = 0. \tag{6.7}$$

Then the roots r_1 and r_2 are

$$r_1 = -\zeta\omega_n + \omega_n\,\mathrm{Sqrt}(\zeta^2 - 1), \tag{6.8}$$

$$r_2 = -\zeta\omega_n - \omega_n\mathrm{Sqrt}(\zeta^2 - 1), \tag{6.9}$$

where Sqrt $(\cdot)$ is the square root of $(\cdot)$.

Depending on the value of damping ratio ζ, three classes of roots arise from values of $\zeta = 1$, $\zeta > 1$ and $\zeta < 1$; the first one is named critical damping, the second is overdamped and the third is called underdamped. We analyse these situations one by one.

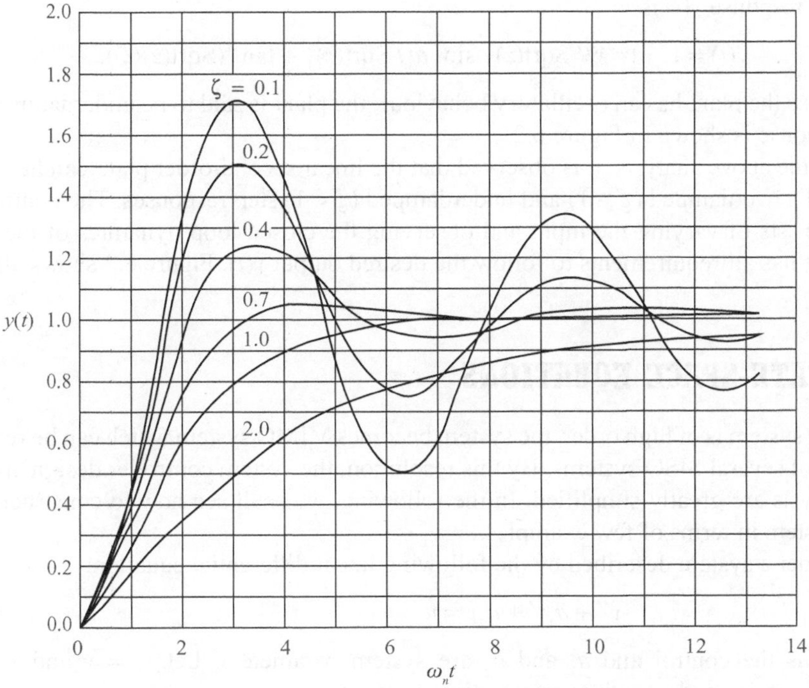

Figure 6.2 Second-order Step Responses

Case 1: Critical damping

For a damping ratio equal to unity ($\zeta = 1$), $r_1 = +j\omega_n$ and $r_2 = -j\omega_n$. Then the time response of the system derives to be

$$y(t) = [1 - e^{-\omega_n t} - \omega_n t e^{-\omega_n t}], \tag{6.10}$$

which is said to be critically damped without having any overshoot. Figure 6.2 shows this mode of operations.

Case 2: Overdamping

For a damping ratio more than 1 (i.e. $\zeta > 1$), the plant is said to have overdamped response. We define

$$z = (\zeta^2 - 1). \tag{6.11}$$

Also Sqrt(z) = square root of (z), then

$$y(t) = 1 - (e^{-\zeta\omega_n t}/(2\text{Sqrt}(z)))\,[(\text{Sqrt}(z) - \zeta)\, e^{-\omega_n t \text{Sqrt}(z)} + (\text{Sqrt}(z) + \zeta) e^{\omega_n t \text{Sqrt}(z)}]. \tag{6.12}$$

In this case, the plant has a sluggish non-oscillatory behaviour (time response); the plant is said to be overdamped. Figure 6.2 shows this time response.

Case 3: Underdamping

The roots are complex conjugate to each other when damping ratio is less than unity (i.e. $\zeta < 1$).

The time response $y(t)$ is

$$y(t) = 1 - \{e^{-\zeta\omega_n t}/\text{Sqrt}(z)\}\ \sin\{\omega_n t\ \text{Sqrt}(z)\} + \tan^{-1}(\text{Sqrt}(z)/\zeta). \tag{6.13}$$

In this case the plant has an oscillatory behaviour; the plant is said to be underdamped. The time response is shown in Figure 6.2.

From the above analysis, it is observed that the linear second-order plant can have critical ($\zeta = 1$), overdamped ($\zeta > 1$) and underdamped ($\zeta < 1$) step responses. The control of a plant consists of varying the input and observing the closed-loop dynamics of the plant. The plant has all requirements to follow the desired output $y(t)$. Figure 6.2 shows all three behaviours.

6.5 STATE-SPACE EQUATIONS

If a SISO system is of high order, the system becomes MIMO system which can be resolved in terms of several SISO systems. By this resolution, the system controller design and system analysis are greatly simplified. In the following, we shall see how to construct SISO linear system in terms of few examples.

Consider a system described by the following linear differential equation:

$$y'' + a_1 y' + a_2 y = u, \tag{6.14}$$

where u is the control and a_1 and a_2 are system parameters. Let $x_1 = y$ and $x_2 = y'$. Differentiating x_1 and x_2 with respect to time t we get

$$x_1' = x_2, \tag{6.15}$$

$$x_2' = y'' = -a_2 x_1 - a_1 x_2 + u, \tag{6.16}$$

$$\text{and } y = x_1, \tag{6.17}$$

where x_1 and x_2 are state variables. Equations (6.15)–(6.17) are state-space equations as derived from Equation (6.14).

Often the two equations are represented by matrix-vector form as

$$\begin{bmatrix} x_1' \\ x_2' \end{bmatrix} = \begin{bmatrix} 0 & 1 \\ -a_2 & -a_1 \end{bmatrix} \begin{bmatrix} x_1 \\ x_2 \end{bmatrix} + \begin{bmatrix} 0 \\ 1 \end{bmatrix} u, \tag{6.18}$$

$$y = \begin{bmatrix} 1 & 0 \end{bmatrix} \begin{bmatrix} x_1 \\ x_2 \end{bmatrix}, \tag{6.19}$$

or

$$x' = Ax + Bu \tag{6.20}$$

and

$$y = Cx, \tag{6.21}$$

where $x = [x_1\ x_2]^T$ is known as state vector. System parameter matrices are

$$A = \begin{bmatrix} 0 & 1 \\ -a_2 & -a_1 \end{bmatrix},\ B = \begin{bmatrix} 0 \\ 1 \end{bmatrix} \text{ and } C = \begin{bmatrix} 1 & 0 \end{bmatrix}. \tag{6.22}$$

The eigenvalues of state-space equations, Equations (6.15)–(6.17), can be calculated from characteristic equation $|sI - A| = 0$, where

$$|sI - A| = \left| s\begin{bmatrix} 1 & 0 \\ 0 & 1 \end{bmatrix} - \begin{bmatrix} 0 & 1 \\ -a_2 & -a_1 \end{bmatrix} \right| = \left| \begin{bmatrix} s & -1 \\ a_2 & s+a_1 \end{bmatrix} \right|, \tag{6.23}$$

$$= s^2 + a_1 s + a_2 = 0, \tag{6.24}$$

with eigenvalues

$$s_1 = [\,-a_1 + \text{Sqrt}(a_1{}^2 - 4\,a_2)]/2, \tag{6.25}$$

$$s_2 = [\,-a_1 - \text{Sqrt}(a_1{}^2 - 4\,a_2)]/2, \tag{6.26}$$

where Sqrt( ) is the square root of ( ). The characteristic Equation (6.24) and the state-space Equations (6.15)–(6.17) are the same and one can be derived from the other.

Reader can refer to Section 6.4 for (i) critically damped, (ii) overdamped and (ii) under-damped cases.

In fact, the state-space theory is the same as theory of solving the differential equation, but the presentations are different. Modern control principles are based on state-space theory. In general, the SISO dynamical system is described by nth-order linear differential equation as

$$y^{(n)} + a_1 y^{(n-1)} + \cdots + a_{n-1} y' + a_n y = b_m u^{(m)} + b_{m-1} u^{(m-1)} + \cdots + b_0 u \tag{6.27}$$

with $m < n$. Let the intermediate variable z be the solution to the differential equation

$$z^{(n)} + a_1 z^{(n-1)} + \cdots + a_{n-1} z' + a_n z = u \tag{6.28}$$

and define n state variables as

$$x_1 = z, \qquad x_2 = z', \ldots, x_n = z^{(n-1)}. \tag{6.29}$$

Then

$$\begin{bmatrix} x_1' \\ x_2' \\ \cdot \\ \cdot \\ \cdot \\ x_{n-1}' \\ x_n' \end{bmatrix} = \begin{bmatrix} 0 & 1 & 0 & \cdots & 0 \\ 0 & 0 & 1 & \cdots & 0 \\ \cdot & \cdot & \cdot & \cdots & \cdot \\ \cdot & \cdot & \cdot & \cdots & \cdot \\ \cdot & \cdot & \cdot & \cdots & \cdot \\ 0 & 0 & 0 & \cdots & 1 \\ -a_n & -a_{n-1} & -a_{n-2} & \cdots & -a_1 \end{bmatrix} \begin{bmatrix} x_1 \\ x_2 \\ \cdot \\ \cdot \\ \cdot \\ x_{n-1} \\ x_n \end{bmatrix} + \begin{bmatrix} 0 \\ 0 \\ \cdot \\ \cdot \\ \cdot \\ 0 \\ 1 \end{bmatrix} u. \tag{6.30}$$

and

$$y = [b_0 \;\; \ldots\ldots \;\; b_m \;\; 0 \;\; \ldots\ldots \;\; 0] \begin{bmatrix} x_1 \\ x_2 \\ \cdot \\ \cdot \\ x_{n-1} \\ x_n \end{bmatrix}. \tag{6.31}$$

In the above, we have discussed the state-space equation of linear systems. In fact, there are many different methods available to get the state-space equations.

Let us consider the following nonlinear differential equation. This is describing an inverted pendulum (one arm of robot) as

$$ml^2\,\Theta'' + mgl\,\sin\Theta = f, \qquad (6.32)$$

where Θ is angular variable (the pendulum output), the external input is f, m is the pendulum mass, l is the length of string and g is the gravity constant.

Let $x_1 = \Theta$ and $x_2 = \Theta'$, the state-space equations of the pendulum can be obtained as

$$\begin{bmatrix} x_1' \\ x_2' \end{bmatrix} = \begin{bmatrix} 0 & 1 \\ \dfrac{-g}{l}\sin x_1 & 0 \end{bmatrix} \begin{bmatrix} x_1 \\ x_2 \end{bmatrix} + \begin{bmatrix} 0 \\ \dfrac{1}{ml^2} \end{bmatrix} f. \qquad (6.33)$$

We shall next discuss Lyapunov stability. There are two approaches, the first method and the second method. Both are equally important, but the second method is universally adopted.

6.6 LYAPUNOV STABILITY

In 1960, Lyapunov stability theory has been formulated and is applicable to nonlinear system. Routh–Hurwitz criterion and Nyquist criterion are not generally applied to nonlinear systems. They are used to test stability of linear time-invariant systems and whether the linear time-invariant system poles are in the right half of the s-plane. If any pole is in the right half of the s-plane, we will declare that the system under test is unstable.

The nonlinear system is described as

$$x' = f(x, t), \qquad (6.34)$$

where $x\ \varepsilon\ \boldsymbol{R}^{n\text{x}1}$ is the state variable vector and $f(x, t)\ \varepsilon\ \boldsymbol{R}^{n\text{x}1}$ is the nonlinear function on vector x and time t.

In order to discuss the stability of the nonlinear system shown in Equation (6.34), the concepts such as state equilibrium point, stability in the sense of Lyapunov, asymptotic stability and exponential stability are first defined. Then, we investigate how to get these stabilities.

The state equilibrium point is x_e and the function $f(x_e, t) = 0$ for all time t.

The stability in the sense of Lyapunov:

Refer the two-dimensional Figure 6.3; it says any trajectory starting from any initial point x_0 will remain within portions between two circles $s(\delta)$ and $s(\varepsilon)$ and does not reach equilibrium point, x_e.

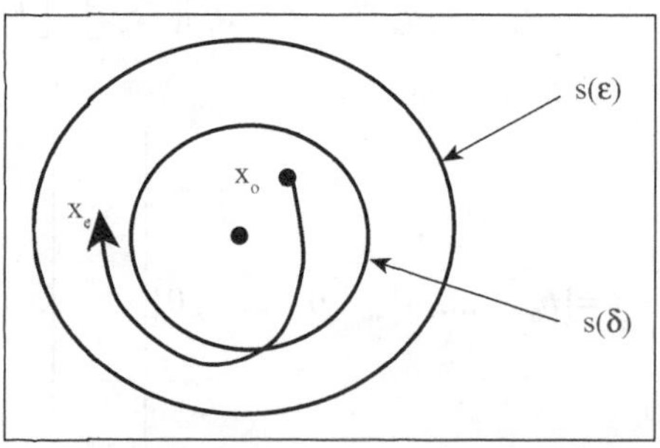

Figure 6.3 Stability in the Sense of Lyapunov

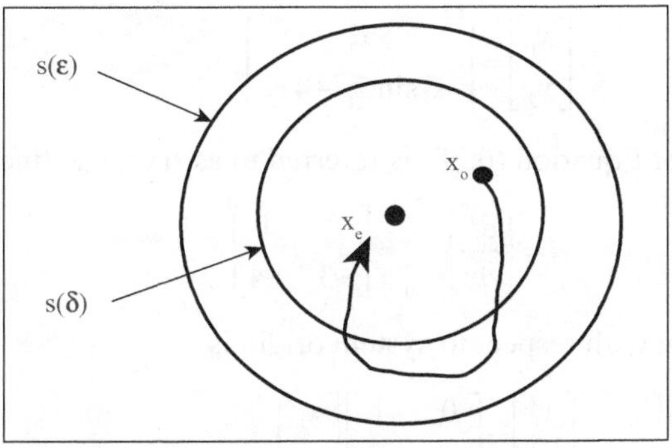

Figure 6.4 Asymptotic Stability

Asymptotic stability:

Refer the two-dimensional Figure 6.4; it says any trajectory starting from any initial point x_0 not only coming out from the circle $s(\delta)$ or entering circle $s(\varepsilon)$ will reach the equilibrium point, x_e, asymtotically.

Exponentially stable:

The state equilibrium point, x_e, is said to be exponentially stable if it is stable in the sense of Lyapunov; for each spherical region $s(\delta)$, the norm $\|x_0 - x_e\| \le \sigma$; where $x_0 = x(t_0)$ has a bounded initial value at $t = t_0$. The norm $\|x - x_e\| \le \alpha e^{-\lambda t}$ as $t \to \infty$ with constants $\alpha > 0$ and $\lambda > 0$. λ is the convergence rate. Typically, the exponential stability is a special case of asymptotic stability.

In conclusion

State equilibrium point:	$f(x_e, t) = 0$, for all t
Stability in the sense of Lyapunov:	$\|x_o - x_e\| \le \delta,\ \delta > 0$
Asymptotic stability:	$\|x - x_e\| \le \varepsilon,\ \varepsilon > 0$
Exponential stability:	$\|x - x_e\| \le \alpha e^{-\lambda t}$ with $\alpha, \lambda > 0$.

6.7 LYAPUNOV FIRST METHOD

A nonlinear system is given by

$$x' = Ax, \tag{6.35}$$

where

$$A = \{\partial f/\partial x\}_{x=0} = n \times n \text{ matrix } \{\partial f_i/\partial x_j\},\ i, j = 1, 2, \ldots n. \tag{6.36}$$

The linear system in Equations (6.35) and (6.36) is generally called linearized system of nonlinear system. For local stability of nonlinear system we have the following concept:

If all the eigenvalues of system in Equation (6.34) lie in the left half of the s-plane, the nonlinear system is stable. Otherwise, the system is said to be unstable.

Consider the following nonlinear system:

$$x'' + 4x' + 3 \sin x = 0. \tag{6.37}$$

Let $x_1 = x$ and $x_2 = x'$, we obtain the following state-space equation; the origin (0, 0) is the system equilibrium point.

$$\begin{bmatrix} x_1 \\ x_2 \end{bmatrix} = \begin{bmatrix} x_2 \\ -3\sin x_1 - 4x_2 \end{bmatrix}. \tag{6.38}$$

The right-hand side of Equation (6.38) is referred to as $f(x)$. The linearized system has

$$A = \left\{ \frac{\partial f}{\partial x} \right\}_{x=0} = \begin{bmatrix} 0 & 1 \\ -3 & -4 \end{bmatrix}. \tag{6.39}$$

The linearized system with respect to system origin is

$$\begin{bmatrix} x_1' \\ x_2' \end{bmatrix} = \begin{bmatrix} 0 & 1 \\ -3 & -4 \end{bmatrix} \begin{bmatrix} x_1 \\ x_2 \end{bmatrix} \tag{6.40}$$

with two eigenvalues in the left half of the s-plane. If atleast one of the eigenvalues of the linearized system of Equation (6.40) has a positive real part, then the nonlinear system is unstable.

6.8 LYAPUNOV SECOND METHOD

It is difficult, and not impossible, to find a linearized model for the nonlinear system. Lyapunov second method introduces a positive-definite function named $V(x, t)$. This is referred to as a candidate of Lyapunov function. Since we do not know how the 'candidate' behaves, we continue to mention this name. This is an energy-based function. Its derivative, $V'(x, t)$, has to be less than zero (negative) for a system to be stable.

Let us consider a nonlinear system described by

$$x' = f(x, t), \tag{6.41}$$

where $x \in \boldsymbol{R}^{n\times 1}$ is the state variable vector, $f(x, t) \in \boldsymbol{R}^{n\times 1}$ is a nonlinear function on x, t is the time and $f(0, t) = 0$ for all $t \geq 0$. This method can also be applied to linear systems. If there exists a scalar function $V(x, t)$ which has a continuous first-order partial derivative in a ball s, with radius r and centred at the system origin, such that

$$\text{(i) } V(x, t) > 0 \text{ and (ii) } V'(x, t) < 0 \tag{6.42}$$

then $V(x, t)$ is said to be a Lyapunov function and the equilibrium point at the system origin is asymptotically stable.

Table 6.1 lists the various attributes of Lyapunov function.

Let us consider a third-order nonlinear system as

$$\begin{aligned} x_1' &= x_1(x_1^2 + x_2^2 + x_3^2 - 1) + x_2 + x_3, \\ x_2' &= -x_1 + x_2(x_1^2 + x_2^2 + x_3^2 - 1) + x_3, \\ x_3' &= -x_1 - x_2 + x_3(x_1^2 + x_2^2 + x_3^2 - 1). \end{aligned} \tag{6.43}$$

System origin is the system equilibrium point. We use Lyapunov second method of stability. Let us define the following candidate of Lyapunov function:

$$V(x) = x_1^2 + x_2^2 + x_3^2. \tag{6.44}$$

Differentiating $V(x)$ with respect to time

$$V'(x) = 2x_1x_1' + 2x_2x_2' + 2x_3x_3'. \tag{6.45}$$

After some manipulations, we get

$$V'(x) = 2(x_1^2 + x_2^2 + x_3^2)(x_1^2 + x_2^2 + x_3^2 - 1). \tag{6.46}$$

Table 6.1 Attributes of Lyapunov Second Method

No.	Attributes	Explanation
1	Physical meaning?	The total energy is continuously decreasing for stable system. That is why, $V(x, t) > 0$ and $V'(x, t) < 0$
2	Any general method?	No general method for finding another candidate of Lyapunov function. One's experience demonstrates the ability of determining one more candidate of Lyapunov function
3	Uniqueness?	More than one Lyapunov function is possible
4	Sufficient condition?	If you cannot find a Lyapunov function, it does not mean the system is unstable
5	Stability of system origin?	Lyapunov second method is a sufficient condition. If the second method shows the convergence condition, the system stability is ensured

This shows that when the trajectory is within the following spherical region

$$x_1^2 + x_2^2 + x_3^2 < 1, \tag{6.47}$$

then the first-order derivative of $V(x)$, $V'(x) < 0$. The system origin is asymptotically stable if the trajectory is within the spherical region as indicated by Equation (6.47).

6.9 CONTROL UNIT

The arm of a robot can be moved (or constraint to move) in any way which is suitable to robotics engineer. The control unit is depicted in Figure 6.5. All the blocks are required for a minimum system intended for a successful functioning of robots in industries. Table 6.2 lists the functions of each block within the control unit.

Table 6.2 Control Unit Items and Their Functions

No.	Items in Control Unit	Their Functions
1	Peripheral interface	Checks the suitability of peripherals in sending suitable data. Program interpretation: a higher level instruction is decoded into a number of lower level instructions
2	System supervision	Checks whether the system is suitable for further processing
3	Coordinate transformation	This transforms the higher level instructions to a lower level
4	Computer memory	For data storage, retrieval and decision making
5	Trajectory computation	For trajectory computation, storage of data and retrieval
6	Decision making and monitoring	Supplying the additional data to trajectory computation
7	Servo control	To send required control signals to servomotors
8	Sensory data processing	To supply the required data to decision making and servo control

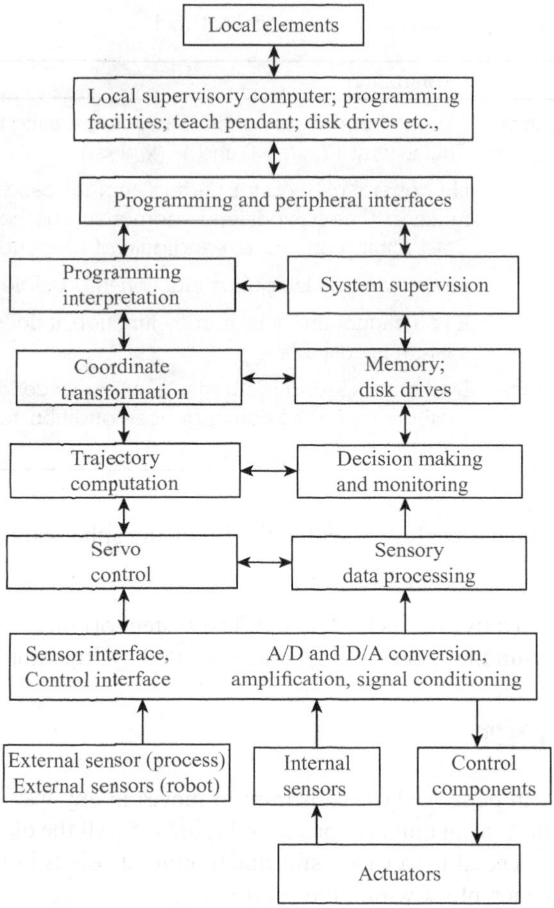

Figure 6.5 Schematic of Control Unit Block

There can be additional processing that are related to the current jobs and are listed in Table 6.3.

Table 6.3 Extra Processing and their Needs

No.	Processing	Their Needs
1	A/D and D/A conversion	To supply the necessary data in terms of A/D and D/A conversion
2	External process sensors	To test the process sensors have all the required data
3	External robot sensors	To direct the robot to finish its task
4	Internal sensors	To test whether the internal sensors have given relevant data
5	Control components	Input and output devices to supply and retrieve data
6	Actuators	Whether actuators such as conveyor systems and the welding rods are ready with all necessary data

6.10 ELECTRIC, HYDRAULIC AND PNEUMATIC DRIVES

Electric drives are now commonly applied to several drive systems. In the past, hydraulic and pneumatic drives were used in the robot systems. The main reason being hydraulic drives are powerful and pneumatic drives are economical. Electrical drives have come in the way between high power and economical. The electric drives can be far better than hydraulic drives and pneumatic drives. We consider all these three drives for comparison in terms of advantages and disadvantages. Table 6.4 shows advantages and Table 6.5 shows disadvantages of these drives.

Table 6.4 Advantages of Electric, Hydraulic and Pneumatic Drives

No.	Electric Drives	Hydraulic Drives	Pneumatic Drives
1	Electric motors are lighter than their counter parts in hydraulic and pneumatic drives	Hydraulic drives are heavier than electric drives	Pneumatic drives are lighter than the hydraulic drives; but some are heavier than electric drives
2	Accuracy and repeatability are better than the other two	Hydraulic drives are not as good as electric drives in repeatability and accuracy	Though these drives are not as good as other drives, they are used in several applications in industries
3	These drives are very acceptable environmentally being quiet and clean	A high power to size ratio is obtained with hydraulic drives and actuators	These are suitable to applications where hydraulic and electric drives are not suitable
4	These drives are easily maintained and repaired; structural components are light weight and cheaper	Hydraulic actuators are used in several applications such as stacking cement bags in a lorry	It is the cheapest form of drive; components, in addition to air, are easily available in factories
5	These drive systems are well suited for electronic control; new motor designs are coming into practice that improve power to weight and power to size ratios	These drives generally have a large load carrying capacity. Large forces can be applied directly at any required location	These drives have a few moving parts; hence, they are well suited for fast work cycles

Table 6.5 Disadvantages of Electric, Hydraulic and Pneumatic Drives

No.	Electric Drives	Hydraulic Drives	Pneumatic Drives
1	The main requirement is that of having a sort of mechanical transmission. This adds mass and unwanted movement consuming additional power	They can be less reliable than electric drives; leakage can cause loss in performance and can be a contamination in work area	Only a limited sequence operating speed is often available. If mechanical stops are used, resetting the system can be very slow

(Continued)

Table 6.5 Continued

No.	Electric Drives	Hydraulic Drives	Pneumatic Drives
2	With the introduction of direct drive motor systems, the above disadvantage can be overcome gradually	Servo control of hydraulic systems is complex and is not widely understood as electric servo control	Pneumatic drives are not suitable for moving heavy loads under precise control due to compressibility of air
3	Newer brushless motors do allow electric robots be used in some fire-risk applications such as spray painting and in tunnelling	For smaller robots hydraulic power is usually not economically feasible since the cost of hydraulic components does not decrease with size	If moisture enters the units and ferrous metals have been used, then the damage due to individual components is larger

6.11 INDUSTRIAL VISION SYSTEM

Figure 6.6 shows a computer-based industrial vision system. This basically has a camera, lighting and a robot. The system has a camera for capturing the images, a lighting provision for directing the light onto the object and a robot for some processing or pick and place operation. In addition, the system may have controllers for robot operation and one or more vision cameras with their controllers. All these elements are required for a useful industrial vision system.

Now let us understand a set of requirements of industrial vision system:

(a) The cost should be as low as possible
(b) The system should be simple to program and easily understood
(c) The system should be easily reprogrammed

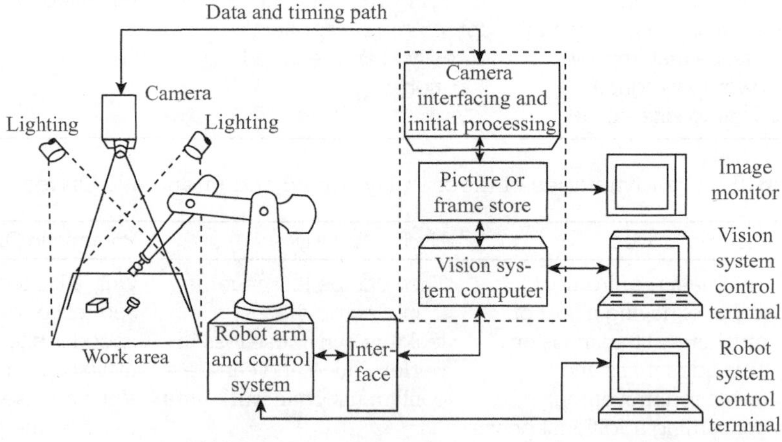

Figure 6.6 Industrial Vision System

(d) It must be fast in operation
(e) The hardware is to be rugged and reliable
(f) The system must have high integrity and less error prone
(g) The system should be able to use lighter scene of illumination

The investigators should be fully conversant with robot configuration that includes drives and control systems. The likely interfacing requirements are to be noted to anticipate the potential complexity of achieving integration. Although the first installed robot is a single robot, the possibility of applying more than one has to be considered. Initially, a survey of working with one robot has to be done. The raw materials and the parts at the dispatch area should be utilized to carry out the required operation. As the normal workflow is traced, the following situations should be looked at as they provide ideal conditions for a factory to be robotized.

The conditions are as follows:

(a) Operations that are simple and repetitive, for example, unloading die casting machines or pick and place operations
(b) Jobs where the cycle time is not too short, probably over 3 s and preferably greater than 5 s
(c) Robot tolerance on parts and tools such that an average precision will be able to carry out the required operation
(d) Heavy jobs with a heavy product will probably require a very robust type of robot
(e) The operation being considered should not have an integral inspection element
(f) Situations where the worker is dissatisfied; this may be due to the work being tedious, dirty or socially isolated
(g) Toxic or inflammable atmosphere that creates hazardous environment
(h) The jobs require little intelligence judgement
(i) The product variety should not be too great; this will minimize the tooling costs
(j) Design and manufacturing costs are to be low
(k) Situations where more than one shift can be worked, thus making financial justification easier
(l) Areas with quality problems. Not suitable for less or no quality production
(m) Situations where parts have been previously positioned and oriented. There can be dangerous situations in which parts can come out from fixtures

6.12 INSPECTION USING INDUSTRIAL VISION

From Figure 6.6, it can be seen that the robot realizes an industrial vision using a camera. The applications of industrial vision include measuring external and internal dimensions of an object and categorization. Artificial intelligence (AI) is an intelligent component with which the robot works. AI categorizes whether the object belongs to a class or otherwise.

Qualitative inspection: This is an attractive area for investment since it is estimated that visual means can give a good satisfaction for robot engineers and technicians in doing these jobs.

Qualitative inspection involves the vision system for checking a product on an accept or reject basis. A crack on a bottle or a flaw in moulding are a few examples for qualitative inspection.

Quantitative inspection: Quantitative inspection involves using the system for measuring the dimensional feature of a product or component. Examples are measuring the width of steel strip in a steel mill, gauging the length of a shaft during or after turning and inspecting a component dimensionality on a conveyor belt.

Identification: Rather than checking an attribute or measuring a feature, the camera-based system is used to identify and classify the components. The vision process may involve identifying an individual item in a bin of jumbled components, reading the characters from a label or identifying a component by its shape. The outcome of identification of a part may be used as an information input to a subsequent processing the system.

Information for decision making: The visual data are used, in this case, to help the robot to decide what to be done next. For example, laser vision is used in seam welding; the visual data give immediate guidance to the robot to direct the welding torch thus ensuring an optimum quality weld. In more sophisticated applications vision can be used in conjunction with AI techniques to determine the order of assembly of components.

The complete robotic artificial vision system normally consists of eight components and is depicted in Figure 6.6.

(a) Camera: without this the robot vision cannot be realized

(b) Camera interface: a hardware that is essential to interface the cameras

(c) Initial image processing equipment: this hardware has to be with the camera or otherwise

(d) Frame grabber: catches an area by which the camera vision has to be addressed

(e) Microcomputer: an initial image processing system

(f) Interfacing equipment: this generates an area with which a computer vision has an address

(g) Scene illumination system: lighting arrangements for the vision camera to identify parts

(h) Robot or other system: to realize a robotic camera-based system

6.13 CAMERA

Two types of cameras are being used in industrial vision-vacuum type and solid-state type. Photodiode cameras are the first generally used as solid state and are still popular. However, relatively new solid-state cameras are being used for its light weight and are thus most economical. Among the most recent solid-state varieties, charge-coupled devices are being used advantageously. Figure 6.7 shows solid-state cameras.

Manufacturers prefer to produce solid-state cameras; these cameras manufactured by them are of the following qualities:

(a) Low weight: Using large-scale integrated circuits, it is possible to include the video amplifier, the pulse generator and the scanning logic circuits–all within one chip. Hence, it is rugged and weighs only a few hundred grams.

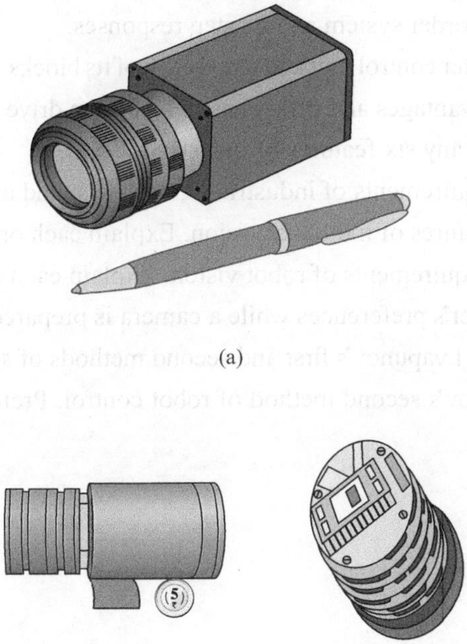

(a)

(b)

Figure 6.7 Solid-state Cameras: (a) Length Compared with Writing Pen and (b) Size Compared with Coin

(b) Small size: A solid-state camera will probably be at least half the diameter and 0.1 times the length when compared with tube-type cameras

(c) Longer life: Solid-state camera has a longer life and is more reliable having longer meantime between failures

(d) Most economical: Solid-state cameras consume only a few milliwatts of power compared with the tube type which consumes a few watts

(e) Risk of damage: There is no risk of damage by high light intensities. On the other hand, it is possible to damage tube-type cameras by exposing them to intense light source

(f) Solid state: Solid-state cameras have a broader temperature operating range at the lower temperature ranges

(g) Blooming: The disadvantage of solid-state cameras is that they can suffer from blooming; that is, if the solid-state diode is exposed to severe light source for longer time, the blooming effect appears

REVIEW QUESTIONS

1. List several varieties of reaching task points in robot control and explain them.
2. What are some varieties of PID control? Explain them.

3. Explain second-order system and its step responses.
4. Draw the computer control unit and name each of its blocks. What are their functions?
5. What are the advantages and disadvantages of three drive systems.
6. List and explain any six features of industrial vision.
7. What are the requirements of industrial vision? List and explain any 10 items.
8. List any four fcaturcs of industrial vision. Explain each one of them.
9. List any eight requirements of robot vision. Explain each one of them.
10. List manufacturer's preferences while a camera is prepared for industrial vision.
11. List and explain Lyapunov's first and second methods of system stability.
12. Explain Lyapunov's second method of robot control. Preferably illustrate using an example.

ROBOT PROGRAMMING AND WORK CELL 7

One's ability to perform a given task competently decreases in proportion to the number of people watching

Mark R. Frank

This chapter covers the following key topics:

Introduction – Language Structure – Operating System – Robot Programming Languages – Current Programming Languages – Application – Robot Motion – SCORBOT-ER – Example – Sensor Integration – Robot Work Cell – Work Cell Lay-outs – Interference Problems – Example – Further on Robot Work Cell

7.1 INTRODUCTION

The sophistication of the user interface has been improved so that the manipulators and other industrial equipment are applied to more and more demanding applications. In fact, most of the challenges of the design and use of industrial robots focus on the aspects and complexity of problems.

The term 'work cell' is used to describe a collection of local machine tools, one or more robots, conveyor systems, part feeders and fixtures. At next higher level, work cells are interconnected so that the factory can control a factory flow. Hence, programming the robots is often considered a variety of interconnected machines in an automated factory work cell.

Teach pendant is humanly handled button box that allows control of robot joints. This involves moving a robot to a desired goal point and recording its positions in a memory that would need during feedback. During the teach phase, the user would guide the robot through an interaction with data of teach pendant; and during play phase, the user tests whether the robot has reached the desired stage.

This chapter discusses two aspects of robotics – programming the robots and forming the robot work cell. Both aspects are required for successful industrial robotics. The first aspect is a method of controlling a robot and the second is of controlling the robot so that it completes the industrial task successfully. The first aspect needs a way of assembling a set of instructions (programming) and the second needs a way of assembling the local equipment (work cell).

7.2 LANGUAGE STRUCTURE

The robot operating system is depicted in Figure 7.1. Some of the elements such as supervisory, compile, edit and execute modes are indicated. These are required for program languages that are to be run. This system is beyond the first generation; let us call this as second generation though some of the first-generation language statements are also characterized in the second-generation language statements.

7.2.1 Operating System

In using textual languages, the programmer looks into a Cathode Ray Tube (CRT) monitor, alphanumeric key board and a teach pendant. There should also be a medium of storing the program such as magnetic tape or disk. Nowadays, the conventional disk has become minidisk so that language statements are stored and executed at a faster rate. There should be some mechanism that indicates a programmer whether to write a new program or edit existing one or to execute the current one entirely to perform a set of new robot operations. This functioning is called operating system. This term is not new to computer scientists to describe a software that will do the execution to the satisfaction of the software experts. The purpose of the operating system is to facilitate the functioning of robot by the programmer. In this way, the operating system helps the programmer to maximize the performance and efficiency of robots and the associated circuits.

The robot language operating system has the following three basic modes of operations:

(a) Supervisory mode
(b) Compile and edit mode
(c) Execute mode

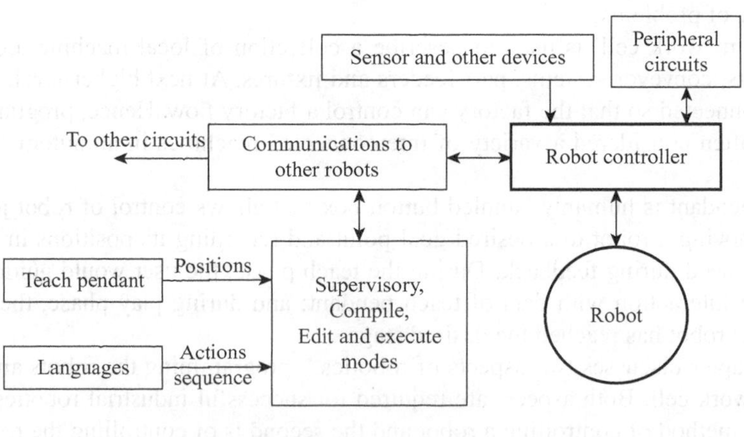

Figure 7.1 Robot Operating System

The supervisory mode has to accomplish the overall supervisory control of robotic system. In this mode, the user can define locations in space using teach pendant, set the speed control, store the relevant program, transfer programs, store the program back in to control memory or move back and forth between other modes of edit and execute.

The compile and edit mode, together, provide an instruction set that allows user to compile and later, or then and there, edit existing program. In doing so, the user can modify or insert changes in the existing program. The operation of editing the program may differ from the one which has been already edited in a different language. This includes modifying the instructions and writing new instructions. The program already edited is processed by the operating system using a compiler. That is, LISP (discussed later) is a robot language processed by an interpreter. A compiler is a program which can be understood by a robot controller that makes a language translation and executed by a robot controller.

Execute mode runs the program. The robot performs the sequence of instructions in the program during the execute mode. The programmer can test a new program (separate languages) by employing debugging procedures that are built within the language. For example, the program may call for the manipulator to exceed its joint limits in moving from one point to another. Both points have been identified (or taught) to the robot. However, due to exceeding joint limits, the robot would stop and the reason for stopping is printed out in the monitor. This condition can be overcome when the points are redefined or by adjusting the program instructions.

In addition to these, there are facilities for teach pendant, sensors and other peripheral interfaces. All these are 'speaking' to robot controller, the main and important portion to execute the control of the robot. There is another communication channel which monitors other robots within the factory.

7.3 ROBOT PROGRAMMING LANGUAGES

Artificial intelligence (AI) is immensely used in programming languages. To implement the problem solving techniques, one must use the programming languages to be discussed in this chapter. Some high-level languages (such as BASIC and FORTRAN) can be used to implement the same AI concepts. These languages are not ideal since they are not symbolic languages. Symbolic languages use arbitrary symbols to represent and manipulate almost anything. Numeric data alone will not be useful in this case.

Some of symbolic languages are listed in Table 7.1. The first is the LISP program, developed by John McCarthy in 1958. LISP is the oldest programming language currently in use. Other AI languages, such as MACLISP, INTERLISP and QLISP evolved from LISP. Knowledge representation, logical reasoning and common sense reasoning evolved from LISP. Others such as PLANNER, CONNIVER, SAIL, PROLOG and Fuzzy were developed to improve the features of LISP which are being applied in various specific areas.

There are reasons for not including a thorough study of AI in this text. Some of the fundamental AI concepts that must be incorporated in various aspects are communication, vision, navigation and tactile sensing. An industrial robot has these features. It can be seen from Table 7.1 that new languages keep evolving continuously. LISP is presently the standard one with the AI language field evolving. An interested reader can refer the books on LISP.

Table 7.1 Summary of Current AI Languages

No.	Languages[a]	Date Developed
1	LISP	1958
2	QLISP	1960
3	MACLISP	1966
4	POP-2	1967
5	SAIL	1969
6	PLANNER	1971
7	CONNIVER	1972
8	PROLOG	1977
9	FUZZY	1977
10	INTERLISP	1978
11	FOL. PROLOG	1979

[a]AI programming languages continue to have more modern concepts.

7.3.1 Current Programming Languages

There are several ways of describing robot programming languages–five out of them are usually pointed out.

Specialized manipulation languages: VAL is a kind of manipulation language for controlling industrial robots. Developed by Unimation Inc., VAL is of a general purpose language which cannot handle floating point numbers or character strings and subroutines cannot pass arguments. A more recent version V-II provided these features. Some statements are as follows:

```
APPRO P1, 50
MOV P1
DEPRT 50
```

Here 50 is a distance in mm. APPRO (approach), MOV (move) and DEPART are executed in joint interpolated motions. The current incarnation of this language, V+, includes many new features. The new features include subroutine package and matrix manipulations.

AL: Another example of a specialized manipulation language is AL developed by Stanford University. Although this language is now relic of the past, some features are still not found. Force control and parallelism are not included in this version.

Robot library for an existing computer language: These robot languages have been developed starting from a popular computer language and adding a library of robot-specific subroutines. Pascal is one kind of general purpose language which is used for frequent calls to the predefined subroutine package for robot specific needs. An example is the language AR-BASIC from American Cimflex, essentially a subroutine library for a standard BASIC language. JARS, as developed by NASA's Jet Propulsion Laboratory, is a kind of language based on Pascal.

Robot library for a new general purpose language: These programming languages have been developed by first creating a new general purpose language as a programming base and then supplying a library of predefined robot specific subroutines. Examples are RAPID developed by ABB robotics, A Manufacturing Language (AML) developed by IBM Robotics and KAREL developed by GMF Robotics.

Task level programming languages: This level of robot programming is embodied in task-level programming languages. These languages allow a programmer to command robot's desired subgoals of the task directly, which is required to specify the details that the robot is to take. In such a system, the programmer is able to include instructions in the application program at a significant higher level than in explicit robot programming language. In the task-level programming, the robot system has to decide and act according to the task being instructed. For example, if an instruction to 'grasp the nut' is issued, the system has to plan a path that avoids all obstacles around and performs the grasping of the nut. In contrast, in an explicit programming language, all the positions of obstacles are to be taught to the robot by the programmer. True task-level programming of manipulators does not exist yet, but it has been an active research topic today.

There are other varieties of robot programming languages. In the following, we list only three of them as the others are not required at this context.

LISP: LISt Processing: The mathematical and logical operations are embedded in the statements. Some example statements are as follows:

```
(PLUS 3 4)                      7
(TIMES 5 7)                    35
(ADD 1 8)                       9
(PLUS (PLUS 2 3)(PLUS 4 5))    28
```

AML – subroutine oriented: There are three statements in the AML language – executable statement, variable declaration statement and subroutine declaration statement.

RAIL: RAIL is developed by Automatrix Inc. and is used in Robovision and Cybervision statements.

7.3.2 Application

Figure 7.2 shows an automated work cell that completes a small subassembly in a manufacturing process. The following steps are to be completed by the manufacturing process:

1. The conveyor is signalled to start; it is stopped when the vision system reports that the bracket (part) or object is detected.
2. Using the output of the vision system, the manipulator grasps the bracket. When there is no bracket, the robot system moves away and the vision is repeated. If the bracket is found and vision is successful, the conveyor is signalled to stop.
3. The bracket is placed on the fixture. At this point the conveyor is signalled to move one step ahead.
4. The bracket-pin assembly is grasped by the robot and is placed in the press.
5. The press is commanded to actuate and it presses the pin the rest of the way.
6. By force sensing, the assembly is checked for the proper insertion of the pin.

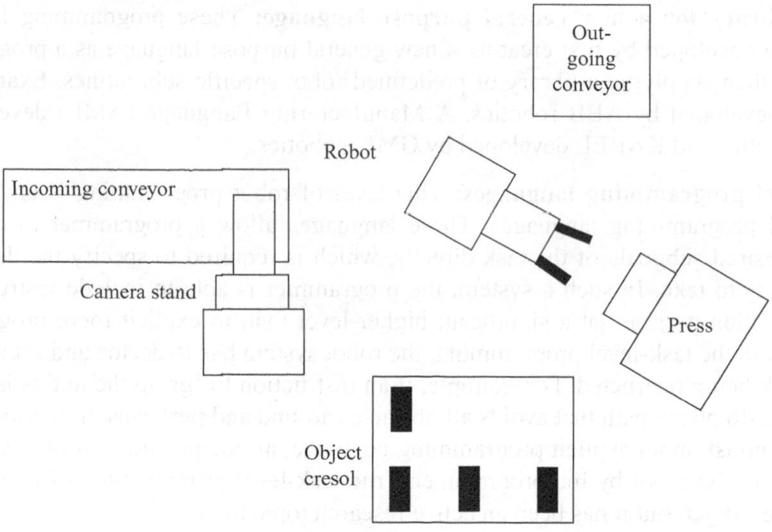

Figure 7.2 Example Illustration

7. If the assembly is good, the robot places the finished part on to outgoing conveyor. If the pallet is full, the robot signals to the operator.
8. Go to step 1

This is a task possible by today's industrial robot. A good programming environment offers the programmer a productivity. A programmer is to continually repeat 'edit-compile-run' cycle of a compiled language. Hence, most robot programming languages are now interpreted so that individual language statements can be run at a time during program development and debugging. Many of the language statements cause movements of physical devices. Only a tiny amount of time is required and this duration is insignificant for the manufacturing line. Typical programming support such as text editors, debuggers and a file system is also required.

7.4 ROBOT MOTION

The basic function of a robot programming is allowing descriptions of desired motions of a robot. Motion statements can describe either joint interpolated motion or Cartesian coordinate straight line motion or both. They allow the user to specify via points and the goal points which one of the above-mentioned motions to use. In addition, the user has the control over speed and duration of the robot functioning.

To illustrate various syntaxes for motion primitives, we consider the following manipulator motions example: move to position location *A* then to location *B* through straight line then move without stopping through via1 and come to location *C*. Assuming all these points have been taught by teach pendant or described textually, this program segment can be written as follows:

In VAL II	*In AL (controlling manipulator 'garm')*
Move *A*	Move garm to *A*;
Move *B*	Move garm to *B* linearly;
Move via 1	Move garm to *C* via via1;
Move *C*	

Most languages have similar syntax for simple motions statements. Differences in the basic motion primitives from one robot program to another become more apparent if we consider features such as the following:

1. Ability to do math structured types such as frames, vectors and rotation matrices
2. Ability to describe geometric entities such as frames
3. Ability to impart constraints on velocity of a particular move
4. Ability to describe to go fast

7.4.1 Example

EXAMPLE 7.1

This is on VAL programming. Consider the following program segment called PALLET: to sequentially move nine objects on (three rows × three columns) pallet to a location called MACH.

Solution:

The VAL program goes in the following way even though there are many alternatives to write and subsequently compile the statements.

```
                                                  SOME COMMENTS
 .PROGRAM PALLET                                    Program PALLET.
      1.        SET HOLE = CORNER
      2.        SETI COL = 0                      Initiate IMMEDIATELY
      3.  100   SETI ROW = 0                      a row and a column
      4.  200   GOSUB PALLET.SUB                  A call to subroutine
      5.        SHIFT HOLE BY 50.00, 0.00, 0.00
      6.        SETI ROW = ROW + 1                This is an increment of a row
      7.        IF ROW LT 3 THEN 200              If row < 3, jump to 200
      8.        SHIFT HOLE BY -150.00, 50.00, 0.00
      9.        SETI COL = COL + 1                This is increment of a column
     10.        IF COL LT 3 THEN 100              If column < 3 jump to 100
   .END                                           END statement of main

 . PROGRAM PALLET. SUB                            Subroutine PALLET.
      1.        OPENI 0.00                        Gripper opening immediate
      2.        APPROS HOLE, 100.00               Straight line approach
      3.        MOVES HOLE                        Straight line move
```

4.	CLOSEI 0.00	Gripper closing immediate
5.	DEPART 50.00	50 mm departure
6.	APPROS MACH, 50.00	Approach in straight line
7.	MOVES MACH,	Straight line moving
8.	OPENI 0.00	Open immediately
9.	DEPART 50.00	Departure 50 mm
10.	RETURN 0	Return from subroutine
.END		END statement of subroutine

FURTHER COMMENTS

The CORNER and MACH locations represent the two precision points associated with this program.

The numerical distances associated with program are in millimetres.

Each pallet is away from the other by 50 mm.

SETI, CLOSEI, OPENI are Immediate Operations.

PUMA 560 takes considerably less time than 1 minute to complete this program depending on the distance between MACH and CORNER as well as the select speed.

7.5 SCORBOT-ER

The Eshed Robotec Ltd has designed the SCORBOT-ER system. The SCORBOT-ER system has been developed both for industrial applications and for educational purposes. This system with minimum configuration is shown in Figure 7.3.

The arm of this robot is driven by DC servomotors. The transmission system includes gears, timing belts and a lead screw. A set of optical encoders are included on all motor shafts. This set tells the controller the position and velocity information.

The teach pendant is an important device to teach the robot controller the current and future motions of the robot manipulator. Figure 7.4 illustrates the teach pendant in detail. Here, teaching means that we are allowed to use or teach the robot to determine the via points and the path to follow. This is required for future basic applications. We have thus full control over each axis movement. The end effector can perform its required operations for a specified application.

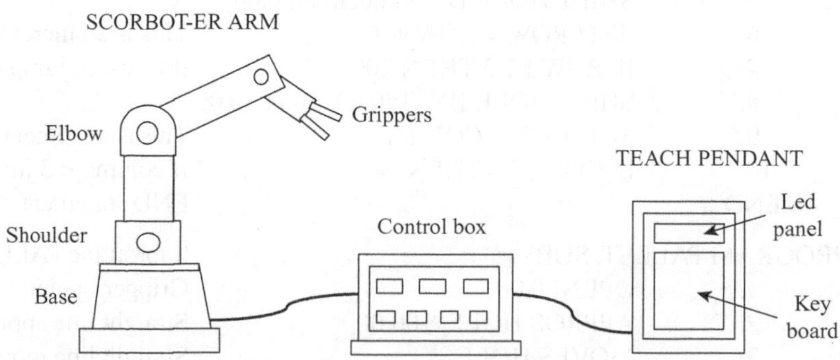

Figure 7.3 Basic Connections of SCORBOT-ER Robot System

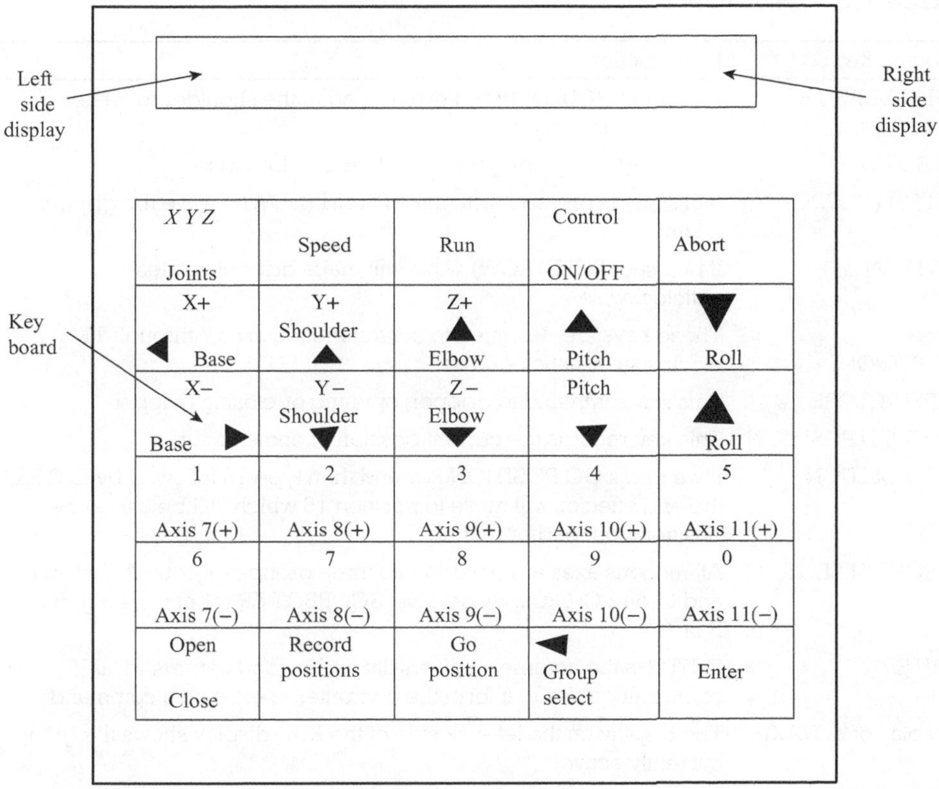

Figure 7.4 Teach Pendant of SCORBOT-ER

Table 7.2 Control Keys and Their Functions

Control Key	Key Function
SPEED	The SPEED key is used to set up the current motion speed. If we press 'SPEED ± number + ENTER' then the SPEED of the axis increases (or decreases) by its percentage number
JOINTS/*X*, *Y*, *Z*	If we press this key, the motion along the *X* or *Y* or *Z* axis is modified
RUN	This key is used to run an existing program saved in the hard disk
CONTROL ON-OFF	This key, if pressed, will make the CONTROL ON; if pressed again it makes the CONTROL OFF
ABORT	By pressing this ABORT key, the complete control to the robot is switched off
SHOULDER/*Y*+	This key makes the SHOULDER to rotate clockwise
BASE/*X*+	By pressing this key, robot base offers a clockwise rotation
ELBOW/*Z*+	If pressed, the elbow joint will rotate clockwise
PITCH (UP)	This position will make PITCH (UP) wave the gripper up
ROLL (CW)	If pressed, ROLL (CW) wave will make gripper rotates clockwise
BASE/*X*–	By pressing this key, robot base offers an anticlockwise rotation

(*Continued*)

Table 7.2 Continued

Control Key	Key Function
SHOULDER/*Y*–	Pressing SHOULDER/*Y*– key will lead to the shoulder to rotate anticlockwise
ELBOW/*Z*–	If pressed, the elbow joint will rotate anticlockwise
PITCH (DOWN)	If this key is pressed, will make PITCH (DOWN) wave the gripper down
ROLL (ACW)	If pressed, ROLL (ACW) wave will make gripper rotates anticlockwise
Axes 7–11 CONTROL	These keys are designed to control motion axes 7 through 11 clockwise or anticlockwise
OPEN/CLOSE	This is a command to gripper; opening or closing gripper
RECORD POSITION	This key records the current position of end effector
GO POSITION	If we press GO POSITION key and then type 15 followed by ENTER, the end effector will move to position 15 which will be a recorded position from earlier motion
GROUP SELECT	All motions axes are divided into three groups – Group A, Group B and Group C. Group A is active; SCORBOT-ER robot has no group B or C
ENTER	ENTER is the common, universally understood command, as confirmation key to inform the computer to execute a command
Display of *A*/*B*/AXIS	The display on the left-half side of the LCD display shows the group currently active
Display of JOINTS/*X*/*Y*/*Z*	When power to control box is ON, the display on the right-half panel shows the coordinate system currently active

7.5.1 Example

EXAMPLE 7.2

Let us consider an ACL application program suitable for SCORBOT-ER robot. This is to simulate the work of an industrial robot which picks up a part from one location and transport to another location. This is depicted in Figure E7.2.1. The robot operation is divided into the following seven steps:

1. Opens gripper
2. Moves the gripper to point *A* from home position
3. Closes the gripper (holds the part)
4. Moves the gripper to point *B* along a given trajectory
5. Opens the gripper
6. Moves to the home position
7. Closes the gripper

Assume that the Cartesian coordinates of points *A* and *B* in the base frame are A (3000, –2500, 0) and B (3000, 2500, 0), respectively.

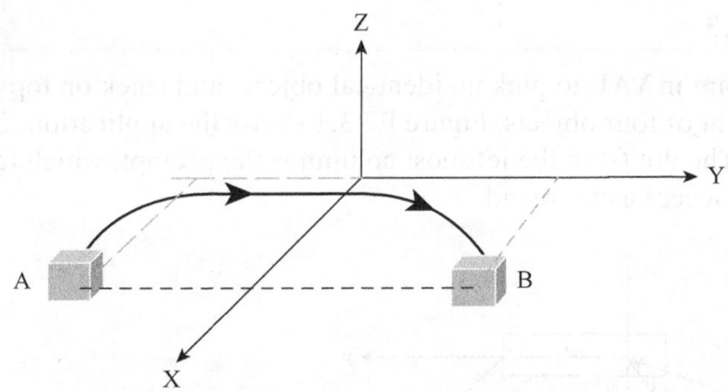

Figure E7.2.1 Trajectory

Given that *z*-axis value is computed from:

$$z = 1250 - ((x - 3000)^2 + y^2)/5000$$

Solution:

```
GLOBAL     Y1       K     Z1     ZV
DIMP       PV[50]
PROGRAM  PATHAB
SET        Y1 = -2600
FOR        K = 1 TO 50
SET        Y1 = Y1 + 100
SET        Z1 = Y1 * Y1
SET        ZV = 1250 - Z1
HERE       PV[K]
SETPVC     PV[K]    X     3000
SETPVC     PV[K]    Y     Y1
SETPVC     PV[K]    Z     ZV
SETPVC     PV[K]    P     -900
SETPVC     PV[K]    R     0
DELAY      5
PRINTK
ENDFOR
END
PATHAB     50
HOME
SPEED      20
OPEN
MOVED      PV[1]
CLOSE
MOVESD     PV       2     50
OPEN
HOME
CLOSE
END
```

EXAMPLE 7.3

Develop a program in VAL to pick up identical objects and stack on top of each other to a maximum height of four objects. Figure E7.3.1 shows the application. One side of each cube is 75 mm. The dot (·) in the leftmost column is the prompt, which tells the user that VAL is ready to accept a command.

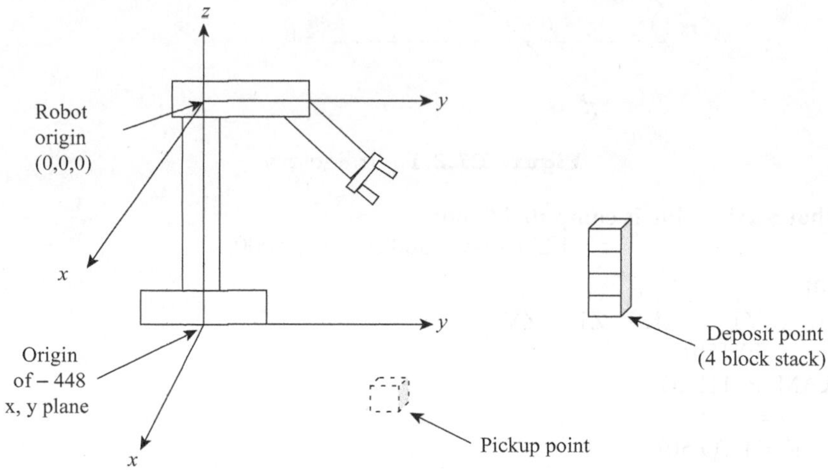

Figure E7.3.1 Work Space for VAL Programming

Solution:

```
.LOAD STACK
.LOCATIONS
.OK
.LISTP STACK
.PROGRAM STACK
REMARK
REMARK                          PARTS FROM A FIXED LOCATION
REMARK                          CALLED PICKUP, THEN DEPOSITS THEM
REMARK                          AT A LOCATION CALLED B. IT IS ASSUMED
REMARK                          THAT FOUR PARTS ARE TO BE STACKED
REMARK                          ON TOP OF ONE ANOTHER.
OPENI                           The Gripper opens immediately and waits
SET B = DEPOSIT                 Deposits the cube at B
SETI COUNT = 0.                 A variable count is initiated
10 APPROS PICKUP, 200.00        A label 10 is associated with PICKUP
MOVES PICKUP                    Moves the gripper to location PICKUP
CLOSEI                          The gripper closes and waits for a sufficient time
DEPARTS 200.00                  Robot moves along its approach vector
APPRO B, 200.00                 Robot moves to 200 mm to point B
MOVES B                         Robot moves in a straight line to point B
OPENI                           Robot deposits the cube
DEPARTS 200.00                  Robot departs 200 mm above location B
SETI COUNT = COUNT +1           Counter increments by 1
```

19. TYPEI COUNT	Displays the count values
20. REMARK	COUNT: NO OF ITEMS STACKED
21. IF COUNT EQ 4 THEN 20	A test to check if COUNT = 4
22. REMARK	MOVE LOCATION OF B BY 75.00 MM
23. SHIFT B BY 0.00, 75.00	Modifies B so that *z* is increased by 75 mm
24. GO TO 10	Goes to 10
25. 20 SPEED 50.00 ALWAYS	Speed is maintained as 50%
26. READY	Moves the robot to its ready position
27. TYPE *** END OF STACK PROBLEM ***	Robot printer works
.END	
.LISTL	
DEPOSIT	–445.03 130.59 –448.44 –87.654 88.890 –180.000
PICKUP	163.94 433.84 –448.38 178.006 88.896 –180.000

```
EXEC STACK
COUNT = 1
COUNT = 2
COUNT = 3
COUNT = 4
*** END OF STACK PROGRAM ***
PROGRAM COMPLETED; STOPPED AT STEP 28
```

7.6 SENSOR INTEGRATION

An important part of robot programming has to do with interacting sensors. By interacting sensors we mean those that are used for placing them at selected locations in addition to guiding conveyor belts. These sensors at the minimum have the capability to touch and force sensors and the conveyor belts to follow *if-then-else* constructs. Integration with a vision system, as seen in the earlier example, allows us to send the coordinates of the object of interest to manipulator system. In the earlier example, the camera frame can be computed relative to station frame so that a desired goal frame can be acquired from the data. Some sensors can be part of other equipment within work cell. Some robots can get signals from belt conveyor sensors so that they can track the belt's motion and acquire objects from belt as it moves.

The interface to force control capabilities comes through special language statements. Such force control strategies necessitate an integral part of manipulator control system. The robot programming simply serves as an interface to those capabilities. In systems that support active force control, the description of such ability is to display force data collected during a constrained motion. The AL language describes six component of stiffness – three translational and three rotational – with a bias force. In this way, the manipulator's apparent stiffness is programmable. For example

```
Move garm to goal
with stiffness = (100, 100, 0, 100, 100 100)
With force = 20*ounces along zhat;
```

'zhat' is an axis that has a bias force of '20 ounces'. The second statement is programmable.

7.7 ROBOT WORK CELL

Industrial robots generally work with other pieces of equipment. These pieces of equipment include conveyors, production machines, fixtures and tools. The robot and other pieces of equipment form the robot work cell. Some times human personnel are also included. They perform tasks that are not easily automated. These tasks might consist of inspections that require sense of touch. A robot generally does not possess sense of touch.

While designing robots, two problems that robot design engineers face are as follows:

(i) Physical design of the robot work cell
(ii) Design of coordinated control.

These problems are to be solved one by one so that the applications of robotics are expanded. An example is provided in the following section.

7.7.1 Work Cell Lay-Outs

There are three possibilities in constructing robot work cell such as

(i) Robot-centred work cell
(ii) In-line robot work cell
(iii) Mobile robot work cell.

(i) Robot-centred work cell

Robot is placed approximately at the centre of other industrial equipment (Figure 7.5). The robot performs operation, either servicing single production operation or serving a machine tool. Robot can also perform a processing operation. Die casting, for example, is a processing operation. Die casting is the earliest function of a robot. Robot serves a die-making function which picks up certain partly or fully finished product and quenches the product in a quench bath before keeping it on an outgoing conveyor. The other operations of the robot are scarping off unwanted impurities from the die, load and unload parts from and to the machine tools and similar operations. Part feeders and delivery chutes can make the outgoing conveyors.

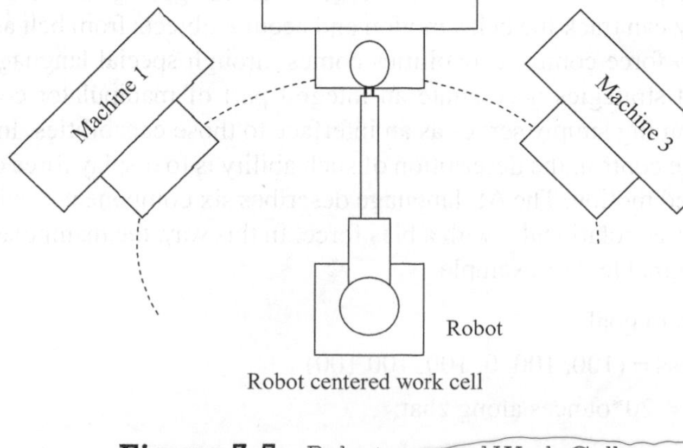

Figure 7.5 Robot-centred Work Cell

Nowadays, more robots with more functions are required to serve more equipment. Sometimes, robot should wait (or idling) till the processing is completed by a machine (e.g. CNC machining). Sometimes, machine has to wait till it gets the robot service (e.g. stamping).

(ii) In-line robot work cell

Figure 7.6 describes this configuration. The robot is located along a moving conveyor and to perform tasks on the product coming to the conveyor. Many in-line robot work cells involve more than one robot. One common example is car body assembly; robots are positioned along the assembly line to spot weld car body frames and panels.

Following of are three categories of transfer system that can be used with in-line robot cell configuration:

1. Synchronous transfer
2. Continuous transfer
3. Non-synchronous transfer.

Synchronous transfer: Intermittent transfer system moves the parts with a start and stop motion from one work station along this line to another work station. Here, all the parts are moved simultaneously and then registered at the next respective locations. The robots are stationary along the line of transfer (conveyor). The advantage is that a part is registered and kept stationary so that robot finishes the process during the robot work cycle.

Continuous transfer system: Here the parts are moved continuously along the line at constant speed. This means that the position (sometimes orientation) of the part is continuously changing. Although this improves productivity, the problem is for the robot to have sufficient time for working on the part. This problem is solved by the robot with two methods such as stationary baseline tracking and moving baseline tracking.

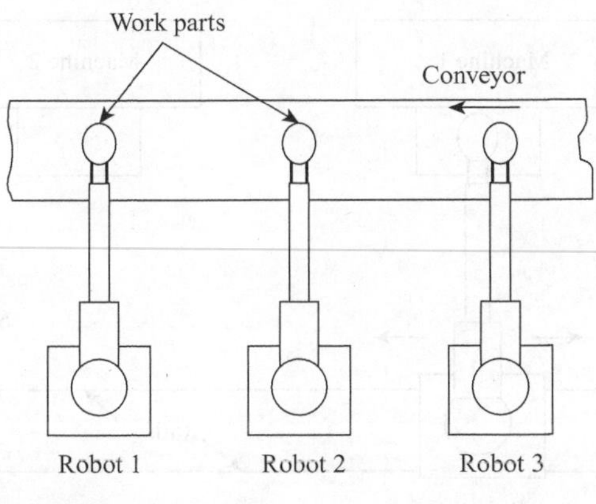

Figure 7.6 In-line Robot Work Cell

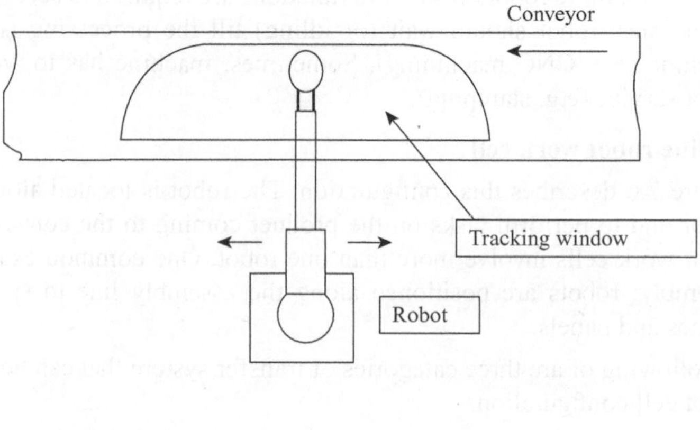

Figure 7.7 Concept of Tracking Window

Stationary baseline tracking: Figure 7.7 depicts this stationary baseline tracking system. Robot is located in a stationary position and it creates a tracking window. Parts are allowed to pass through this tracking window. When a part enters the tracking window, the robot starts its work on the part. It has to complete its work before the part leaves the window.

Moving baseline tracking: This method is about moving the base of the robot along the conveyor to some calculated distance. This requires a cart (with wheels) on which the robot is to be mounted. This creates an extra degree of freedom for the robot. Due to this solution, sufficient time is available for a robot to do a job on the part coming along the production line. The problem of possible collision between neighbouring robots is to be taken care. The tracking window is the intersection of work volume of robot with the production line. Figure 7.8 illustrates this moving baseline tracking. When the robots are placed at sufficient distances for collision-free operations, additional floor space is required; the time of production is also increased

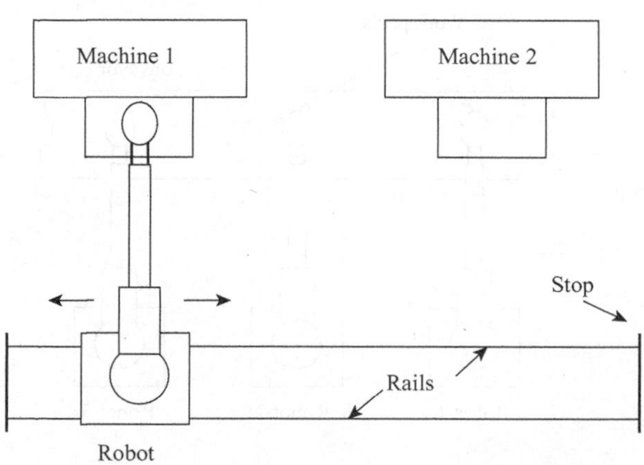

Figure 7.8 Mobile Robot Work Cell (Floor Mounted)

as the parts travel additional distances. Sensor-based AI control can avoid robot collisions as a robot has to know where exactly the other robot is extending its arm.

One problem is that the robot has to continuously manipulate its joints so that it does the required job on the part without interruptions to the part when it is moving on the conveyor. Real-time controlling of robot has to be very effective. Sensing the presence of part within the tracking window is required. Robot starts its functioning as the part enters the tracking window. Another problem is the velocity of parts coming along the line. Velocity has to be known and fixed. Variations in velocity will deter the effectiveness of robot cell performance.

Non-synchronous transfer: This is also known as 'power and free system'. Each part moves independently along a conveyor in a stop-and-go fashion. Each part is placed on a cart that moves along the conveyor; at the work station, the cart is brought away from the conveyor nearer to the robot so that robot takes its time for performing a task on the work part. This system is complex and is to be provided with sensors to decide where exactly the cart is. Sometimes, the conveyor may be stopped if the duration of robot's task on the part is substantially long.

(iii) Mobile robot work cell

The robot is capable of moving to various equipments within the cell. A mobile base is required. This can be accomplished by having rails either on the floor or overhead so that the robot can reach the equipment easily. Overhead system is expensive but reduces floor space. A very effective but complex design is to have more than one robot operating in a network of rail system overhead.

7.8 INTERFERENCE PROBLEMS

Some robot work cells offer a difficult choice of picking robots for servicing the machines. By servicing the machines we mean to load and unload the machines with partly finished parts so that the final operation is to be given to the robot. In this case, the in-line robot or robot centred work cells usually give this difficult choice. In either case, the machine waiting for the robot to service and/or robot waiting for the machine to complete the process can occur. We shall observe both of these cases here.

The first type of interference can occur when the machine process cycle is too long and unequal between machines. The robot cannot wait for such a long time. We cannot modify the machine process time. In such situations, we can only modify the number of machines. Machine waiting for a robot is termed as *machine interference*. This condition can occur in any machine–robot setup. The machine interference can be measured as a ratio or percentage. This is the ratio of total idle time of all the machines to one robot cycle time.

The second type of interference is physical that the robot waiting for the completion of processing. We can only increase the number of robots so that all will work for the same product process completion. Here, there is a danger of robot arms collision. This is easily avoided by separating the distances between robots. It is not economical to increase the number of robots due to its cost. However, there are some applications in which it is quite desirable for two or more robots to share the same working space. An example is placing a work part in one location and the other robot picks up the part. It is too difficult manually to design this situation. The alternative approach is to coordinate the robot working space so that one or some minimum number of robots can work together. Here, care must be taken not to allow the arms of the robots to interfere with other arms of other robots. Such a waiting time is usually measured in *robot waiting time*. This is the ratio between robot waiting time in one robot cycle or its percentage. The following example indicates how to compute the waiting times.

This is a machine cell with one robot and three machines. The cycle times of machines are indicated in Table 7.3.

Table 7.3 Machine Run Times and Service Times

Machine No.	Run Time (s)	Service Time (s)
Machine 1	10	5
Machine 2	5	5
Machine 3	5	5

The Process Chart

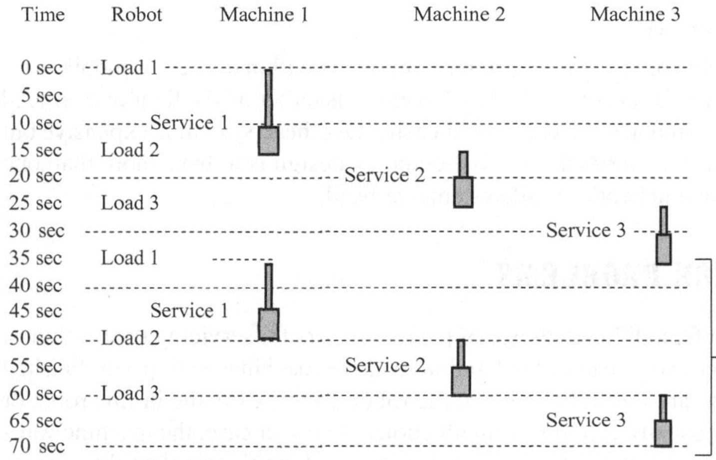

Robot cycle = 35 sec

Machine interference = 60/35 = 1.714 ie. 171.4 %

Robot waiting time = 0/35 = 0 %

7.9 FURTHER ON ROBOT WORK CELL

There are several other considerations in the design of robot work cell. The following are considered important:

Changes to other equipment in the work cell

Special fixtures and control devises must be made to the equipment to work as independent mechanism. Examples include work-holding nests and conveyor stops to position and orient the parts for the robot. Changes in the robot arm to gain access to the equipment are necessary. Sometimes limit switches are required to access various components and equipment in the cell.

Part position and orientation

Raw work parts are delivered to the cells and it is important that the robot has a precise pickup location. They should be in a known position and orientation so that the robot will grasp and hold this part accurately.

Part identification

This important problem can be solved by optical means. Vision systems and limit switches will be useful in part identification.

Protection of robots

In certain applications, such as spray painting and hot metal-working operations, a means of robot protection from adverse effects of its environments is required.

Control of work cell

The activities of robot are to be coordinated with other equipments in the cell. This work cell is to be referred by robot engineers to work cell controllers.

Robot and human safety

A means of protecting human personnel from harm in and around robot working must be provided. This can be accomplished by means of fences or other barriers to protect humans.

REVIEW QUESTIONS

1. Draw a block diagram of robot operating system. Explain the functions of each of the blocks.
2. What are three basic modes of robot operating system? Explain them.
3. Give an example of industrial task other than those indicated in this book. Elaborate the steps and a program listing (VAL and LISP).
4. What are interacting sensors?
5. Robots, in general, do not possess the sense of touch. Suggest sensors to incorporate the sense of touch?
6. List and explain any three robot programming languages.
7. Prepare the symbolic instructions program on any one of industrial tasks. The example need not be available in this book. Use VAL or LISP statements.
8. List and explain robot work cells.
9. Explain robot work cell that has three categories of material transfer operations.
10. What is the concept of tracking window?
11. What are the other considerations of robot work cell? Explain them.

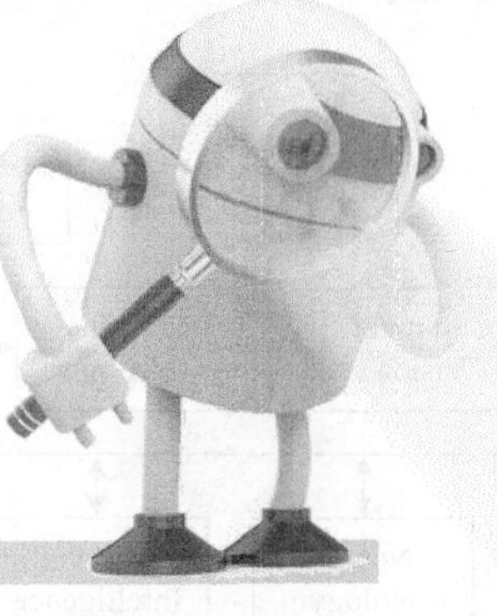

ROBOT VISION 8

Learning is the discovery that something is possible

Fritz Perls

This chapter covers the following key topics:

Introduction – Robot Vision – Lighting Devices – Analogue-to-Digital Conversion – Example – Image Storage – Illumination – Feature Extraction – Example – Object Inspection – Object Recognition – Procedure of Robot Vision

8.1 INTRODUCTION

In this chapter, we will discuss five aspects regarding vision – robot vision, computer vision, machine vision, image acquisition and image processing. They all synonymously indicate the same thing. But each of the mentioned aspects has its own description and definition. Robot vision means vision engineering as applied to robotics to incorporate vision-based control. Computer vision points the use of computers in image processing in terms of developments of complex algorithms connected to computer vision. Machine vision focuses on vision such as vision-based measurement and inspection. Image acquisition and image processing are two aspects towards processing images in getting out information and using them for specific applications. All books and journals which have these terms have significant overlaps irrespective of the techniques and applications they cover. All these are used for imaging and image applications and control of some automation, such as robots.

Some applications of robot vision are vision-based servo (guidance and control of robot arm), complex inspection of products, feature identification and extraction, part recognition and part orientation.

8.2 ROBOT VISION

Robot vision has mainly two parts – hardware and software. The most important of them is the 'frame'. The part of memory where the digital image is stored, retained and retrieved is known as frame. The hardware in which the digital image is captured (grabbed) is known

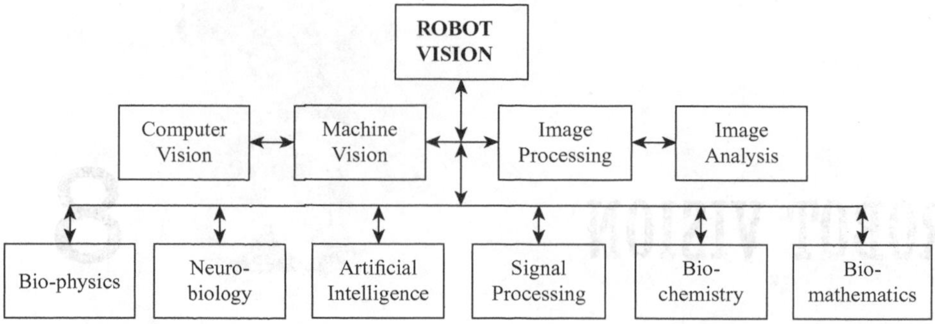

Figure 8.1 Various Components of Robot Vision

as 'frame grabber'. This hardware is capable of capturing images of 100 or more frames per second. The digital frame captured is a matrix of data 'seen' by camera.

The elements of this matrix are called as 'picture elements' or 'pixels'. The maximum number of pixels is determined by a sampling process performed on each image frame. A single pixel is the projection of a smallest portion of the image which reduces that portion to a single value. The value is a measure of a light intensity for that element of the image. Each pixel value is converted to a digital value. The digital value is between 0 and 255 (both numbers are included) with no decimal point.

Some components, aspects and requirements of robot vision are shown in Figure 8.1 and Table 8.1.

The first aspect involves acquiring and digitizing images. The images will be stored and subjected to processing and analysis functions for data reduction and interpretation. These steps are required since the camera is required for this purpose and for further applications. Usually, an image uses threshold to produce binary images with various feature measurements to reduce the data representation of the image. This data reduction can change the representation of a frame from several thousand bytes of raw image data to a few hundred bytes of feature value data, the second being easier to analyse the image.

The second aspect is image processing and analysis. A computer with suitable software is needed and for this, an interface (hardware and software) to robot is also necessary. The reduced image happens to be several hundred pixels derived from original image or directly from binary image. The features are matched against previously computed values stored in the computer.

The final aspect is of robot control. The objectives include the robot control for specific applications such as measurement of dimensions of a product. The product may be finished or partly unfinished. The result includes inspection, part identification, location and orientation.

Table 8.1 Aspects and Requirements

No.	Aspects	Requirements
1	Acquiring and digitizing images	Camera (mainly digital camera)
2	Image processing and analysis	Computer and interface for processing
3	Application in robot control	Robot for specific application

Another way of classifying vision systems is in accordance with number of grey levels (light intensity levels) used to characterize the image. In binary image, the grey scale values are divided into either of the two categories – black or white (0 or 1). In contrast, the binary 0 can be changed as 1 and 1 as 0.

8.3 LIGHTING DEVICES

Good illumination of the scene is needed owing to its effect on the level of complexity of image processing algorithm required. In bad illumination, the task of interpreting a scene is more difficult. In contrast, proper lighting techniques should provide a high contrast and minimize specular reflections that create shadows. This will happen when the lighting is specifically designed into the system.

The basic types of lighting sources with their explanations are provided in Table 8.2.

Table 8.2 Names of Light Sources with Explanation

No.	Name of Light Source	Explanation
1	Diffuse surface devices	Examples of diffuse surface illumination are from typical fluorescent lamps. Some light tables can also be included in diffuse surface devices
2	Condenser projectors	A condenser projector transforms an expanding light source into stream of condensed light source
3	Flood or spot projectors	Flood light and spot light can be used in illuminating surface areas
4	Collimators	Collimators are used to provide a set of parallel beams of light on the object
5	Imagers	Slide projectors are imagers. Object enlargers form image on the target at the object place

Two basic illumination techniques are used in robot vision – front lighting and back lighting. Front lighting means that the light source is on the same side of the product as that of camera. Accordingly, the reflected light is used to get the image viewed by the camera. In the back lighting, the light source is directed towards the camera. Backlighting is enough for applications in which the silhouette of object is sufficient for recognition.

8.4 ANALOGUE-TO-DIGITAL CONVERSION

The analogue-to-digital (A/D) conversion technique involves taking an analogue voltage signal, producing an equivalent output in digital form and storing in the memory of computer. The A/D conversion has three phases – sampling, quantization and encoding.

Sampling: An analogue signal can be sampled periodically to get a series of discrete time analogue signals. This process is illustrated in Figure 8.2. By setting the sampling time as Δt, we get approximate sampled voltage points and sampled voltage levels. How well we approximate the analogue signal is determined by the sampling rate of A/D converter. The sampling rate adopts the concept of sampling theorem. The sampling theorem states as follows:

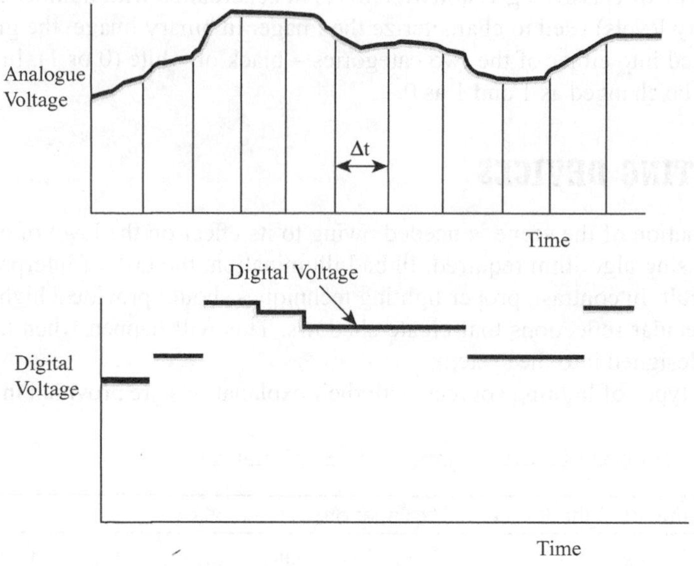

Figure 8.2 Sampled and Digitized Analogue Voltages

If we wish to reconstruct the entire analogue signal, the fastest sampling rate is not to be less than twice the highest frequency of the analogue signal.

Quantization: Each sampled discrete time voltage level is assigned to a number of defined amplitude levels (grey scale). These amplitude levels correspond to a grey scale used in the system. Predefined amplitude levels are characteristic to a particular A/D converter and consist of a set of discrete values of voltage levels. The number of quantization levels is defined by 2^n where n is the number of bits of A/D converter output. For example, an 8-bit converter would allow us to quantize at $2^8 = 256$ different values whereas a 4-bit value would allow only $2^4 = 16$ different quantization levels.

Encoding: The amplitude levels that are quantized are to be changed into digital code. This process of encoding involves representation of amplitude levels by a binary digital sequence. The ability of encoding process to distinguish between various amplitudes is a function of the spacing of each quantization level.

Given the full-scale range of an analogue signal

$$\text{Quantization level spacing} = (\text{full-scale range})/2^n.$$

Then, the quantization error is

$$\text{Quantization error} = \pm \tfrac{1}{2} (\text{quantization level spacing}).$$

8.4.1 Example

EXAMPLE 8.1

A camera has a 8-bit word, 0–5 V with an A/D converter and then one grey level = 5/256 V = 0.0195 V. This is the quantization level spacing. The encoded values areas shown in the following table:

Voltage Range (V)	Binary Number	Grey Scale
0.0000–0.0195	0000 0000	0 pure black
0.0195–0.0390	0000 0001	1 dark −1
.	.	.
.	.	.
4.9610–4.9805	1111 1110	254 light grey
4.9805–5.0000	1111 1111	255 pure white

Quantization error = ± (0.0195/2)

8.5 IMAGE STORAGE

The A/D converted data are stored in a 'frame buffer', a part of frame grabber. This should be done in real time, that is, data are to be stored as it is acquired. Data acquisition is sometimes faster than that of frame buffer. A combination of row–column counters is used to locate data in a frame buffer. For example, if $F(10\times10)$ matrix is allotted to the frame buffer, each datum has a location (x, y) with $x = 1, 2\ldots10$ and $y = 1, 2\ldots10$. Frame buffer, F, can be represented as (the decimal number is actually a binary coded decimal representing the light intensity):

$$F = \begin{bmatrix} 36 & 104 & 200 & \ldots\ldots & 72 \\ 19 & 200 & 46 & \ldots\ldots & 42 \\ \cdot & & & & \\ \cdot & & 156 & & \\ \cdot & & & 142 & 98 \end{bmatrix}$$

$f(x, y)$ (arrow pointing to 156)

8.6 ILLUMINATION

Good illumination on the object space is important to identify and interpret various objects and their sections/parts, so that a robot can manipulate or process the object. Five kinds of illumination are given in Table 8.3.

Table 8.3 Lighting Units with Modification

No.	Lighting Units	Modification
1	Omni-direction lighting	A diffuser is used to make lighting uniform
2	Coaxial lighting	Using a circular polarizer to the camera
3	Linear polarizer	Polarizer to cameras and lamps
4	Back light	To use flat fluorescent lamps to light the other side of objects
5	Imager	Slide projection on object so as to identify size and shape

8.7 FEATURE EXTRACTION

In robot vision, it is often necessary to distinguish one object from another. This is usually accomplished by means of features that uniquely characterize the object. A list of features of robot vision is given in Table 8.4. The list shows only two-dimensional features cases that can be roughly categorized as those that deal with area features and those that deal with boundary features.

Table 8.4 Basic Features

No.	Features	Equations
1	Area	–
2	Grey scale	Maximum, average and minimum
3	Perimeter length	–
4	Diameter	–
5	Minimum enclosing rectangle	Length and breadth
6	Centre of gravity	$CG_x = (1/n)\ \Sigma x$; $CG_y = (1/n)\ \Sigma y$ where n is the pixel
7	Eccentricity	(Maximum chord length A)/(Maximum chord length B) Denominator is measured perpendicular to numerator
8	Aspect ratio	The length-to-width ratio of a boundary rectangle which encloses the object. Search for the minimum aspect ratio
9	Compactness	$(\text{Perimeter})^2/\text{Area}$
10	Thinness	Diameter/Area
11	Holes	Number of holes in the object
12	Moments	Given a region, R, and coordinates of points (x, y) within or on the boundary of region, the pqth order moments of image of the region is given as $M_{pq} = \Sigma_{x,y}\ x^p\ y^q$

8.7.1 Examples

EXAMPLE 8.2

Figure E8.2.1 shows an example of a product to be finished. Find the minimum features that fit the product.

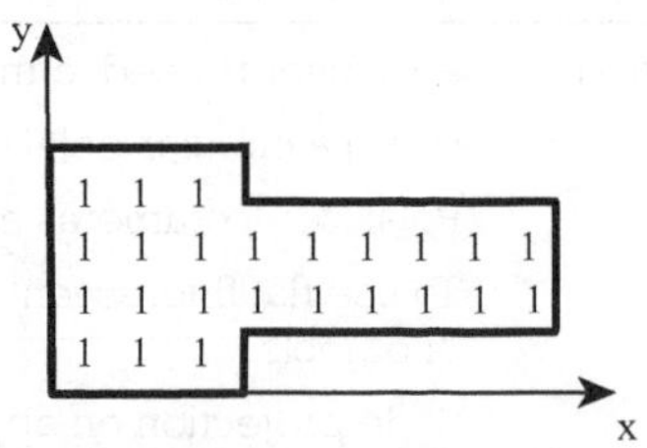

Figure E8.2.1 Example Figure

Solution:

(a) Moment (M_{00}) = $x^0 y^0$
(b) Eccentricity = (Max x length)/(Max y length) = 2.25
(c) Perimeter = 2* (9 + 4) = 26
(d) Area = (3*4) + (6*2) = 24
(e) Diameter = 9
(f) Thinness = (Diameter/Area) = (9/24) = 0.375
(g) Compactness = ((Perimeter)^2/Area)) = (26^2/24) = 28.17
(h) No. of holes = 0

An object that satisfies all the above features belongs to one category.

8.8 OBJECT INSPECTION

In this section, the use of robot vision and robotic application towards object inspection, identification, and visual servoing and navigation are discussed.

Object inspection is carried out by robot vision system which has a camera, especially digital camera, suitable components and equipment. The inspection is considered as main role to support the application. Here, the objectives are checking gross surface defects, discovering flaws in labelling, verifying the components in assembly, measuring dimensional accuracy and checking for holes and other features in a completed part. When these kinds of inspections are performed manually, there is a tendency of human error. Also the time needed in most of manual inspection operations requires that the procedure be accomplished on a sampling basis. But in robotic application, every part, finished or unfinished, is checked for correct assembly.

The second category, the object identification, is concerned with the applications in which the purpose of machine vision system is to identify and to classify an object rather than to inspect it. Inspection implies that the part be either accepted or rejected. Identification involves a process in which the part itself or its position and orientation are determined. This is usually followed by a subsequent decision and action taken by the robot. Identification includes part sorting, palletizing and depalletizing, and picking parts that are randomly oriented in a conveyor or in a bin.

In visual servoing and navigational control, the purpose of robot vision system is to direct the actions of robot (and other devices in the robot work cell) based on its visual inputs. The generic example of robot visual servoing is where the machine vision system is used to control the trajectory of end effector towards an object in the work space. Industrial examples of this application include part positioning, part retrieving, reorienting, assembly, bin picking and seam tracking. Clearly, the visual data are just an important input in this type of task. A great deal of intelligence is required in the robot controller to use these data for navigation and collision avoidance.

8.9 OBJECT RECOGNITION

Another important part is that of recognizing the object whether it belongs to current classification; if so whether to accept it or not. Usually, two methods are applied.

(i) Template matching technique: The method of chain code is very common. The chain code allows one to categorize and adopt either 4-directional chain codes (Figure 8.3) or 8-directional chain codes (Figure 8.4). Each has certain advantages over the other. If the object has ±90° turns, it is faster to recognize. Otherwise, if the object has ±45° turns, the second category is better; however, the recognition of shape is slower than the first.

Figures 8.5 and 8.6 show two applied examples. The operator has to initially test the chain codes that belong to correct category. Then, he/she has to lead the tested chain codes along the directional ones and to assure them. The other one (i.e. Figure 8.6) is left as an exercise for students. Table 8.5 shows the assured chain codes.

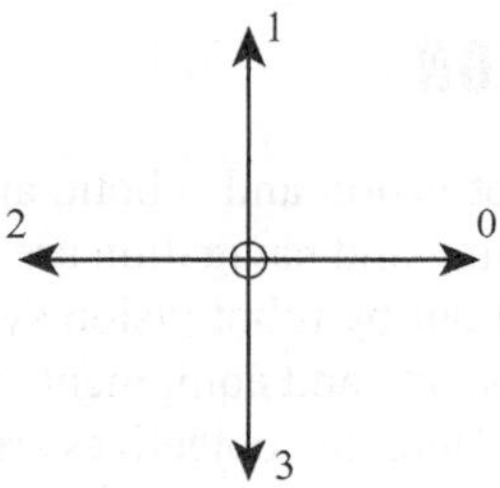

Figure 8.3 4-directional Chain Code

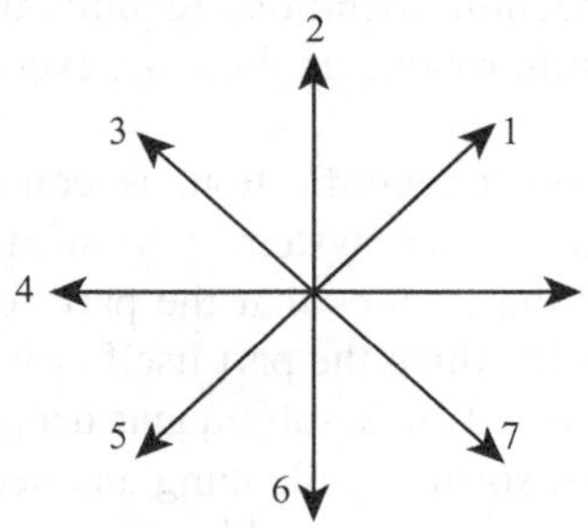

Figure 8.4 8-directional Chain Code

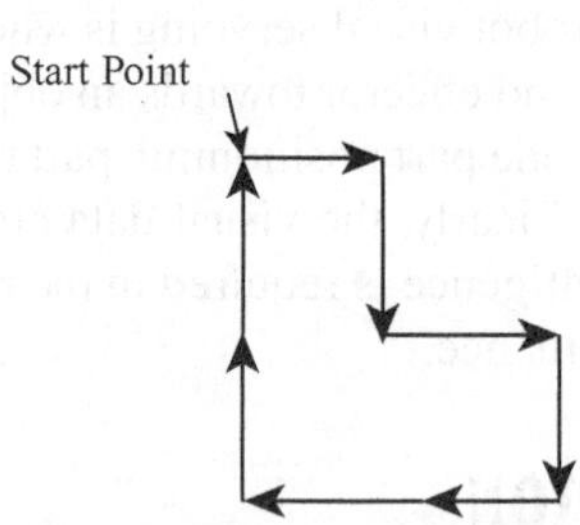

Figure 8.5 Example 1

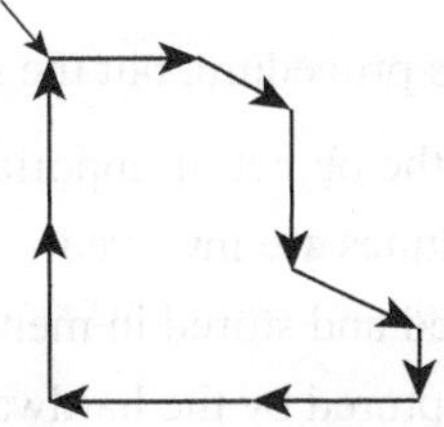

Figure 8.6 Example 2

Table 8.5 Comparison Between 4- and 8-point Inspections

Sl. No.	4-point Inspection	8-point Inspection
1	Start with any start point and follow the lines as indicated by arrows	Start with any start point and follow the lines as indicated by arrows
2	Compare with 4-point inspection and mark number for lines	Compare with 8-point inspection and mark number for lines
3	Take the minimum of 4-point inspection	Take the minimum of 8-point inspection
4	This is the shape number of 4-point object	This is the shape number of 8-point object

The chain code of 4-point object: 03032211
We get the differences as:

Difference between 0 and 3 is 3 (refer 4-directional chain code)
Difference between 3 and 0 is 1
Difference between 0 and 3 is 3, and so on
Difference between 1 and 1 is 0
Difference between 1 and 0 is 3
Difference code is 31330303.

Take the minimum decimal value of difference:
Shape number: 03033133

This shape number is for the object, uniquely recognized and independent of rotation by ±90°.

(ii) Structural techniques: Structural techniques of object recognition consider relationships between edges (features) of an object. If the image of an object can be subdivided into four straight lines or primitives connected at their end points at right angles, the object is a rectangle. This kind of technique is known as syntactic pattern recognition. This is the most widely used techniques. Structural techniques are different from decision theoretic techniques. In the second case, the techniques deal with a pattern on quantitative basis. This ignores most part of interrelationships among object primitives. A detailed description of this technique can be found in complete robotic books.

8.10 PROCEDURE OF ROBOT VISION

There are several steps for this procedure, but the most important steps are as follows:

- Camera is focused on the object of importance.
- Special lighting techniques are involved.
- Image frame is captured and stored in memory.
- Image is frequently captured by the hardware (frame grabbing).
- Rate of frame grabbing can go even 100 frames/s nowadays.
- Frame consists of matrix of pixel data.
- Pixel is a projection of light intensity.
- Each pixel intensity (colour) is converted to a digital value.
- The digitized image matrix is stored as an image frame.
- The image matrix is processed and analysed for data reduction, feature extraction and interpretation.
- Feature is an important aspect derived from the image for any application.
- Robot is servoed (controlled) based on the interpretation.
- More grey scale produces a near natural monochrome image.
- We can also divide the black to white spectrum into two parts:
  0 representing pure black and 1 a pure white; or 0 representing a pure white and 1 a pure black.

We refer to this image as binary image. In most of the industrial applications binary image is sufficient.

REVIEW QUESTIONS

1. Explain five aspects of robot vision.
2. List and explain various components of robot vision.
3. What are the aspects and requirements of robot vision?
4. List and explain some of basic lighting devices.
5. Explain three phases of A/D conversion.
6. How are images stored in the memory of camera?
7. Explain five aspects of illumination.
8. What are the basic features of illumination? Do they belong to 3D features? If yes, why? If not, why not?
9. Distinguish between 4-directional and 8-directional chain codes. What are their applications?
10. What are the procedures of robot vision?

ROBOT APPLICATIONS

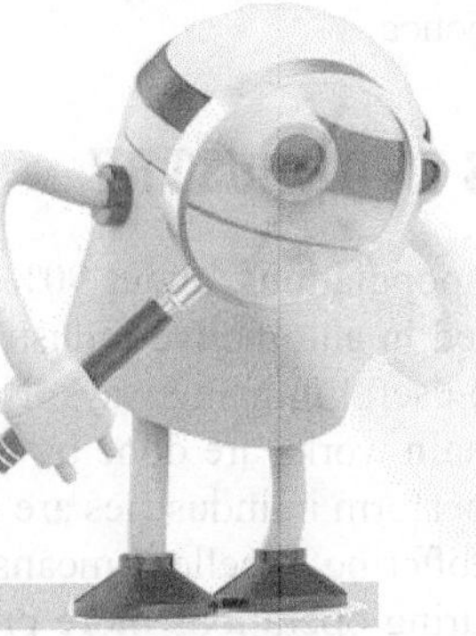

9

To teach is to learn twice
J. Joubert

This chapter covers the following key topics:

Introduction – Robots in Industry – Robots in Handling – Machine Loading and Unloading – Materials Transfer – Palletizing – Welding – Arc Welding – Spot Welding – Compliance – Assembly – Injection Moulding

---

9.1 INTRODUCTION

Robotics applications are limited only by human intelligence. Over the last 30 years or so, especially in industries, robots are emerging with various applications. That is, almost every day we come across a new application of robot. One of key features is its versatility. The mathematical models which have been developed in the previous chapters are to calculate its motions, its control, to determine its trajectories and to frame transformations to perform specific tasks for which it is purchased for. Robotics is going to be a prominent component of manufacturing industry affecting human labour at all levels from managers of production shop floor to unskilled workers.

This chapter is intended to introduce industrial applications of robots. A programmable robot with a number of degrees of freedom and varied configurations can perform specific and diverse tasks. This will perform the required tasks, as per program statements, with varied end effectors and tools; it can be reprogrammed and adapted to changes in process or production line. After all, the robots as they are now and are going to be for decades to come are dumb machines, which are supposed to obey the computer commands. They are not 'super workers' and it is important to understand that they are all tools or machines working at our commands.

It is important to remember that the robot is not conscious of what it is doing. It can only move their end effectors to well-defined positions in well-defined manner and perform the tasks such as pick and place, drilling, paint spraying, scraping, welding, assembly, inspection and whatever industrial tasks it is commanded for.

9.2 ROBOTS IN INDUSTRY

In today's robot populations, about 90% are working in industries and out of them about 50% are deployed in automotive industries. Robots in industries are known as *industrial robots*. They are useful in many ways. Nowadays, the industries require more competitive robots since certain works are done by robots are precise than humans. Most of the tasks that robots can perform in industries are not advisable for deploying humans.

Robots help offering excellent means of utilizing modern technologies. They help to make manufacturing operations more profitable and competitive. However, the technologies are new to industrial robots and in general to industrial scenes. Robots applications in industries are confined to

(i) Pick and place
(ii) Spray coating
(iii) Assembly
(iv) Inspection

All of the above are depicted in Figures 9.1–9.4, respectively. Pick and place operations can have one or a few obstacles. Automotive industries, in addition to other industries, need robots for pick and place operations. Peg-in-assembly task requires a robot which has the power of lifting, sometimes, heavy objects. Inspection task needs a robot that has at least

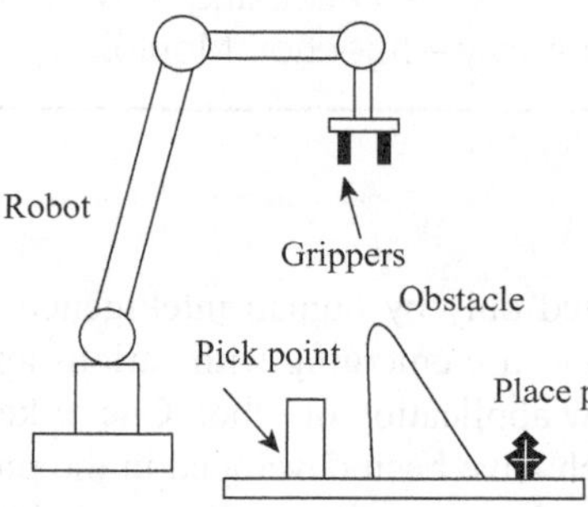

Figure 9.1 Pick and Place

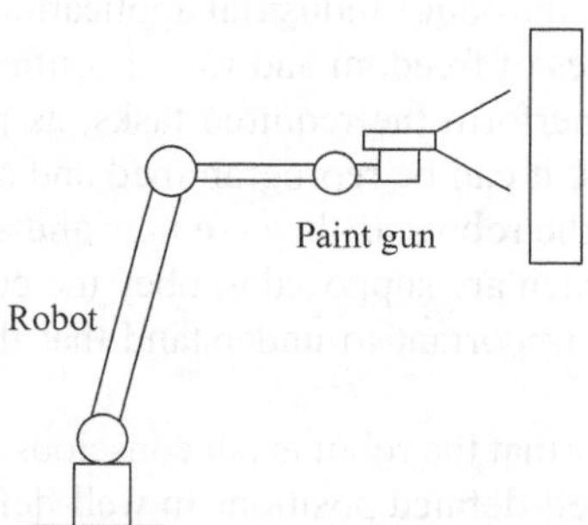

Figure 9.2 Spray Coating

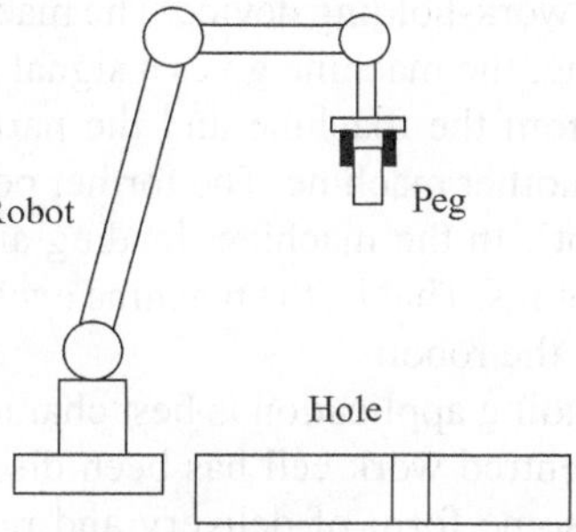

Figure 9.3 Assembly

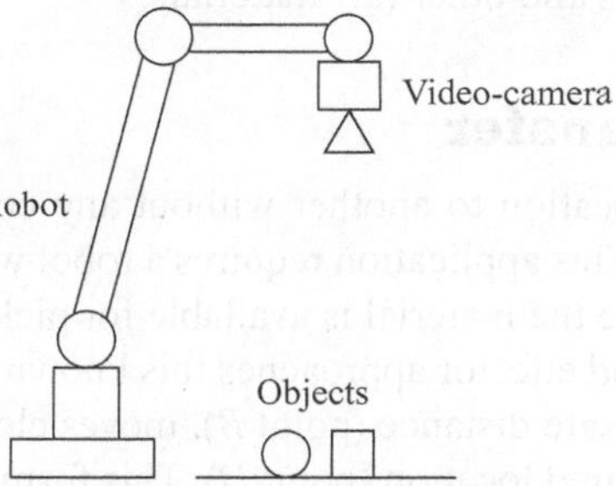

Figure 9.4 Inspection

one video camera plus a lighting device. The other cameras, when needed, are placed on the sides of robot each with its own lighting devices. Spray coating requires a lighter robot. It is proved that robot spray coating is performed more economically compared with humans.

9.3 ROBOTS IN HANDLING

This is the most basic operation in which the robot is to pick an unfinished object and to place it in other positions suitable for processing. Several tasks performed by the robot require this basic handling operation.

Some examples of operations are as follows:

(i) Machine loading and unloading
(ii) Materials transfer
(iii) Palletizing
(iv) Welding

These applications are discussed in Sections 9.3.1–9.3.4.

9.3.1 Machine Loading and Unloading

The loading of parts into a machine and unloading it from the machine are suitable tasks for a robot. The robot picks up the part from a specific location and places it in a desired position and orientation into a work-holding device. The machine now starts the processing. After completing its processing, the machine gives a signal to the robot to unload the part. The robot unloads the part from the machine and the part is delivered to another point which may be a conveyor or another machine. The former point is known as 'loading point' and the latter 'unloading point'. In the machine, loading and unloading are to be coordinated with that of machine timings. That is, the machine cycle time is to match with loading and unloading the material by the robot.

Machine loading and unloading application is best characterized by robot-centred work cell of the robot. The robot-centred work cell has been discussed in Chapter 7. The work cell has other items such as some form of delivery and removal systems. The work cell has other machines so that the robot has to work with all such machines. More complex machine group, in a robot work cell, is possible with more robots and more number of machines. Machine loading and/or unloading is not restricted to parts only. Robots can be used to load and/unload tools and other raw materials.

9.3.2 Material Transfer

To move a part from one location to another without any complex constraints is termed material transfer operation. This application requires a robot with a few degrees of freedom (Figure 9.5). The place where the material is available for picking by the robot is known as pick-up point. The robot's end effector approaches this known location (point *A*), grasps by its gripper, moves away to a safe distance (point *B*), moves close to another point (point *C*) and places the part at the desired location (point *D*). This forms a typical work cycle known as (*A-B-C-D*). The orientation of the part will be done at location point *B* or point *C*.

The pick-and-place operation can have many variations – simple operations to complex operations. On the simpler side, the part can be dropped at the desired position. The part can be dropped from location *C*, and the work cycle is only *A-B-C*. More complex work cycles of material transfer operations include picking the part from a moving conveyor pickup point, changing motion patterns of conveyor from cycle to cycle, changing delivery point location based on some attributes of the part or moving the part through multiple points in the work cycle to avoid obstacles.

9.3.3 Palletizing

Pallet is a place which has several materials uniformly stacked. Palletizing is the process of staking of materials side by side or one over the other. Removal of stacked materi-

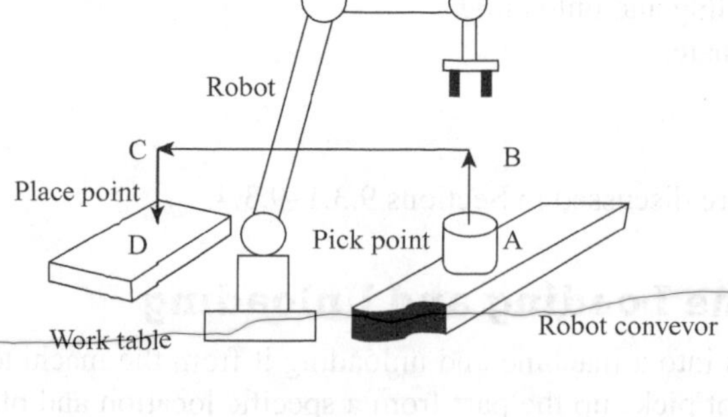

Figure 9.5 Pick and Place Operation

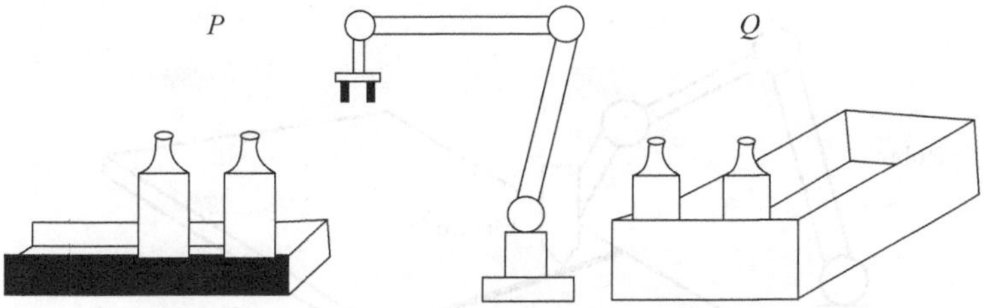

Figure 9.6 Palletizing Function

als is named depallatizing. Both processes have their own problems. Palletizing requires robot operations to systematically stack materials without gaps in between them. Stacking medicinal pills, coke bottles, gas cylinders, electric switches and the like are examples of palletizing. Depallatizing is a process that will remove the staked materials in a sequence one by one.

In Figure 9.6, points *P* and *Q* are indicated. Individual parts are picked from a fixed place (pick up point, *P*) and moved to a comfortable point (delivery point, *Q*) for robot. Palletizing operation is generally a pick-and-place operation with a major difference; that is, the pick-up point is to have its ringing indication whether the pallet is full. If so, a ring will be heard to change the pallet and the operator changes the pallet. Nowadays, these operations are also automated. Points *P* and *Q* need not be stationary; but one of them or both can be moving. The next location can be easily computed from the part count, size of the pallet, size of the part and location of the corner of pallet.

Robots can be easily programmed to perform palletizing operation. The program computes the delivery point for each part when keeping the count of number of parts. This is to detect when the pallet becomes full. The robot repeats the cycle of pick and place with a different placing point for each cycle. We are talking about two-dimensional pallets but three-dimensional pallets are also possible. These operations are complex since each pallet is to be known where it is stacked, not only horizontal but also vertical.

9.3.4 Welding

Many robot applications such as when the end effector has a tool instead of gripper are more than just material handling operations. These robots are classified for processing applications. With the tool, the robot manipulator performs some manufacturing processes. Sometimes, the tool is temporarily attached to the end of robot. This provides greater flexibility for tool changing.

Some of the processes where the robots applied are (i) arc welding and (ii) spot welding. The robots are applied in industries for many other processing applications. Drilling, polishing, water jet cutting and the like are these applications. Artificial intelligence is applied in some form of robot applications. Chess coin movement is a way of such applications.

Arc Welding

A human welder using the arc welding unit generally performs the arc welding. The arc welding (or fusion welding) unit consists of an electrode, a low voltage, high current converter and electric cables. The arc is initiated between the two metal plates to be joined and the electrode; this arc is sufficient to melt metal to join the pieces together. The welder

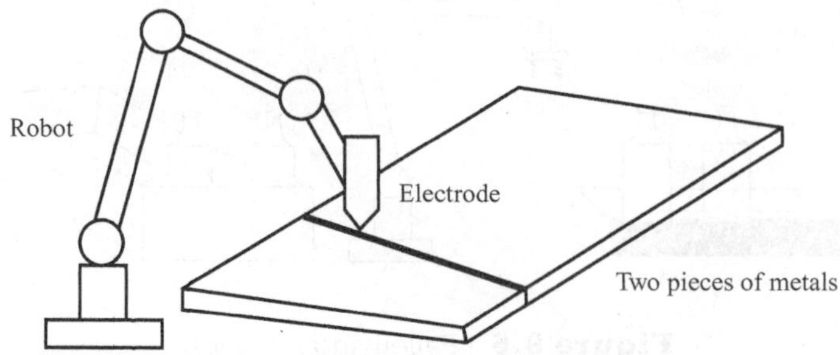

Figure 9.7 Two-dimensional Robot Arc Welding

creates and maintains this pool of molten metal by electrode touching metal pieces. The welding path may be straight or curved not only in the two dimensions but also in three dimensions. This distinguishes an expert welder from the first time welder. The accuracy and quality of the weld depend on skill of the expert welder.

A typical diagram of a two-dimensional plate welding is shown in Figure 9.7. When a consumable electrode is used, it contributes to the metal as filler. The electrode size is reduced requiring an additional movement of electrode towards the parts to maintain the arc. Many times, a replacement of electrode is also needed. The quality of arc weld depends on the speed of electrode movement. If the speed is low, more metal will deposit and the weld is weak. If the speed is more, less metal will deposit and the weld is weak. In both the cases, the weld is weak and hence the speed has to be just sufficient. The arc produced emits ultraviolet radiation which is injurious to human eye and can cause permanent vision damage. The arc welding has other hazards such as high temperature, molten metal, flying sparks and toxic fumes. When a robot is employed and programmed for arc welding, all these problems are overcome and the human is not affected.

There are a few reasons why robot is used in the arc welding:

1. The arc welding process is usually of low quantity.
2. The arc welding parts (or components) to be welded have variations.
3. The edges of parts are very irregular.
4. For a robot, negotiating a straight line path is as difficult as negotiating curves.
5. If the speed of electrode is low, more metal will deposit and the weld is weak; if the speed is high, less metal will deposit and the welding joint becomes weak.

All these problems are circumvented by skill and judgement of human welder.

Spot Welding

This type of welding is a process in which two metals are permanently joined through localized coalescence, a phenomena that occurs at the weld joint. This results from a combination of three factors such as heat, pressure and metallurgical properties. Under such conditions, it is possible to achieve welding with

(i) high temperature and no pressure
(ii) normal temperature and no pressure
(iii) varying degrees of pressure and temperature.

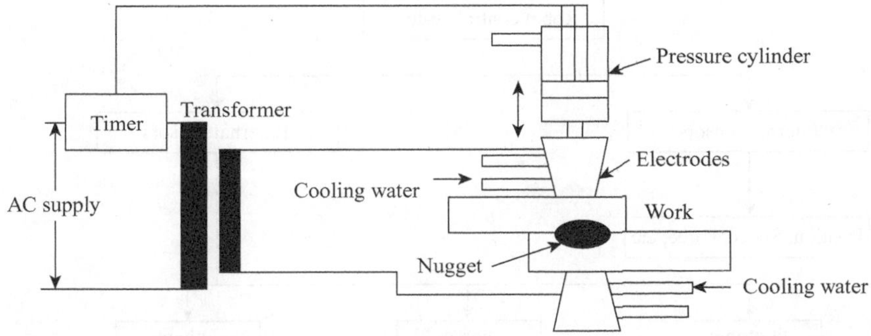

Figure 9.8 Spot Welding Unit

In spot welding, heat and pressure are used to produce coalescence, whereas in arc welding, heat and metallurgical properties are used to melt the material to be joined. The additional portion of material is used to fill the joint.

Spot welding is the most common application of industrial robots. This is mainly due to the fact of widespread use in the automotive industry. Motor car manufacturers have applied robots in spot welding not only to improve productivity but also to release human operators from tiring, awkward and unpleasant tasks. Spot welding is the simplest form of resistance welding.

An arrangement of electrodes and work for spot welding is shown in Figure 9.8. Note that both the pressure cylinders and one of the electrodes are from the robot's finger; the other electrode is from the other finger. The electrodes are connected from a low voltage, high current transformer. High electric current is passed through the electrodes which produces heat. This heat is due to the electrical resistance of the material. A specific amount of low pressure is applied at the beginning. This creates a temperature and then increases to aid the coalescence. The electrodes are water cooled. Pressure is increased via a pneumatic or hydraulic cylinders. Their areas are reduced at the tips to produce weld from 1.5 to 13 mm in diameter. The pressure is applied in a controlled manner to produce weld. The robot, then, disconnects the electrodes and the welding gun is moved to a new location. The guns are massive to the operator but for the robots it is easily manageable. A current of 1500 A is carried by the cable and the size of cable creates no problems to robot. The manipulation of spot welding gun to achieve the welds in the appropriate position and hold the gun in correct orientation demands a robot with full six degrees of freedom.

In conclusion, spot welding is a task ideally suited to industrial robots that are servo controlled. The non-servo controlled robot will not create any acute problems in spot welding. This spot welding is a process with a complexity, but the technical requirements are easily handled by existing system. Reprogramming of robot is fully utilized. The spot welding removes an unpleasant and difficult task to human workers.

9.4 COMPLIANCE

Robot compliance is an important aspect in robot application. It is defined as follows:

> *Robot compliance means initiated or allowed part movement for the purpose of alignments between mating parts.*

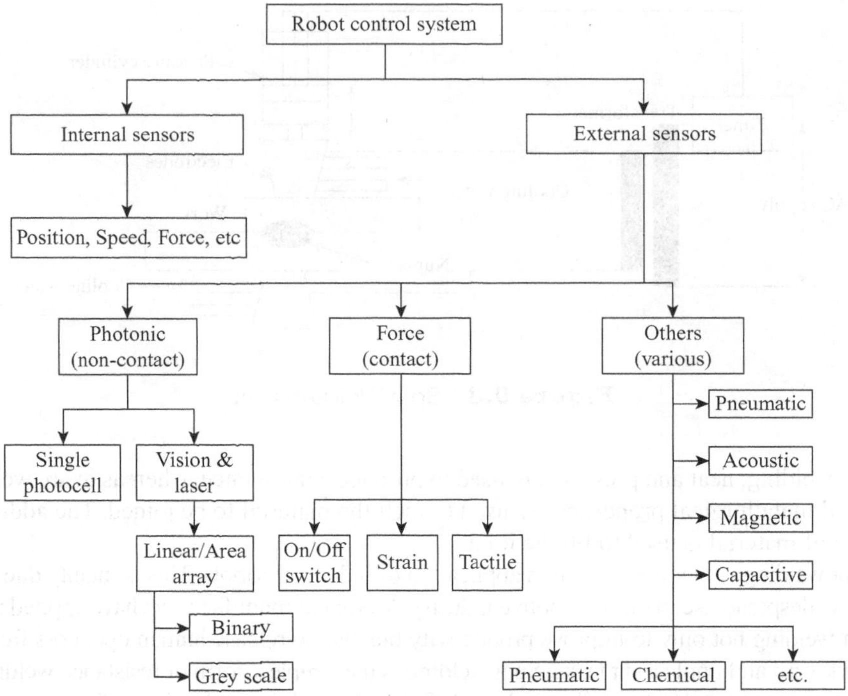

Figure 9.9 Two Ways of Providing Compliances

Figure 9.9 shows robot compliance divided into two parts – active compliance and passive compliance. Active compliance systems measure the force and torque (*F-T*) present when the robot is performing the programmed tasks. They are often called (*F-T*) sensing systems. Force sensing allows the robot to detect changes in the dimensions and adapt the programs to correct them; the torque sensing allows a robot to detect any variations in torque. In (*F-T*) sensing systems, the robot moves are based on a required measured force instead of a programmed point. Active compliance is required when a gear assembly requires compliance. The (*F-T*) sensors are available with varieties of specifications, both for forces and for torques.

Passive compliance is basically a mechanical device. It allows the principle of mechanism of mating parts. Passive compliance, as illustrated in Figure 9.10, shows the principle of remote centre compliance (RCC) device. The theory of operation of the RCC is a little abstract; however, understanding the principle of operation of the device is simple. The system that includes the part (the hole) and gripper (that carries peg) is considered to be mated. The location of the centre of compliance is determined by the design of RCC and depends on the location of compliance elements and their orientation to the RCC.

RCC device provides all three types of compliances such as axial, lateral and rotational. RCC device is attached between the wrist and the gripper. The working of an RCC device is shown in Figure 9.11. The device has four lateral and four rotational springs as compliance elements; Figure 9.11 shows only four springs (two lateral and two rotational).

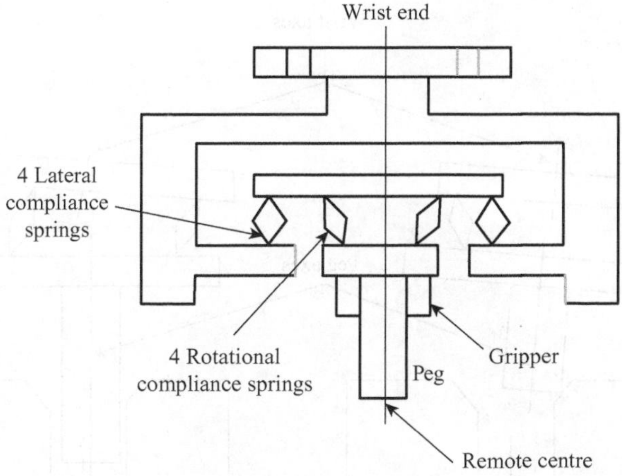

Figure 9.10 RCC Device

The gripper has the peg to be inserted into the hole. The axes of peg and the hole may be parallel or may cross each other; but, they do not coincide. If they are in coincidence, then they can be shifted a bit one way or other for a successful insertion. When the robot tries to push the peg, the contact point between hole and peg generates a lateral force. This is due to the edges of hole which have been chamfered. Under the lateral force, the lateral compliance springs deform and allow the peg to move laterally without the movement of the wrist. This stage is shown in Figure 9.12. Consequently, the peg is moved horizontally by the robot and slides into the hole.

There can be a misalignment between the axes of the peg and the hole. In this case, the axes are not parallel. Let us assume that the robot is pushing the peg at an angle as

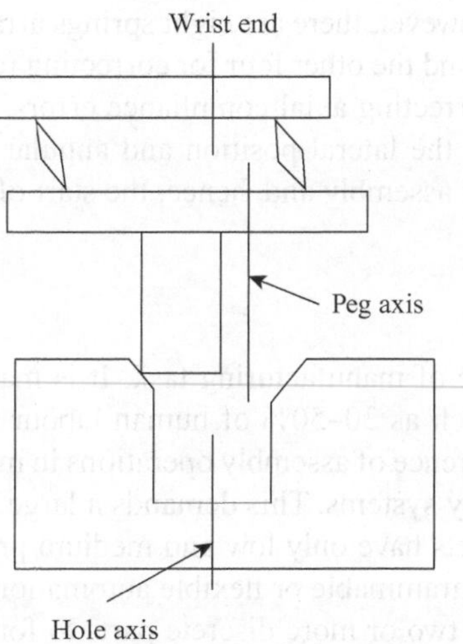

Figure 9.11 Use of RCC to Correct Lateral Error

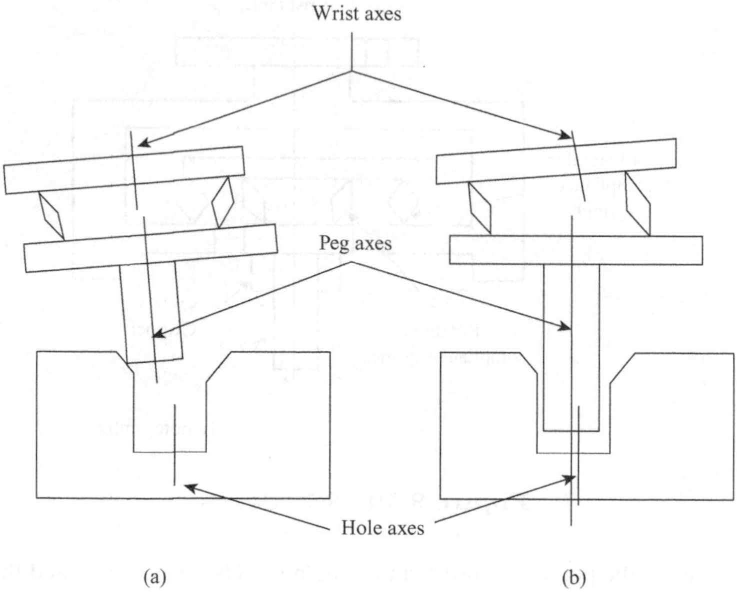

Figure 9.12 RCC Position in Correcting Rotational Error: (a) Initial Stage and (b) Final Stage

illustrated in Figure 9.12. When the peg comes in contact with the corner of the hole, angular moment is created and the angular compliance of RCC occurs which deforms the respective springs. The deformation of angular elements results in making peg axis parallel to the hole axis.

The vertical components of forces generated by the contact of peg with hole deform or compress all the eight spring elements providing an axial compliance. Note that Figure 9.10 shows only four springs; however, there are eight springs arranged in a circular form – four for correcting lateral error and the other four for correcting rotational error and all the eight springs are required for correcting axial compliance errors. This axial compliance pushes the peg into the hole once the lateral position and angular position errors are corrected, accomplishing a successful assembly and, hence, the start of next robotic action.

9.5 ASSEMBLY

Assembly is the final stage of manufacturing task. It is manual and labour intensive. An estimation says that as much as 30–50% of human labour is required to get the finished product. High rate of occurrence of assembly operations in manufacturing industry requires a set of automated assembly systems. This demands a large investment. A welcoming fact is that a majority of products have only low and medium production. This is justifiable to employ robots or other programmable or flexible automation systems.

Assembly means fitting two or more discrete parts to form a new part or subassembly. The final assembly requires a complex way of assembled parts which require considerable amount of handling, positioning and orienting parts. In addition, the subassembly requires a calculated and controlled force to mate both or more parts together properly. A removable screw driver with a screw or a removable spanner with a nut to tighten is temporary assembly.

Assembly, further, means an interaction between two or more parts being assembled. So, the assembled parts maintain their relationship with one another. Assembly operation can be mating of two parts, placing a third part together and fastening all of them with a nut. This indicates that welding is a sort of assembly operation.

In the assembly task of peg-in-hole, if for some reasons the hole is not present at the expected location or the peg is grasped wrongly, a robot would be most of the time unable to perform the assembly successfully. Of course, a human is required to help the robot to find the hole or to grasp the peg correctly. This uncertainty still persists in this physical world to make the task somewhat most difficult.

9.6 INJECTION MOULDING

The most widely and profitably used process is the injection moulding process. This is mainly for plastic moulding producing plastic components. The most common way of injection moulding process is shown in Figure 9.13. Granulated or powder made plastic is fed into the hopper and this in turn is fed into the heated cylinder in which a reciprocating screw plasticizer works. This screw acts in two ways – plasticizing and injection. After passing through plasticizing zone, it passes into plastic injection zone. Here, the moulded plastic is heated to a high temperature. The plastic becomes semisolid and it is injected into the plastic moulding machine.

The moulding machine works in two stages – inputs the desired shape of the finished part and allows to cool the desired finished part. This opens along the mould parting line to allow the part to be ejected. A sprue is a part within the mould. Within the mould, the molten plastic flows into the die through the sprue. It then travels along a runner system and through a gate into an actual component cavity. The sprue, runner and gate system are usually cooled, by a cooling water jet, along with the component and thus have to be removed from the finished article either by hand or by some mechanical means.

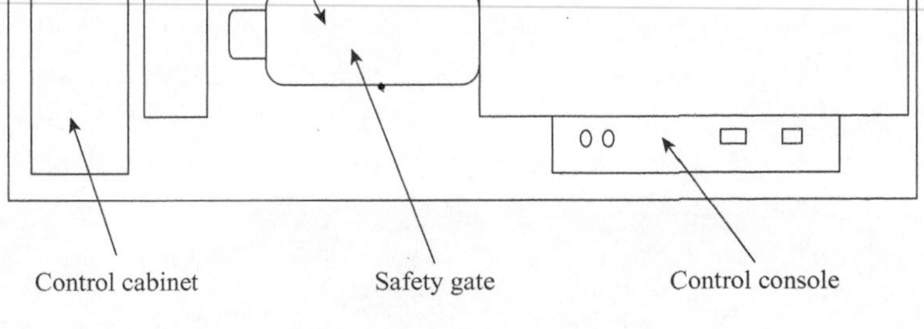

Figure 9.13 Injection Moulding

The advantages of applying a robot to injection moulding are as follows:

(i) Decreased moulding cycle time
(ii) Higher speed of production
(iii) More consistent cycle time
(iv) Better quality components
(v) Removal of human operators

Human workers, if robots are not utilized, have to work in this hazardous environment. This type of hazardous environment is created from high clamping pressure (nearly 5 K tonnes) and high temperature (approximately 300 °C). This high temperature produces noxious fumes. Any unexpected escape of molten metal from this high pressure and high temperature will certainly create unpleasantness.

REVIEW QUESTIONS

1. List a few more industrial applications that are not discussed in the book given to you. Explain them.
2. List and explain any four basic industrial applications of a robot.
3. Explain with figures: robots in handling objects for (i) machine loading and unloading, (ii) materials transfer, (iii) palletizing and (iv) welding.
4. Explain with figures (i) arc welding and (ii) spot welding.
5. Define compliance.
6. Using a block diagram explain two ways of providing compliance
7. Expand RCC. Explain how does RCC overcome errors.
8. What is the assembly of a final stage of manufacturing?
9. What are the advantages of injection moulding?
10. Draw a diagram of injection moulding and explain the internal activities of each part.
11. List and briefly explain a few more robot applications that are not given in this book.

ROBOT TRAJECTORY PLANNING 10

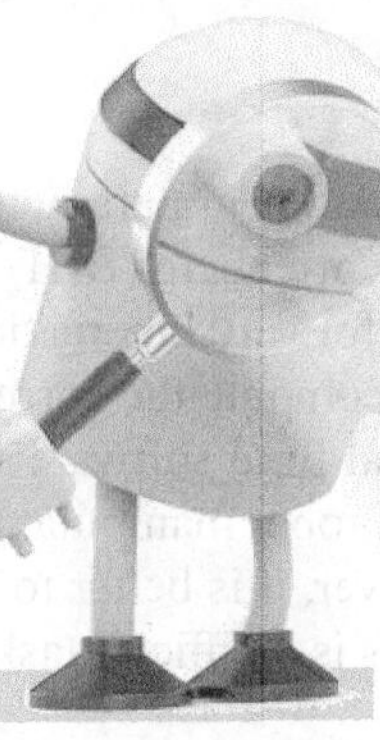

Only a mediocre person is always at his best
William Somerset Maugham

This chapter covers the following key topics:

Introduction – Trajectory Planning Terminology – Steps in Trajectory Planning – p-degree Polynomial Trajectories – Linear Function with Parabolic Blends – LFPB-1 – LFPB-2 – Issues on LFPB Trajectories – Bang-Bang Trajectory – Cartesian Space Versus Joint Space – Examples

10.1 INTRODUCTION

An industrial robot to follow a pre-planned path is the largest problem of motion. This is termed as trajectory planning for the robot. During the trajectory planning, the object (or the work piece) has to be moved in accordance with a planned trajectory. The goal of the trajectory planning is to describe the requisite motion as a time sequence of joint-link end-effector locations and their higher order derivatives. These are generated by interpolation or approximation by properly fitting a polynomial function. The use of higher order derivatives than third-order derivatives is not required. The trajectory planning serves as giving reference inputs or control set points to the robot's control system. The control system, in turn, assures that the robot executes the planned trajectories.

In trajectory planning, the work piece has to be moved from the start (initial) point to a final (end) point; the movement has to be in accordance with the planned trajectory. When there is no obstacle in the path, the robot can have its own trajectory by which the object can be moved. This can be a straight line path or a curved one which has constraints on the robot's joints. Once there are some obstacles on robot path, the robot cannot be moved in any direction. This has to be in accordance with a specific path already planned. The start positions, orientations, linear velocities, accelerations and finally the end positions are to be in line with the planned path.

Forward kinematics of robot manipulators is the study of position and orientation of robot arm (tool, gripper or end effector) with respect to a reference coordinate system with joint variables and arm parameters. The reference coordinate system usually is the robot base. Denavit-Hartenberg (D-H) method provides algorithms which result in a matrix method to derive the forward kinematics solutions for robot manipulators. A real-time software is then used to compute the corresponding variables.

Another set of real-time software to compute the joint-based controllers is called the *inverse kinematics* of robot manipulators. This gives the desired position and orientation of the robot arm. However, it is better to be careful as the computation of inverse kinematics of robot manipulators is a difficult task because of unavoidable nonlinearities and multiple solutions.

With the background of the forward and inverse kinematics, when there is no obstacle in the work space, the following two methods can be considered: firstly, planning a path in Cartesian space and secondly planning a path in joint space.

Cartesian space seems to be attractive since the path in this space can be easily visualized. In addition, the Cartesian space requires a two-time inverse kinematics – one at the start point and the other at the final end point. This is required for one-stage trajectory. Obtaining these two-time kinematics is easier than many-time requirements. However, Cartesian space approach requires a large look-up table to store the data which may be enormous for a quality path. On the other hand, if data can be computed using inverse kinematics in real time, the robot working efficiency can be greatly affected.

Joint space is the second path planning which requires many-time applications of inverse kinematics. Here, some advantages can be seen. One-stage trajectory requires the applications of inverse kinematics for many times. For example, after using the inverse kinematics to determine the initial and final joint positions, the robot inverse kinematics is not required again. The path can then be easily planned in the joint space. Care must be taken to see that the path does not have discontinuity.

If a trajectory is planned in the work space with obstacles, arbitrarily chosen trajectories will not be useful. Therefore, one of the trajectories having *via points* has to be chosen. We shall see in later discussion that a polynomial with a certain number of via points will have to be fit to accommodate the planned path.

This chapter introduces a typical trajectory planning issue. With regard to the inverse kinematics, a solution to joint space is the same as that of Cartesian space and the solution to Cartesian space is the same as joint space. There should be one solution; thus, problems are solved in Cartesian space due to its advantages. Every robot has a capability to move from an initial location to a final location in Cartesian space.

10.2 TRAJECTORY PLANNING TERMINOLOGIES

Let us discuss some terminologies involved in trajectory planning. These are listed in Table 10.1. These terminologies are important as we design a robot to have the set of desired trajectories.

10.3 STEPS IN TRAJECTORY PLANNING

There are three steps in trajectory planning. All of them, either in parts or in total, are required to solve the trajectory planning problem.

(i) **Task description:** This is the first step in the motion planning problem. This task can be grouped into the following three different categories:

Table 10.1 Terminologies and Their Meanings

No.	Terminology	Meaning
1	Path	A path is a locus of points to be traversed by the robot in executing the specified task. A path should have at least two points – the start and end points along with proper orientations
2	Trajectory	A trajectory is a path with specified qualities of motion, that is, a path on which a time history is specified in terms of start position, orientation, velocity, acceleration and finish point
3	Via points	Via (path) points are the set of intermediate locations between the first start point and the final goal point. The robot is expected to pass through these points to reach the final goal point. In this chapter, other terms such as destination point, final point and finish point are used interchangeably
4	Spline	This is a smooth time function that passes through the set of path points
5	Joint space trajectory planning	This is an additional description. However, in this planning, each via point is specified in joint space in terms of position and orientation of end-effector frame relative to a base frame. Each of this point is converted to a set of desired joint positions by application of inverse kinematics. A smooth time function is then determined for each joint when a robot passes through these points
6	Cartesian space trajectory planning	In this planning, the path is explicitly given in Cartesian space. The path constraints (start positions, orientations, velocities, accelerations and end points) are specified in Cartesian coordinates. Later, these constraints can be represented in joint space through inverse kinematics
7	Trajectory generation	This is an act of computing the trajectories as a time sequence of values in real time, using a trajectory planning algorithm. This algorithm takes care of spatial and temporal constraints
8	Path update rate	It is the rate at which the trajectory points are computed. This rate will indicate which next point the robot has to move

The first category is the pick and place application. This task is specified as initial and final end-effector locations. No particular specification on intermediate location of end effector is given and the planner is free to formulate any convenient path. The user specifies a goal point for an initial point, both points are known in Cartesian space.

In the second category, if in addition to start and finish points, a specific path is to be traced by the end effector in Cartesian space. This is known as continuous path motion and continuous trajectory. For example, in arc welding, the arc welder specifies the type and parameters of the path to be traced.

The third category of task description is where more than one set of points are specified. For example, a pick and place operation in a place where obstacles are present. The first point is called the initial point while the last point is called the goal point. The intermediate locations are via points. However, the task performed by tool or gripper in the point-to-point motion along the continuous path motion is not part of trajectory planning.

(ii) **Employing a trajectory planning technique:** Various trajectory planning techniques fall into one of the two categories – the joint space technique and Cartesian space technique. We shall consider Cartesian space technique. In terms of point-to-point motion, with or without path points, Cartesian space techniques are employed. By adopting inverse kinematics, joint variables and their derivatives can be obtained.

(iii) **Computing the trajectory:** The final step is to compute the time sequence values and their derivatives attained by any function generated by trajectory planning techniques. These values are computed at a particular path update. The path update rate in real time lies between 20 and 200 Hz in a typical industrial robot system.

10.4 p-DEGREE POLYNOMIAL TRAJECTORIES

We shall study the use of a p-degree ($p = 3$) polynomial as the interpolation function when a set of path points is given. For a smooth motion between two points, the selection of a single polynomial for entire joint path depends on the number of constraints such as

(i) initial position, (ii) initial velocity, (iii) final position and (iv) final velocity.

A third degree ($p = 3$) polynomial (cubic spline) with four coefficients can be used as

$$q(t) = a_0 + a_1 t + a_2 t^2 + a_3 t^3, \tag{10.1}$$

where a_i, $i = 0, 1, 2$ and 3 are four coefficients which are related by their constraints. However, for a simple pick and place task, just specifying start and finish points are not enough for satisfactory performance. While picking up an object, the motion of the end effector must be directed away from the supporting surface of the object. This will avoid crashing of the end effector with the supporting surface.

Thus, a 'lift-off position' along with outward normal away from the surface has to be considered. On the basis of similar consideration, a 'set-down position' is selected to specify the smooth and correct approach, which gives following four position constraints:

(i) Initial position
(ii) Lift-off position
(iii) Set-down position
(iv) Final position.

In a typical pick and place operation, the robot begins moving at constant acceleration, away from initial position in an appropriate direction. For example, in a vertically upward direction from a horizontal surface, the initial position is complete. After it has moved to a safe distance, lift-off is complete. From then onwards, it moves at a constant velocity with zero acceleration. This constant velocity and zero acceleration motion continue until it reaches the set-down position. From this point it begins to vertically decelerate reaching the final position. Figure 10.1 illustrates this trajectory. This is known as linear function with parabolic blends (LFPB) and has six constraints – four as listed above, one acceleration constraint in the beginning and one deceleration constraint in the final stage.

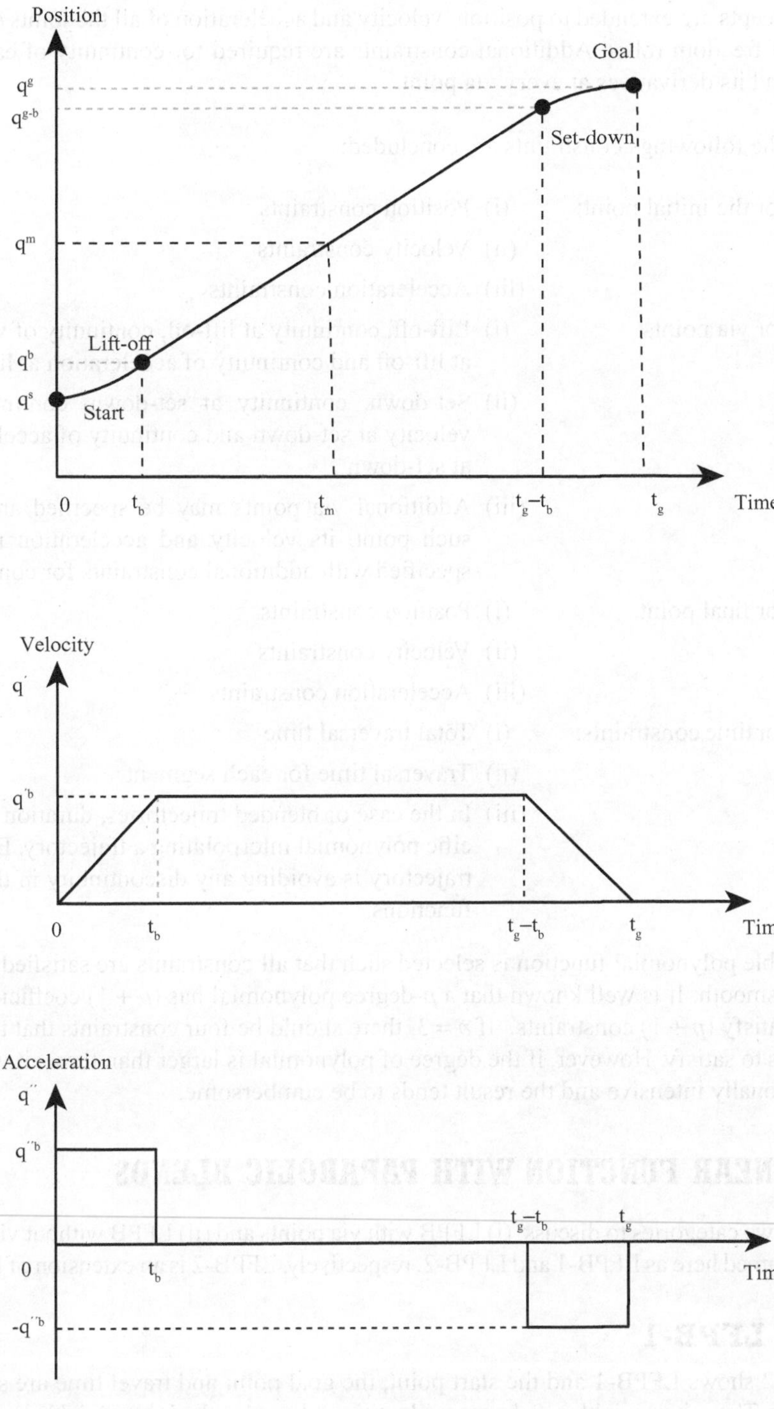

Figure 10.1 Position, Velocity and Acceleration Profiles

These concepts are extended to position, velocity and acceleration of all the joints of multi-degrees of freedom robot. Additional constraints are required for continuity of each joint variable and its derivatives at every via point.

Thus, the followings constraints are concluded:

(a) For the initial point:
   (i) Position constraints
   (ii) Velocity constraints
   (iii) Acceleration constraints

(b) For via points:
   (i) Lift-off, continuity at lift-off, continuity of velocity at lift-off and continuity of acceleration at lift-off
   (ii) Set-down, continuity at set-down, continuity of velocity at set-down and continuity of acceleration at set-down
   (iii) Additional via points may be specified and each such point, its velocity and acceleration may be specified with additional constraints for continuity

(c) For final point:
   (i) Position constraints
   (ii) Velocity constraints
   (iii) Acceleration constraints

(d) For time constraints:
   (i) Total traversal time
   (ii) Traversal time for each segment
   (iii) In the case of blended trajectories, duration of specific polynomial interpolating a trajectory. Blended trajectory is avoiding any discontinuity in the time functions.

A suitable polynomial function is selected such that all constraints are satisfied and trajectory is smooth. It is well known that a p-degree polynomial has $(p + 1)$ coefficients and this will satisfy $(p + 1)$ constraints. If $p = 3$, there should be four constraints that the polynomial has to satisfy. However, if the degree of polynomial is larger than three, it would be computationally intensive and the result tends to be cumbersome.

10.5 LINEAR FUNCTION WITH PARABOLIC BLENDS

There are two categories to discuss: (i) LFPB with via points and (ii) LFPB without via points. Both are named here as LFPB-1 and LFPB-2, respectively. LFPB-2 is an extension of LFPB-1.

10.5.1 LFPB-1

Figure 10.2 shows LFPB-1 and the start point, the goal point and travel time are specified by the user. The joint variable, q_i^j, has two identities where i is the joint variable and j is the via point. The motion of end effector starting at $t = 0$ and ending at $t = t_g$ is the complete time history of a joint.

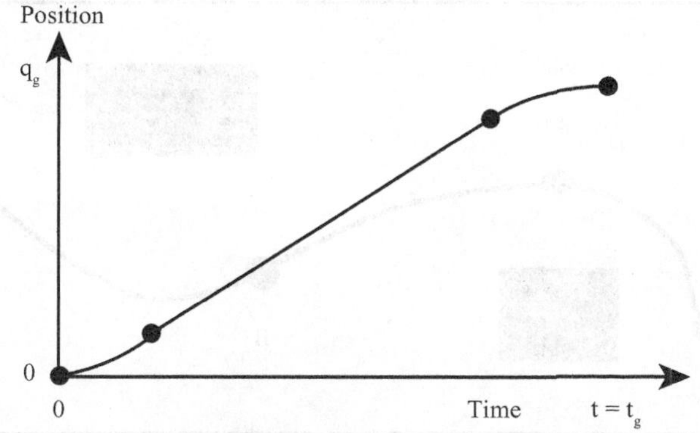

Figure 10.2 LFPB-1

The two identities i and j are ignored since there is only one joint variable for which no identities are necessary. Let q^s and q^g be the start and goal points, respectively, of the joint variable $q(t)$. Further, $q(t)$ is zero for all $t < 0$ and $t > t_g$. This along with the continuity requirements offers four constraints:

$$q(0) = q^s;\ q(t_g) = q^g;\ q'(0) = 0;\ q'(t_g) = 0. \tag{10.2}$$

There are four constraints and hence the polynomial $q(t)$ has to be of third order.

We shall consider Equation (10.1) as an illustration. Then, the parabolic velocity profile is

$$q'(t) = a_1 + 2a_2 t + 3a_3 t^2 \tag{10.3}$$

and the linear acceleration profile is

$$q''(t) = 2a_2 + 6a_3 t. \tag{10.4}$$

Substituting the constraints of Equation (10.1) into Equations (10.2)–(10.4), with some modifications, results in the following set:

$$a_0 = q^s, \qquad a_1 = 0,$$
$$a_2 = (3/t^2_g)\,(q^g - q^s), \qquad a_3 = (-2/t^3_g)\,(q^g - q^s). \tag{10.5}$$

Thus, the cubic polynomial to interpolate the path connecting the initial joint position to final joint position is

$$q(t) = q^s + (3/t^2_g)\,(q^g - q^s)\,t^2 - (2/t^3_g)\,(q^g - q^s)\,t^3. \tag{10.6}$$

10.5.2 LFPB-2

It should be noted that the problem of LFPB-2 has a possibility (accessibility), as shown in Figure 10.3 (a) where a number of obstacles have to be overcome. Here two obstacles are indicated and they are avoided by two via points A and B. In this case, we can plan two polynomials, one for each via point. $P(0)$ and $P(t_f)$ are the start and the stop points.

It is the usual practice that each via point is specified in terms of desired end-effector position and orientation. Similar to the single goal point problem, each via point is transformed into a set of joint variable values by inverse kinematics. Assume that q_j ($j = 1, 2$) are two cubic polynomials which are defined in the interval $0 < t < t_f$. P_j ($j = 1, 2$) are two portions as shown in Figure 10.3 (b). The total time interval for the path is

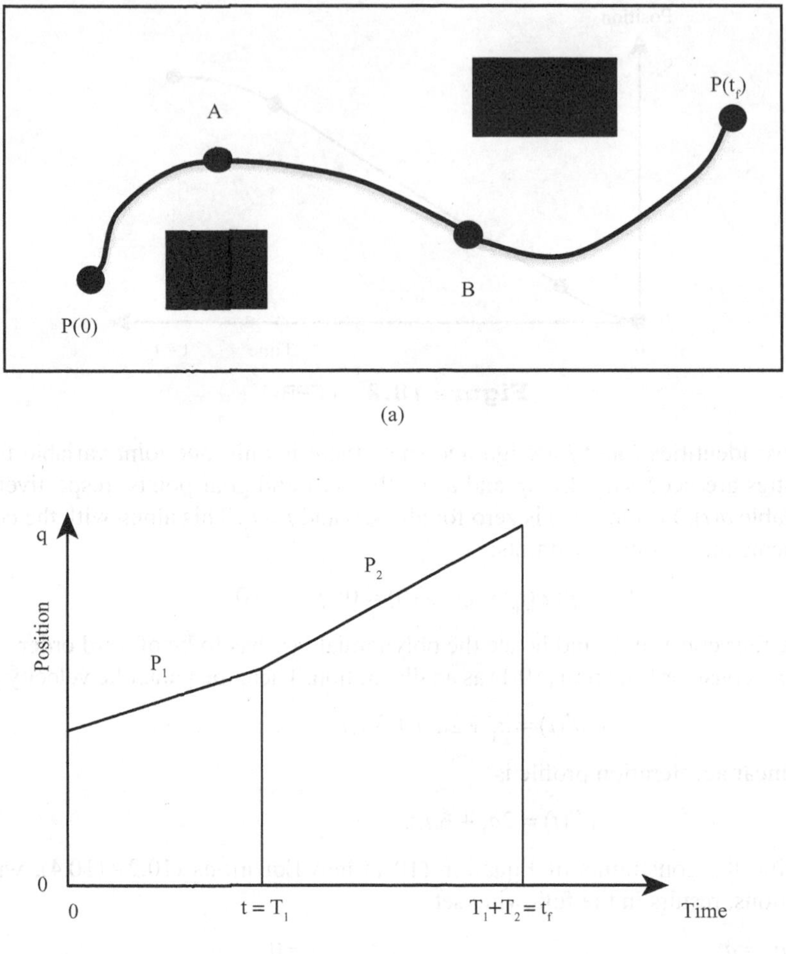

Figure 10.3 (a) Obstacle Avoiding with a Special Trajectory and (b) LFPB-2

$$T = t_f = T_1 + T_2. \tag{10.7}$$

Additional constraints are required for smooth trajectory at each via point. Once these constraints are specified, two cubic polynomials are determined as before. There are several accessibilities of specifying the additional constraints at each via point. However, two accessibilities are considered here.

Accessibility 1: The set of constraints imposed on two cubic polynomials between path points is given by

$$P_1(t) = q^1 \qquad P_2(t) = q^2. \tag{10.8a}$$

The derivatives of $P_1(t)$ and $P_2(t)$ are

$$P'_1(t) = q'^1(t), \qquad P'_2(t) = q'^2(t). \tag{10.8b}$$

The four constraints in Equation (10.8) represent a set of four equations from which four unknown coefficients of cubic $P_j(t)$ and $P'_j(t)$ can be obtained where $j = 1, 2$.

Accessibility 2: Joint velocities at via points are computed using a heuristic criterion. In this case, the constraints imposed on the cubic polynomials are the same as those used in the previous case, but the method of specifying via point velocities are different. The joint variables $q(t)$ at the via points along travel times T_1 and T_2 are specified by the user. The joint velocities at via points are assigned using a criterion which preserves the continuity of $q(t)$. Two such criteria are as follows:

$$q'^1(t) = 0, \qquad q'^2(t) = 0. \tag{10.9}$$

It is to be noted that this solution creates velocity as zero both at the start and finish points. The acceleration, though linear, has discontinuity at initial and final positions. Due to this the motion has jitters at the start and goal points. When the user does not bother the jitters in the acceleration profile, he/she can consider this accessibility. If he/she is concerned, then has to use another method which avoids jitters.

10.6 ISSUES ON LFPB TRAJECTORIES

There are three issues that must be observed: (i) the calculation of blend time, (ii) the calculation of final time and (iii) the constraints on final time; velocity constant is also included in parts (ii) and (iii).

The blend time, t_b: When the constant acceleration a and the velocity v, both in acceleration segment and velocity segment, respectively, are known, the computation of blend time is direct as

$$t_b = v/a. \tag{10.10}$$

We only know initial and final positions (Θ_o, Θ_f), constant velocity (v) and final time (t_f). First, from Equation (10.10), acceleration can be written as $a = (v/t_b)$. Then, the expression for first segment is

$$\Theta(t) = \Theta_o + (v/t_b)\, t^2/2 \qquad (0 \le t \le t_b). \tag{10.11}$$

Due to the continuity of time response in LFPB problems, at $t = t_b$, the following equation holds:

$$t_b = (\Theta_o - \Theta_f + v\, t_f)/v. \tag{10.12}$$

Computation of final time, t_f and constant velocity, v: If Θ_o, Θ_f, v and a are known, then final time t_f can be computed.

First, using Equations (10.10) and (10.11), we get

$$(v/a) = (\Theta_o - \Theta_f + v\, t_f)/v. \tag{10.13}$$

Then, the final time is

$$t_f = (\Theta_f - \Theta_o)/v) + (v/a). \tag{10.14}$$

Equation (10.14) can be expressed in a following quadratic form:

$$v^2 - a\, t_f\, v + a(\Theta_f - \Theta_o) = 0. \tag{10.15}$$

The solution to the quadratic form is

$$v = (a\, t_f \pm \text{SQRT}(a^2\, t_f^2 - 4\, a(\Theta_f - \Theta_o))/2, \tag{10.16}$$

where SQRT (.) is the square root of (.).

Constraints on final time, t_f and velocity, v: The LSPB trajectory, having a blend time t_b, satisfies the following inequality:

$$0 < t_b \leq (t_f/2). \tag{10.17}$$

Substituting Equation (10.13) into Equation (10.17), we get

$$0 < (\Theta_o - \Theta_f + v\, t_f)/v \leq (t_f/2) \tag{10.18}$$

or

$$(\Theta_f - \Theta_o)/v < t_f \leq 2(\Theta_f - \Theta_o)/v. \tag{10.19}$$

When the initial position Θ_o, the final position Θ_f and the velocity constant v are given, the final time cannot be arbitrarily chosen. This must satisfy the constraints as given in Equation (10.19).

Similarly, when Θ_o, Θ_f and final time, t_f, are given, velocity v must satisfy the following constraint:

$$((\Theta_f - \Theta_o)/t_f) < v \leq (2(\Theta_f - \Theta_o)/t_f). \tag{10.20}$$

Thus, we get information required for a bang-bang trajectory.

10.7 BANG-BANG TRAJECTORY

The bang-bang trajectory is otherwise known as minimum-time trajectory. This is a special case of LFPB-1 and LFPB-2 trajectories. This occurs when $t_b = t_f/2$, where time t_f is the switching time and t_b is the general time function. Figure 10.4 illustrates the bang-bang function which has only two segments. This has two portions, one with constant acceleration, a, and the other with constant deceleration, $-a$. Switching occurs at $t = t_b$ from a to $-a$. The bang-bang trajectory has its maximum velocity, v_{max} and the deceleration is v_{min}. The bang-bang trajectory always offers maximum acceleration as provided by the actuator. Thus, the bang-bang trajectory is an optimal trajectory with minimum final time, t_f.

It can be seen from Figure 10.4 that

$$v_{max} = at_b \tag{10.21}$$

'a' is any constant to be determined and

$$t_f = 2t_b. \tag{10.22}$$

Substituting Equation (10.22) into Equation (10.12), we get

$$t_b = (\Theta_o - \Theta_f + (v_{max}\, t_f)/v_{max}) = (\Theta_o - \Theta_f)/v_{max} + 2t_b. \tag{10.23}$$

Therefore

$$t_b = (\Theta_f - \Theta_o)/v_{max} \tag{10.24}$$

and

$$v_{max} = (\Theta_f - \Theta_o)/t_b. \tag{10.25}$$

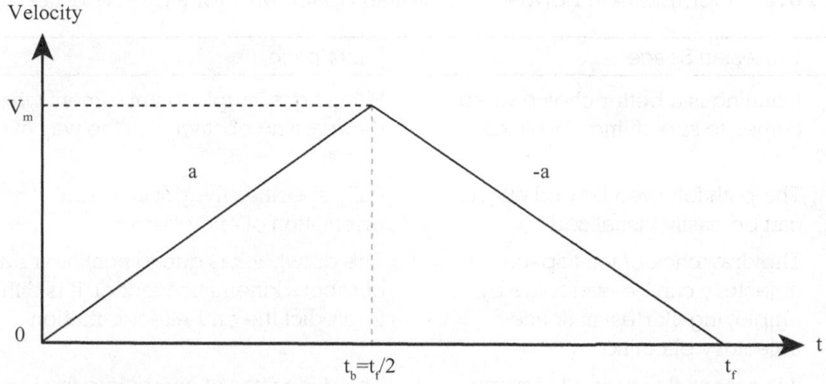

Figure 10.4 Velocity Profile of a Bang-Bang Trajectory

Substituting Equation (10.21) into Equation (10.25), we get

$$t_b = \text{SQRT}\ ((\Theta_f - \Theta_o)/a). \qquad (10.26)$$

Substituting Equation (10.26) into Equation (10.25), we can get the maximum velocity as

$$V_{max} = \text{SQRT}(a(\Theta_f - \Theta_o)) \qquad (10.27)$$

or

$$t_f = 2t_b = 2\ \text{SQRT}\ ((\Theta_f - \Theta_o)/a). \qquad (10.28)$$

Equations (10.25) and (10.28) indicate that the maximum velocity and the final time of a bang-bang trajectory are entirely determined by initial position (Θ_o), the final position (Θ_f) and the acceleration (a).

10.8 CARTESIAN SPACE VERSUS JOINT SPACE

All computations were performed with respect to Cartesian space. Table 10.2 compares Cartesian space with joint space.

10.9 EXAMPLES

EXAMPLE 10.9.1

The SCARA robot has four joints and four limbs. Its second joint is required to move from 30° to 150° in 5 s. Determine the cubic polynomial trajectory to generate smooth trajectory for the joint. What are the maximum velocity and acceleration for this trajectory?

Solution:

The given parameters of the trajectory are as follows:

$$\theta^s = 30°, \quad \theta^g = 150°,$$
$$v^s = 0, \quad v^g = 0,$$
$$t^0 = 0\ \text{s}, \quad t^g = 5\ \text{s}. \qquad (E10.1.1)$$

Table 10.2 Comparison Between Cartesian Space and Joint Space Trajectories

Sl. No.	Cartesian Space	Joint Space
1	Planning is a better choice when it comes to specifying obstacles	Planning is simple to implement when there are no obstacles in the way of end effector
2	The path followed by end effector can be easily visualized	Fully specifies the position and orientation of end effector
3	The drawback of joint space trajectory can be overcome by employing Cartesian space trajectory planning	The drawback is due to nonlinear nature of robot's kinematics model. It is difficult to predict the end-effector motion
4	It is essential to resort to inverse kinematic model and determine the joint variables at every instant of time. Hence, high computational effort is required	The difficulties of computing the joint space schemes are not felt here; the planning is done directly in joint space
5	Overall motion is faster. This demands a higher computational effort	Overall motion is slower. Hence, the computational effort is neither higher nor costlier

The four coefficients of a cubic polynomial are computed as follows:

$$\begin{aligned} a_0 &= 30, \\ a_1 &= 0, \\ a_2 &= (3/t_g^2)\,(\theta^g - \theta^s) = 14.4, \\ a_3 &= -\,(2/t_g^3)\,(\theta^g - \theta^s) = -1.92. \end{aligned} \tag{E10.1.2}$$

Thus, we get the following polynomials interpolating position, velocity and acceleration profiles:

$$\begin{aligned} \theta(t) &= 30.0 + 14.4\,t^2 - 1.92\,t^3, \\ \theta'(t) &= 28.8\,t - 5.76\,t^2, \\ \theta''(t) &= 28.8 - 11.52\,t. \end{aligned} \tag{E10.1.3}$$

As expected, the velocity profile is parabolic and acceleration profile is linear. The maximum values of velocity and acceleration are 2.5°/s and −11.52°/s², respectively.

EXAMPLE 10.9.2

A single cubic trajectory given by

$$q(t) = 30 + t^2 - 6t^3 \tag{E10.2.1}$$

is used for a period of 3 s. Determine the start and goal positions, velocity and acceleration of the end effector.

Solution:

The given function is

$$q(t) = 30 + t^2 - 6t^3. \tag{E10.2.2}$$

Differentiating once

$$q'(t) = 2t - 18t^2 \qquad \text{(E10.2.3)}$$

and differentiating again, we get

$$q''(t) = 2 - 36t. \qquad \text{(E10.2.4)}$$

Thus, at $t = 0$, start and goal positions are

$$t = 0;\ q(0) = 30,$$

$$t = 3;\ q(3) = -123. \qquad \text{(E10.2.5)}$$

Velocity and acceleration are

$$q'(3) = -156$$

$$q''(3) = -106 \qquad \text{(E10.2.6)}$$

EXAMPLE 10.9.3

Design a single polynomial trajectory which starts from the initial position of $\Theta(0) = 10°$, passes through a via point $\Theta(1) = 5°$ and then stops at final angular position $\Theta(2) = 50°$. The velocities of start and stop positions are 0.

Solution:

There are five constraints for the trajectory:

$$\Theta(0) = 10°;\quad \Theta'(0) = 0;\quad \Theta(1) = 5°;\quad \Theta(2) = 50°;\quad \Theta'(2) = 0. \qquad \text{(E10.3.1)}$$

We shall use the following fourth-order polynomial with five parameters as the candidates of the trajectory function:

$$\Theta(t) = a_o + a_1 t + a_2 t^2 + a_3 t^3 + a_4 t^4. \qquad \text{(E10.3.2)}$$

The first and second derivatives are as follows:

$$\Theta'(t) = a_1 + 2a_2 t + 3a_3 t^2 + 4a_4 t^3, \qquad \text{(E10.3.3)}$$

$$\Theta''(t) = 2a_2 + 6a_3 t + 12a_4 t^2. \qquad \text{(E10.3.4)}$$

Using the given five constraints, we get

$$a_o = 10,$$

$$a_1 = 0,$$

$$a_o + 2a_1 + 4a_2 + 8a_3 + 16a_4 = 50,$$

$$a_1 + 4a_2 + 12a_3 + 32a_4 = 0,$$

$$a_o + a_1 + a_2 + a_3 + a_4 = 5. \qquad \text{(E10.3.5)}$$

Then

$$a_o = 10,$$

$$a_1 = 0,$$

$$a_2 = -70,$$

$$a_3 = 90,$$

$$a_4 = -25. \quad \text{(E10.3.6)}$$

Therefore, the fourth-order polynomial trajectories are

$$\Theta(t) = 10 - 70t^2 + 90t^3 - 25t^4, \quad \text{(E10.3.7)}$$

$$\Theta'(t) = -140t + 270t^2 - 100t^3, \quad \text{(E10.3.8)}$$

$$\Theta''(t) = -140 + 540t - 300t^2. \quad \text{(E10.3.9)}$$

REVIEW QUESTIONS

1. List and explain some terminologies of robot trajectory planning.
2. List any three trajectory planning techniques and explain.
3. What is cubic spline polynomial? What are its advantages?
4. What are initial, intermediate and final point constraints? In addition, what are the necessary time constraints?
5. What are LSPB-1 and LSPB-2? Explain in terms of figures.
6. Explain some issues of LSPB trajectories.
7. Explain 'Bang-Bang' trajectory. What are its advantages?
8. Compare Cartesian space and joint space trajectories.
9. What are the applications of (i) inverse kinematics and (ii) forward kinematics in developing trajectory planning techniques?

ECONOMIC ANALYSIS

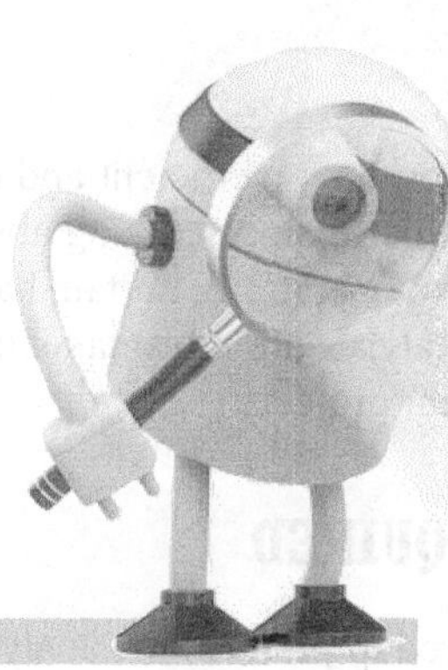

11

Robotics severely challenges artificial intelligence by forcing to deal with real objects in the real world. Techniques and representations developed for purely cognitive problems often do not extend to meet the challenge

Michael Brady

This chapter covers the following key topics:

Introduction – Data Required – Methods of Analysis – Simple Payback Period – Production Rate Appraisal Payback – ROI Evaluation – Robot Installation – Quality of Working Life – Attitudes Towards Robots – Effect on Employment – Current Capabilities of Robots – Future Capabilities of Robots – Example

---

11.1 INTRODUCTION

A question still remains for a robotic project alternative – Will the robot project justify economically? The economic analysis for any engineering project is as important as the robot itself. The management usually decides whether to install the robot project on the basis of analysis. In this chapter, we consider answering the question, even though robot project is still under economic scrutiny. To perform the economic scrutiny of a proposed robot project, certain basic information such as cost of robot and its installation, cost of man and materials in a production cycle, more on the saving and mostly on benefits resulting from the project are required.

The two categories of robot installation that are commonly encountered are as follows:

(i) Involves new application facility
(ii) Replaces the current installation.

The first operation is about a new application facility. Robot installation represents a possibility that must be used to satisfy this need. Various versions are compared and the best one is selected assuming it meets the company's investment criteria.

The second alternative is the robot installation to replace the current method of production. The current method typically involves a production operation being performed manually or by other non-robot methods. The robot can be used as a substitute for human labour. In this situation, the economic justification of a robot installation is a welcome attitude.

This often depends on how inefficient and expensive the manual production is and this is what is compared rather than discussing absolute merits of robot installation.

In either of the two operations, certain basic cost information is required. This information offers which analyses are advantageous in the way of production and to analyse the alternative investment projects.

11.2 DATA REQUIRED

The cost data required to perform the analysis of a robot project separates themselves into two types – investment costs and operating cost. The investment costs include purchase cost of a robot. The engineering costs are also included and they are associated with installation of the robot work cell. Table 11.1 lists and explains the investment cost, whereas costs that can be saved are listed and explained in Table 11.2.

Table 11.1 Investment Cost of a Robot Project

No.	Investment Costs	Explanation
1	Robot purchase cost	The price of a robot is highly variable. For example, the range of price can be anything between ₹40,000 for a simple pick and place variety and ₹800,000 for a sophisticated robot. In general, higher the cost, higher the payable jobs. They are also embedded with artificial intelligence and can be applied to jobs where intelligence is required
2	Maintenance and periodic overhaul	It is the costs of planning and design, by technical staff, to install the robot in a robot work cell. In order to keep the robot in good operating condition, there is a need for regular maintenance. There is also a need for one sweeping overhaul and a need for correcting unscheduled downtime incidents. Maintenance costs in a foundry are greater than those experienced in plastic moulding. This costs approximately 10% of acquisition cost
3	Installation costs	Sometimes utilized fully to a robot project but is often considered as overhead. Includes labour cost and material expenses needed to prepare the installation site. Even if the robot system is to be operated manually, the installation costs are absorbed in robotizing the robot work cell
4	Special tooling	Gripper for a robot is included in the areas such as die cast machine. Gripper can also be included with part positioners, fixtures and other special tools to operate the work cell. The other special tools include indexing conveyor, weld guns, transformers and clamps. A supervisory computer for a sophisticated robot is also considered as a special tool
5	Miscellaneous costs	These costs include additional investment not included by any of the above categories. Robots like any other equipment will exhibit a useful life; it is a practice to depreciate the investment over this useful life. Since a robot is a general purpose equipment, there is an ample evidence of a life span of 8–10 years

Table 11.2 Costs that can be Saved

No.	Operating Costs	Requirement
1	Labour displaced	The prime issue in justifying a robot is labour displacement. The key motivation is the saving of labour cost by supplying a human worker with a robot. A single labour can operate for more than one shift and then multiply the labour saving potential This is also suitable for indirect labour costs
2	Quality upgrade	The mood of human worker changes as the work lies in a hazardous environment or is physically demanding or simply mind-numbering. The robot is more fitting in these jobs and is more advantageous in producing higher quality output. The robot is dumb in such a case but smart enough to do it better. These include costs of supervision, setup, programming and others
3	Maintenance	This covers the anticipated costs of maintenance and repair of robot and its cell. It includes not only indirect labour or the maintenance crew, but also materials and service calls by the manufacturers
4	Cost of utility	This includes costs of utilities to operate the robot cell, for example, electricity, air pressure and gas are the utilities which can be a saving in reducing either one or more than one
5	Inhouse training cost	Inhouse training might be considered to be an investment because much of the training required for installation will occur as the first cost of installation. However, inhouse training has to be a continuing activity and so it can be included as an operating cost
6	Increase in throughput	Higher quality naturally means more net output when a robot works fast enough to match a human worker output. However, there are often circumstances when a robot works faster to increase gross or overall output. The increased throughput is valuable in its right, but improved utilization of capital assets can supplement the economic benefits compared with human worker

It is often convenient to identify the cost saving that will result from the use of robot as compared with an existing method rather than to separately identify the operating costs of alternative methods. Material savings, scrap reduction and advantages resulting from more consistent quality are examples of these savings. All items listed in Table 11.2 are interpreted to allow costs savings between the alternatives.

At the beginning of the project, the investment costs have been paid with no immediate returns. Also, there is a compensating cash flow that represents revenue which should exceed the amount of the operating cost. The difference between operating costs and the revenue is the net flow for the robot installation. These difficulties often immediately prevent the net cash flow reaching the steady-state value anticipated for the robot project. If a robot project is a good investment the net flow will allow the company to recover its investment costs in the project in a relatively shorter period of time.

11.3 METHODS OF ANALYSIS

The methods of analysis include payback methods. Payback methods are the simplest forms of project appraisal. This depends on providing answers to twin questions on the following:

(i) How much does it cost?
(ii) How soon can we recover the investment?

In projects that are philanthropic in nature outlook, initial investment can never be recovered. In normal industrial scene, the financial advisors favour those projects which pay for themselves in a relatively short time. The time is measured in years. Many factors depend on payback period, that is, in years. One or two years is not a problem to financial advisors, but beyond payback period of three or four years should find support.

Following are the three subdivisions to calculate payback periods:

(i) Simple payback period
(ii) Production rate appraisal on payback
(iii) Return on investment (ROI) payback

The first one has been discussed in Chapter 1. However, it is discussed here also but with amended parameters. The second and third methods are new to this chapter.

11.3.1 Simple Payback Period

Here is the scenario for simple payback example. A robot is to work 250 days in a calendar year (not 365 or 366 days). The robot is planned to replace a human worker whose wages and fringe benefits amount to ₹100.00 per hour. This payment is sufficient to run and maintain his/her family. Capital investment for the robot and accessories would be ₹500,000.00. The expenses of maintaining the robot and its accessories amount to ₹15.00 per hour. The company has two options: (i) to run 8 h per day for one shift or (ii) to run 16 h per day for two shifts when production demands so.

Figure 11.1 shows the calculations and illustrations using the simple payback method. If only one shift is undertaken, that is, 8 h per day for annual 250 days, then the payback period amounts to approximately about 3 years. This may satisfy the company financial advisors. If sales forecasts and production plans indicate that two shifts are needed. Then the payback period has to be extended to two shifts per day that reduces it to approximately about 1.5 years. This would justify going ahead with the project with the same nature.

11.3.2 Production Rate Appraisal on Payback

A robot can work in single shift or double shift in the work station. It can work faster than human worker or slower than him. Following are the four possibilities:

(i) Faster than human worker in single shift
(ii) Faster than human worker in double shift
(iii) Slower than human worker in single shift
(iv) Slower than human worker in double shift.

These possibilities will be shown in the denominator with their respective sign – a +ve sign for faster and –ve for slower. In addition, the term in ( ) is also amended with proper data.

Simple payback formula: $P = \dfrac{I}{L - E}$,

where P = Payback period, in years

I = Total capital investment in robot and accessories

L = Annual labor costs replaced by the robot

E = Annual expenses of maintaining robot

In this example:

I = ₹500,000.00

L is at the rate of ₹100.00/hour including fringe benefits

E is at the rate of ₹15.00/hour

There are 250 working days/year, consisting of one or two shifts each of having 8 h or 16 h per day, respectively.

Case 1: Single shift operation

$$P = \frac{500{,}000}{100\,(250 \times 8) - 15\,(250 \times 8)} \approx 3.00 \text{ years}$$

Case 2: Double shift operation

$$P = \frac{500{,}000}{100\,(250 \times 16) - 15\,(250 \times 16)} \approx 1.5 \text{ years}$$

Figure 11.1 Simple Payback Method

Let us assume that the total capital value of robot, associated machinery and equipment is ₹500,000.00. The company takes an annual write down percentage of 15% depreciation of operating cost. This gives an annual depreciation of ₹75,000.00. The actual calculations are given in Figure 11.2.

Results are adjusted to one decimal place, because the accuracy of estimate does not justify more significant figures.

11.3.3 Return on Investment Payback

Changes in inflation applicable to project financing are generally ignored in the first payback analysis. Special purpose automation to produce one workpiece is prone to early obsolescence. Payback period is suitably short compared with the expected life of equipment. Factors such as inflation or annual interest rates do weigh heavily. This is very true when the payback periods are of only one or two years.

Robots are not special purpose automation and their flexibility means that they can be reprogrammed when a product line changes. Their reliability indicates a long working life, at least 8 years. It is reasonable to regard robots as general purpose equipment. Changes in the value of money and the rate of return of money invested provide factors that can be evaluated. We can decide a robot project as go or no-go.

The (ROI) calculation has the same robot used in other two payback methods. All figures and operating conditions are the same but one additional factor has been introduced; the total value of investment in robot buying cost and accessories is to be considered as

Production rate payback: $P = \dfrac{I}{L - E \pm q\,(L + Z)}$

where P = Payback period, in years

I = Total capital investment in robot, associated machinery and equipment

L = Total annual labour saving

E = Annual expense of robot upkeep

Z = Annual depreciation costs of associated machinery and equipment

q = Production rate coefficient

$\pm$= + when robot is faster, – when robot is slower.

In this example,

I = ₹ 500,000.00 invested for robot and accessories

L is derived from a labor rate of ₹ 120.00 per hour including fringes taken over 250 days in full year each day comprisisng either one 8 h or two 16 h shifts

E is derived from a rate of ₹ 15.00 per hour, over the same period as L

Z is ₹ 30,000.00 being 15% of the total capital cost of ₹ 200,000.00 paid for associated machinery and equipment per full year

q is either 20% faster or 20% slower than human operator

Total L equals to ₹ 120.00*250*8 = ₹ 120,000.00 per shift or

₹ 120.00*250*16 = ₹ 240,000.00 for two shifts.

Total E equals to ₹ 15.00*250*8 = ₹ 15,000.00 per shift or

₹ 15.00*250*16 = ₹ 30,000.00 for two shifts.

Case 1(a): Single shift; the robot is 20% slower than human operator

$$P = \frac{500{,}000}{120{,}000 - 15{,}000 - 0.2(120{,}000 + 30{,}000)} \approx 4.90 \text{ years}$$

Case 1(b): Single shift; the robot is 20% faster than human operator

$$P = \frac{500{,}000}{120{,}000 - 15{,}000 + 0.2(120{,}000 + 30{,}000)} \approx 4.00 \text{ years}$$

Case 2(a): Double shift; the robot is 20% slower than human operator

$$P = \frac{500{,}000}{120{,}000 - 15{,}000 - 0.2(240{,}000 + 30{,}000)} \approx 10.00 \text{ years}$$

Case 2(b): Double shift; robot is 20% faster than human operator

$$P = \frac{500{,}000}{120{,}000 - 15{,}000 + 0.2(240{,}000 + 30{,}000)} \approx 3.00 \text{ years}$$

Figure 11.2 Production Rate Appraisal Method

₹ 500,000.00/8 = ₹ 62,500.00 in eight instalments; the life expectancy is of 8 years. ROI calculations are given in Figure 11.3. By any standards, the predicted return rates are impressive. Even single shift working looks good.

Data in the earlier example:

Cost of robot and accessories:	₹ 500,000.00
Yearly depreciation (straight line method) for robot and accessories:	₹ 500,000/8 = ₹ 62,500.00
Annual maintenance costs:	₹ 32,000.00
Annual value of increased output:	₹ 120,000.00
One man has been replaced (i.e. one shift of 8 h operation):	8 h/shift * 250 shifts = 2000 h
Replaced labor rate including direct overhead (i.e. one shift of 8 h operation):	₹ 80.00/h

One shift (8 h) operation

Then, the annual rate of return in percentage

= [Net savings (or income)/Total investment on robot and accessories] * 100

$$= \frac{\text{[(Annual rate of increased output) + (Cost of labour saving) − (Annual depreciation) − (Annual maintenance cost)]}}{\text{[Total investmenton on robot and accessories]}} * 100$$

$$= \frac{(120{,}000) + (80*2000) - (62{,}500) - (32{,}000)}{500{,}000} * 100$$

≈ 37.0%

The rate of return of 37% is quite good. If the invester borrows money at 20% annual interest, he will still get the benefit of net ROI of about 17% per annum.

Two shift (16 h) operation

When robot is operated by two shifts, the annual rate of return is as follows:

Annual rate of increased output:	₹120,000*2 = ₹240,000.00
Cost of labour saving:	₹160*2000 = ₹320,000.00
Annual depreciation:	₹ 62,500.00

Assume: Annual maintenance cost: ₹45,000.00 (you can adjust this cost to more percentage)

Hence, [(240,000 + 320,000 − 62,500 − 45,000)/500,000] * 100

≈ 90.00%

Figure 11.3 Return on Investment

11.4 ROBOT INSTALLATION

Planning for robot installation requires the nature of robotic applications. Choice of applications depends on careful planning and evaluation of economy. A general and a set of specific guide lines are given in the following.

Some considerations for robotic instalment planning are as follows:

(1) General analysis

Repetitive operation (such as pick-place, loading and unloading)
Hostile environment (filled with flammable gas)
Heavy operations (cement bag palletizing)

Multishift operations (one shift or double shift industrial jobs)
Cycle time (slower or faster than manual)
Tiresome (fatiguing and boring) operations

(2) Economic analysis
What is the use of a robot? (replacing human labour, faster/slower than human labour)
Can the use of robot increase the cost of product? (no, creates a reduction in cost)
Can material handling cost be reduced? (naturally reduced material handling cost)
Does the robotic application reduce direct and indirect manufacturing costs? (yes, robot reduces the direct and indirect manufacturing costs)
Can economic analysis be satisfactory? (of course, satisfactory)
Can robotization bring down the rate of return?
(yes, robotization brings down the rate of return)

(3) Feasibility analysis
Is robot really required? (yes, when the job is complex to labourers)
Is the job too complex? (if the job is too complex, robotization is a must)
Is the finish time cycle set prior to programming? (yes, it has to be)
Should the sensors be integrated? (integration of sensors are really needed)
Should the material feeding systems be adequate? (feeding system has to be adequate)
Is there any integration plan with other equipment? (yes, other systems such as machine tools)
Is inspection required? (at regular intervals)
Is there an adequate number of staff? (if not adequate, then robotization is a must)

(4) Selecting the robot
Degrees of freedom (maximum six, mobile robot has additional degrees of freedom)
Coordinate systems (maximum six axes)
Payload capacity (as much as possible)
Speed, accuracy and repeatability (speed has to be high as possible, accuracy has to be small, repeatability has to be as low as possible)
Sensors (as much as possible, including vision)
Programming languages (many varieties)
Integration of interfaces (has to be easy and possible)
Reliability (as high as possible)

(5) Selecting a vendor
Consult experts in robotics
Vendor–user discussions
Finalize a robot (a team of robots)

(6) Preparing the site
Work stations
Drawings on layouts
Materials feeding systems
Materials outgoing systems
Safety to human systems

(7) Get management approval to install a robot/set of robots
(8) Prepare manpower through inhouse training
(9) Install a robot/robots and test the above criteria
(10) Submit a proof to the management on successful working

11.5 QUALITY OF WORKING LIFE

Work environment is surrounded with unwanted noise, vibration, smell, smoke, excessive heat, intense cold, oil spray, flying chips, monotony and risk of personal injury that can be cured in a short time or leaves a permanent disability. For example, a robot can be easily taught to operate the die casting machine. The human worker who works with die casting machine can be transferred to another work where he no longer has to breathe die lubricant fumes. Hot molten zinc no longer spurts at him. He does not have to wear asbestos dress anymore. Now he can work as a man and not as 8 h per day or 16 h per day working robot. Such is our society nowadays.

The worker at the die casting machine has to permanently work there unless he asks for a change. Social improvements usually conflict with lack of funds. Robots are usually exceptions to this rule. Die casting and press operation are the only two examples among many jobs that a robot can perform. There can be many other risk jobs in a factory. Developments make robots become more and more adept and a larger segment of humanity will be released to do more work in intelligent ways.

11.6 ATTITUDES TOWARDS ROBOTS

People, in general, are suspicious of robot's working. Whether a plan is for new road or introduction of automated plant, somewhere someone is going to object it. Even a new employee is going to be regarded as ignorant of robots until he has proved that he can fit in with the rest of the team mates. Every worker has to be assured that change will not occur in his working conditions in reducing his work. When the management fails to inform its workers of forthcoming technological changes, workers should learn themselves if they want the changes. Otherwise they will be left alone to be worried. But, the robot is a special case in this respect and it shows some resemblance to human beings It can autonomously perform some human actions. Table 11.3 lists the cases to be handled and their respective details. The reasons from actual experience which follow social acceptance are provided in the table.

Table 11.3 Cases to be Handled and Their Details

No.	Cases to be Handled	Details
1	Appreciation	Robot's efforts are to be appreciated. A few circuit modifications have changed robot's morale. A new layout had simplified the job of robot. Two months later, the robot performed as required. The production rate was achieved and even exceeded. When it finally made the rate it earned, it received a standing round applause from every man in the department

(Continued)

Table 11.3 Continued

2	Get well cards	Routine maintenance by plant personnel evoked no response from robot. By the time the specialist had diagnosed the illness and a treatment was prescribed to the robot, a full week had elapsed. The robot was, then, working to the satisfaction of all concerned. The department organized a get well party and heaped cards and flowers at robot's pedestal
3	Blue collar support	This is about a blue collar support for robot productivity. Two decades ago there were five robots in a factory. Union leadership demanded the programming of robot by a member of union. But the management did not approve it. It could not trust anyone from the union to program the robots since it believed in engineering skill. A roboticist was called and he/she could not accept the management's decision on programming. However, one day the roboticist accepted the program from the union member. From then onwards it is agreed that to program a robot, engineering background is not essential, but anyone with a strong science background can do the programming.
4	Affection	There are evidence of man's compassion towards robots. The industrial robots are deaf, dumb and blind which create amused affection in human beings. Factory automation do not need such senses, whereas human beings really need such senses. The robot is ready to work as a slave. A robot slave can never be guilty of an insult and it offers no challenge. The roboticist is aware of the humble nature of robot. Robot is therefore truly likable
5	Workers attitude	Three decades ago a US-based company sent a team to visit various representatives of robot manufacturing operation. The team consisted of two members; one of them was a manufacturing engineer whose exposure was on how robots were manufactured and other was an accountant to compare the new and old manufacturing applications. The team was successful in identifying the robot's ability and human's disability. This shows management's hesitancy in blocking a robot installation than rejecting a human worker

11.7 EFFECT ON EMPLOYMENT

The robot's economics clearly proposes that the driving force behind adoption of robots in manufacturing plants is cost effective. Labour displacement is the central benefit. Robots contribute productivity primarily by labour displacement of human workers: the benefits are clear. Human jobs have been eliminated by robots and the machines have taken up works of labour force. By the end of 1979, some 6000 robots were installed worldwide. If we consider two shifts per day, this displacement may total to 9000 robots. The net employment impact of formation of a robotics industry may actually be positive in creating jobs. We can divide the robot labour force into two parts. One part of labour can become manufacturers and the other part can become suppliers to this new industry.

Robotics has no history worthy of extrapolating a sociological cost. This can be the cost of job displacement by robots. Industrial automation over the past century can be evaluated on a cost/benefit basis. One can favour robot technology despite the emotional declarations. Gains are always for productivity which is good and better over the coming years.

Forty thousand robots are entering the work force per year. This is a big number but not disruptive one. This represents about 0.06% of the blue collar work force in the industrialized countries. These countries will employ such a big number of robots. Of course, there will be other automation influences on the productivity. These will probably conceal the impact of robotics since automation is already so much a part of manufacture.

In short, robotics will contribute mainly to the material well-being of mankind, that too without painful dislocation of individual workers. Ten years from now, work week in these days, air and water will be clean again and industrial life will be ever so desirable.

11.8 CURRENT CAPABILITIES OF ROBOTS

The robots are compared with human workers in terms of their attributes and general behaviour. They are just fine to take up jobs which are dirty, repetitive, boring and dangerous in factories. They can do their jobs with positive economic advantages. Yet, robots are stupid, insensitive and limited devices when compared with human workers. No robot can compete with human worker with his acute senses, ability for free thought and judgement, artistic appreciation, capacity for reproduction, efficient conversion of food into energy and capability to recover from illness and injuries. The gap between human and robot will always remain but as technology improves the gap will become narrower. In the following, a set of robot attribute is explained .

1. Teach and play back facilities
   Realizing the fast, automatic programming
2. Positioning accuracies
   Repeatable within 0.1 mm
3. Weight handling capacity
   More than 150 kg
4. Point to point control and continuous path control
   Both intermixed; the cement bags are loaded from manufacturing conveyor
5. Palletizing and depalletizing capabilities
   Both are needed; palletizing capabilities at one station and depalletizing capability at another station are possible
6. High reliability
   Not less than 400 h of mean time between failures (MTBF)
7. Work space command
   With all six articulations from base of robot to the end of hand extremity
8. Synchronizations with moving workpieces
   When the workpiece is coming along the conveyor

9. Interface with a computer
   Computer can make several external interfaces to workpieces coming along the conveyor
10. All the above capabilities are available for a price

This price allows accepted rules of economic justification of any new equipment.
The last item listed is by no means the least and the robot recruitment to its work force is capable of doing the job.

11.9 FUTURE CAPABILITIES OF ROBOTS

In the following, a list of features which are very desirable goals for future robots is given. In general, the more sophisticated the device, the costlier is the device. But trends towards increasing the labour cost encourage robotization in a factory. This is going to happen sooner rather than later.

1. Vision: A human being can look into a basket of parts and pick them out one by one. He should make the object orientation an easy process by robot. This work is included in artificial intelligence community. In academia, this work is named as 'occlusion problem' or 'bin picking problem'. This problem has created most intense research on robots and this intense research and development efforts are entirely justified.

2. Tactile sensing: If tactile sensing technology moves faster than vision technology, then the former can become more important. This is true like a blind person who can be most effective in a number of activities that depend on tactile sensing capability. A robot with tactile sensor can use its sensor to recognize parts just like the way human beings detect what thing is in a dark room. We all come to know what happens when putting a nut into a bolt in a dark place. We almost instinctively back thread until we feel a 'click'; then, we run the nut in the forward direction. This is a tactile sensing experience and we wish to add the sensory perception repertoire to a robot.

3. Mobility: The roving robot in a factory premises which is on wheels and the ability of strolling can make the robot mobile. For bulk factory robot jobs, the robot stands in a single station just as his human predecessor. In fact, a factory job having an operator moving busily outside a 2 m radius is probably inefficient. But there are jobs for mobile operators who must tend stations that are widely separated. The stations need service periodically. For this, mobility is required and to date this mobility has been delivered in a heavy-handed fashion. What is required is a robot that can literally stroll from one station to the next.

4. Minimum spacial intrusion: A human being does not need large working space in factories. But, most robots require substantially more space than their human counterparts. This can actually eliminate the potential use of robot because the cost of laying out equipment is more. Thus, providing factory floor space can kill the 'economic justification' for using the robot. We design the factory floor for many future years so that the human being is comfortable. However, we produce the robots to match human physique. This may literally be easier than to produce manufacturing plants matching the current spacial demands of current day robots.

5. Energy conserved musculature: An elegant research and development statement is to be extended to robotic community. Robots today use more energy to accomplish their work, whereas human operator uses less energy. A calculation says 1/20 h. p. is sufficient to accomplish his job. It is not uncommon for a robot that needs 5–10 h. p. for the same activity. The ratios speak out tremendously in favour of further technological contribution. The robot power drain and human power drain are a compelling argument for furthering material improvement.

6. General purpose hands: Human hand can do several jobs. Till now no robot activity has attempted to imitate the human hand. Rather special purpose hands have been developed. These are to match special tasks. Hand designs are extensive and a library of robot can have such a series of hands. Usually, when job changes, hands also must be changed. There are jobs where single hand is just not appropriate for entire robot task. The solution is simply to give robot an opportunity to select a hand. However, if time is precious, it is distressing to note a robot returns to a home position every time in order to exchange hands. Invention may be necessary and the call will not be unheeded.

7. Voice communication: To date, man–robot voice communication has been something of a joke, because programming is not all that laborious and robot intelligence has not yet advanced to the point where a few modest suggestions will motivate the robot to take desired action. The robots are gaining adaptive features such as rudimentary sensory perception. It may be attractive to allow the human boss to use plain English in instructing the robot as to do its ongoing work. In addition, the robot which is likely to be highly sophisticated can respond to the human boss with a synthesized language which can be understood by both machine and man. Its speech can be used to explain robot internal ailments which need service attention.

8. Self-diagnostic capability: As it becomes easier to assign factory tasks to robots, they become ever more internally sophisticated. Whatever the level of robot sophistication, it is crucial that the machine exhibits job reliability competitive to that of human worker. Thus, the robot user must have a long MTBF and a short mean time to repair. If the machine is an elegant one, then the repair will be intellectually demanding. What is needed is a self-diagnostic software package. The software pin points a deficiency under any failure condition. The software directs the human service staff a set of efficient methods for recovering the performance.

9. Inherent safety: This is one of Asimov's laws of robotics which has been already mentioned. This law becomes more important as the robots become more competent. Nowadays, the robots are utilized in more intimate relationship with other human workers. Safety to man and machine has to be inherent if robots and humans work shoulder to shoulder with the robots doing the drudgery tasks and humans contributing to judgement. This development task is not easy; but fortunately it is also not impossible.

10. All the above capabilities available for a price: Robots must offer these additional capabilities and qualities but still be economically justified. If a robot is designed and devised to give a dazzling visual capability, when the system is a multimillion rupees research accomplishment, it has no place in industries. The robot plus all its accessories must be easily economically justified. The other simple pieces of cost-saving equipment through research completed by academics except for market place remain intractable.

11.10 EXAMPLES

11.10.1 EXAMPLE

A robot installation has an initial (investment) cost of ₹850,000.00. The operation and maintenance costs are estimated to be ₹240,000.00. The project is expected to generate revenues of ₹600,000.00 per year for 4 years. At the end of 4 years the project retires. The estimated salvage value of robot at the end of 4 years is ₹400,000.00.

(a) Determine the payback period for this project.

(b) Determine the equivalent uniform cost for the project, using a 25% rate of return in the interest calculations.

(c) Determine the expected ROI to be derived from the investment.

Solution:

(a) Payback period = [Investment cost/Net annual cash flow]
= 850,000/[600,000 – 240,000]
= 2.36 years

(b) Equivalent union accessories cost (EUAC)
= – 850,000 (A/P, 25%, 4) + 600,000 – 240,000 + 400,000 (A/F, 25%, 4)
= – 850,000 (0.42344) + 600,000 – 240,000 + 400,000 (0.17344)
= 69450.20

(c) ROI, try $i = 35\%$
EUAC = – 850,000 (A/P, 35%, 4) + 600,000 – 240,000 + 400,000 (A/F, 35%, 4)
= – 850,000 (0.5007) + 600,000 – 240,000 + 400,000 (0.1507)
= – 5340.00

Interpolating between 25% and 35%:
$(35 - x)/(35 - 25) = 534.2/7479.4$
$x = 0.3429$
Therefore ROI = 34.29%.

REVIEW QUESTIONS

1. 'Will the robot project justify economically?' How can you answer this question to the satisfaction of industrial managers?
2. What are two categories of robot installation that are commonly encountered in the factories?
3. What are the costs that have to be met with the new robotic installation?
4. What costs can be saved?
5. What are the questions that come in the minds of managers? Explain them.
6. What data are required when a robot engineer thinks about new robotic project based on the following:

(a) Simple payback period

(b) Production rate appraisal on payback

(c) Return on investment evaluation.

7. List and explain six investments on a new robotic project.
8. What are the considerations for selecting a robot for a new project?
9. What type of robots or modifications would you suggest in the following unsuitable, unhealthy situations that are exposed to the workers: environment that offers unwanted noise, vibration, smell, smoke, excessive heat, intense cold, oil spray, flying chips, monotony and risk of personal injury that can be cured in a short time or leaves a permanent disability.
10. What are human attitudes towards robots? Explain in detail.
11. What are current capabilities of industrial robots?
12. What modifications are required for the current robots for obtaining future capabilities?

ARTIFICIAL INTELLIGENCE 12

The whole is equal to the sum of its parts
Euclid

This chapter covers the following key topics:

Introduction – AI Research Goals – AI Techniques – Knowledge Representation – Further on Knowledge Representation – Search Techniques – Problem Solving – Travelling Salesman – LISP– LISP Statements – Examples

---

12.1 INTRODUCTION

LISt Processing (LISP) and PROLOG softwares are suitable for developing artificial intelligence (AI). Software companies which specialize in production rule-based expert systems can take, with proper advantages, one of the softwares as standards.

Compared with LISP, PROLOG is a higher level language which means that it makes many assumptions about the way things to be done – about the kind of inference to be used. The assumption is that PROLOG applications do 'forward chaining' or analogical inference. In this application, the user has to work with language's underlying execution protocol instead of letting it work for him/her.

However, LISP is a general purpose programming language more widely used for producing the entire generation of software. To know the difference in the level of these two languages, one can easily imagine writing LISP interpreter in PROLOG is relaxed; on the other hand, PROLOG interpreter in LISP is more difficult.

Among the recent engineering and scientific applications, if one is interested in building highly integrated production rule expert systems with flexible user interfaces, then LISP is the good choice. Here we discuss LISP in detail to include major statements that help us to construct some easy software package.

Vision and tactile sensing along with the modern concepts of robot control have created the AI. The AI research further requires a programming language. LISP is a suitable programming language for AI. An object-oriented language can assemble basic set of instructions for a higher level, for example, 'GO TO STATION C' within a network of roads. The robot creates a unique set of roads to go to 'STATION C'; the uniqueness can be anything that includes the distance between stations. This is the basic set of AI. Searching for 'STATION C' can be stand-alone or may be intermixed with other activities.

This chapter introduces the AI as one of the very important topics in computer science. Some of the goals are introduced and these goals help for discussion and conclusion of this chapter. Let us describe the AI as

Artificial Intelligence is a part of computer science concerned with the characteristic we associate with intelligence in human behavior – understanding language, learning, reasoning, solving problems and so on.

AI attempts to aim at developing systems (include robotics) that appear to behave intelligently. Often, this is pronounced as designing machines that 'think'. Typically, this work is approved as a subdivision in computer science, although this consists of elements in psychology, linguistics, mathematics and so on. Since there are problems in defining intelligence, AI research is best described by deliberating its goals.

12.2 AI RESEARCH GOALS

The general aim of AI research is development of machines which exhibit intelligent behaviour. Its statement is too broad and ambiguous to be meaningful. Nowadays, the AI area is pursued as a separate area of research. This area of research has certain goals as listed and explained in the following:

(a) Problem solving: There are many examples of problem solving such as playing chess. Given a set of rules and several plans, a robot will be capable of playing chess in an adept way. In future, a robot can be programmed on how to play the chess game. On assembling the chessmen on a board, it can find the initial way to start the game. The robot will start the game and expects the hurdles and obstacles to overcome. It will start the game that goes smoothly till the opponent finds an opportunity to put the hurdles. The robot and its opponent will find the ways and means to cut-off robot's or opponent's chess men. Finally, the game is over and one of the players (mostly robot) will be the winner of the game ending the chess game successfully. The movements of each chessmen should be explained to the robot and the opponent. The robot can 'construct' its assembly strategy by a number of rules.

The robot requires a set of sensors that includes tactile and vision required for the chess play. In addition, the robot understands the chessboard movements and its men through other sensors.

(b) Natural language: In spite of the use of computers even in department stores, many of us are still feel uneasy and uncomfortable with computers. The reason being the computer does not know English language and it may misspell with unwanted meaning. The user has to talk to the computer or instruct it in its language and not in 'natural' language. The problems associated with natural language are numerous. The robot computer should understand the 'human' language and not its natural language. VAL is one of the languages that can be used in place of natural language. The main aspect is that the robot should understand its syntax of the language so the relationship of the given words is understood. Imagine a possible meaning of the word sequence 'kicks out' to indicate dies or deserts.

(c) Expert system: This area of research pertains to the developing systems that appear to behave as human experts in specific areas. The vast data collected from human experts and improving 'the knowledge base' with experience are named as expert system. By systems we mean robots in this context. Expert systems have evolved through a dialogue with human experts. In this way recommendations are to be performed and ask appropriate questions. This questioning continues until the answers are appropriate. Currently, expert systems are used in designing medical diagnosis, configure computer systems, planning electronic circuits and so on. Some of the problems involved with designing of expert systems include the problems of dealing vast amount of data and explaining the systems 'reasoning' on the way to reach a conclusion.

(d) Learning: Learning is one of the important attributes of AI. This is the ability to learn from experience. A system is to have experience by repeatedly using a concept over and over. For example, determining the dimensions of plates with a set of holes is an experience. Tolerance of dimensions is also an attribute to learn. By tolerance we mean the set of allowed variations in dimensions. As the plates with holes are coming on conveyor, robot is directed to measure the dimensions of plate and holes. The robot should have vision as one of the sensors. The image of plate is taken, the dimensions of holes are measured and finally the plate is accepted/rejected.

(e) Vision: Vision is the foremost attribute in developing AI and without this aspect AI research is meaningless. Most of the vision aspects are in determining the scenes. Scene analysis is one of the outcomes of AI research. Objects within the scene are identified and sorted in the scene analysis. The objects can be in solid state such as trees and buildings, colour of liquids in containers and so on. Solids such as banyan trees and liquids such as mercury are alike.

AI research also includes other areas such as automatic programming, deductive reasoning and hardware development. These descriptions are out of the scope of this book.

12.3 AI TECHNIQUES

'Data' is a noun that has a verb following 'data'. 'Knowledge' is the other name of data. The techniques are to be developed for two basic tasks – (i) data representation and (ii) data manipulation. We can bring in knowledge in place of data without modification of the sentence: knowledge representation and knowledge manipulation. We shall also use these in some specific manner. We will not discuss the actual program for doing this; only the general techniques will be discussed which are usually employed by AI programs.

Data representations and data manipulations are knowledge processing. We will describe the approaches of knowledge processing as the general technique that might be employed in AI research.

12.3.1 Knowledge Representation

Before discussing various representations of knowledge and its processing, we must first describe the various types of knowledge which may require representations. Table 12.1 lists various types of knowledge that may require representations.

Table 12.1 Attributes and Explanation

Types of knowledge	Explanation
Objects	More exactly the facts about objects such as 'animals does not have wings' or 'robotics students work tirelessly'. 'Animals' and 'students' are objects (or nouns)
Events	Not only the event such as 'the student broke her arm' but also the cause–effect relation of the event such as 'the student broke her arm yesterday and the teacher made her pay for this'. 'Yesterday' is the event here
Performance	Suppose that AI system is designed to control a robot, which is also one of the activities of AI. Then, it must have data to measure the performance of the robot arm, that is, robot kinematics (how many links the robot has), dynamics (overshoot, settling time), what bits to manipulate (first bit, last bit), any hardware modifications (amplifier circuits) and so on
Metaknowledge	This is knowledge on human knowledge. This includes human knowledge of the origin, its relative importance and its reliability. The information from history student is 'the study of robotics is painless', then any one can give a little more prominence since the wordings come from history student

12.3.2 Further on Knowledge Representation

At this juncture, let us look at some of further techniques for representing knowledge.

(a) Logic: Formal logic was developed by statisticians, philosophers and logicians. This is to provide a method for making a set of inferences from the facts. Formal logic requires facts to be represented in a specific syntax. By applying defined rules of inference to these facts allows conclusion to be drawn. We are familiar with the following example.

Given two statements such as

(a) All roboticists play games

(b) Rajan is a roboticist

We conclude by using rules of inference as

(c) Rajan plays games

Imagine that a computer is gifted with all the possible facts about a subject. When all the applicable rules of inferences are also gifted, then it should be able to develop new facts about the subject which may be inferred from the original set. As the number of facts become more and more, the number of combinations with rules of inference that apply may also increase. This results in generating a problem too large for the computer to solve in a reasonable length of time.

(b) Procedural representations: In addition to storing facts about the subject, it is also necessary to save the information on how to use the facts. In a logic-based system, every rule of inference applied to prove a new point. If there is a possibility on how to use the information in specific areas, then there is an opportunity of a more efficient system to evolve.

(c) Semantic networks: These networks have nodes and arcs (links). They are representations of information. Nodes typically represent objects, concepts or situations while the arcs connect nodes and represent relationship between them. The nodes and links are labelled with simple language descriptions.

Figure 12.1 illustrates a typical semantic network. The information for this network are as follows:

Rajan is a student,
Rajan is a roboticist,
Roboticists play games,
Roboticists may be students,
Students may be roboticists.

The above said inferences can be derived from investigating the semantic network describing Rajan, students and roboticists.

As a means of knowledge representation, networks are extensively used. Unfortunately, they also have their own drawbacks. How does a network deal with ideas and large amounts of knowledge? As with all representation techniques it cannot be pushed to an extreme.

(d) Production systems: This is a storage system called as 'productions'. All the productions have the form of

IF <some expression is true> THEN <some actions>.
Then 'Rajan' and 'Roboticist' can be presented as follows:
IF <Rajan is a roboticist> THEN <he plays games>
IF <the person reads this book> THEN <he is a roboticist>

If we were able to catch 'the person reads this book', then we would know 'he is a roboticist'. Production systems are used to break the problem into successively more manageable and simpler tasks. IF-THEN rules are typically for 'construction of expert system'. They provide uniformly designed system using IF-THEN construction. They will introduce the 'modularity' in which each rule has no direct effect on other rules. However, as the system becomes larger, it becomes inefficient; this can be included along with other problems as well.

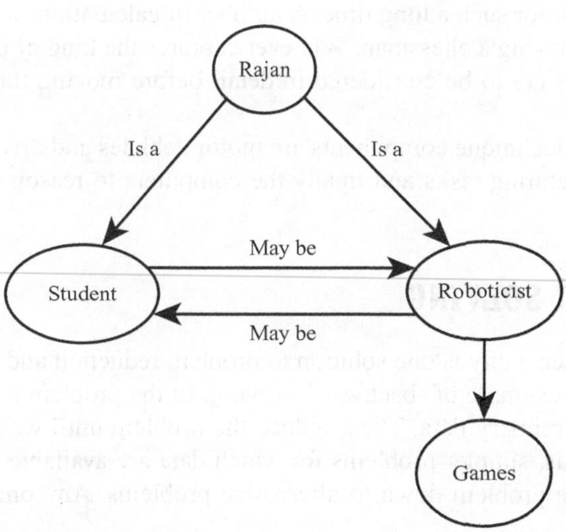

Figure 12.1 Semantic Network Describing Rajan

(e) Frames: One more representation technique is the use of 'frames'. They can be considered as predefined structures with empty slots. The slots are filled to represent specific facts. The slots can be filled by references to other frames or by procedures defining how the slots are to be filled.

For example, the general frame for a student might look as follows:

Student Frame:

Name:	Actual name
Height:	in metres
Weight:	in kilogram
Studies:	Robotics or Machines

Rajan Frame:

Name:	Rajan
Height:	2.2 m
Weight:	90 kg
Studies:	Robotics

(f) Others: This is regarding other representation techniques. Choosing a correct representation technique can reduce the complexity of the problem and also the solution. One way to represent an object is through semantic network node and the action is through its branches. A video view can be through a matrix array with brightness of the pixel as its measure. The type of problem solution is the correct way of deciding the way of representation. Various techniques are applied to different circumstances. In some cases additional representation schemes should be found.

12.4 SEARCH TECHNIQUES

As an example, let us investigate search techniques in chess play. Everyone knows chessboard which has equal white and black squares. There are, respectively, eight white chessmen and eight black chessmen which are to be moved by two opponents. White is generally to start first. There are 10^{120} set of moves and is termed as 'combinatorial explosion'. Chess players need not wait for such a long time. A number of calculations and combinations are to be made before moving a chessman. Whoever captures the king of opponent is the winner. Such techniques are to be considered in detail before moving the chessman by each player.

The other search technique components are motor vehicles and drive paths, robots have to do some manufacturing tasks and finally the computers to reason out and to continue further.

12.5 PROBLEM SOLVING

The right way to reach a city is one solution to problem reduction and problem solving. In this case, we see an example of 'backward' solving. In the problem reduction, we present the goal city as the primary data. Then, reduce the problem until we have a set of primitive problems. That is, simpler problems for which data are available. Simplification may involve breaking the problem down to alternative problems. Any one of these problems

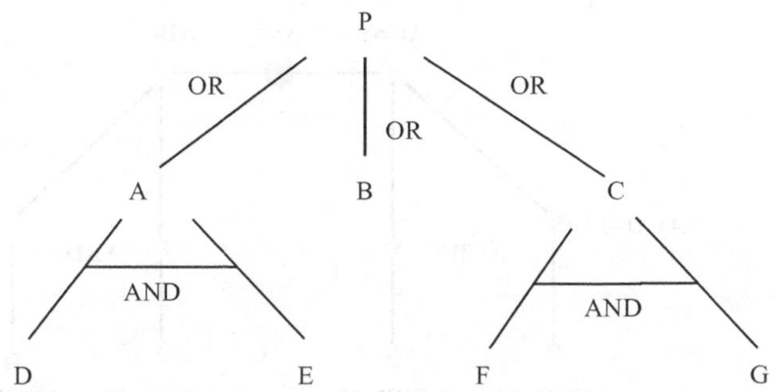

Figure 12.2 AND–OR Graph

can be solved with available data. The scheme is graphically represented as an 'AND–OR' graph. In an AND–OR graph, branches that are connected by a horizontal line are called 'ANDed' and branches which are not connected by a horizontal line are 'ORed'. Figure 12.2 illustrates this simple form.

(P) be solved by solving (A) OR (B) OR(C)
(A) be solved by solving (D) AND (E)
B be solved directly (B is a primitive)
(C) be solved by solving (F) AND(G)
D, E, F and G are primitives
(B, D, E, F and G have all the required data)

12.6 TRAVELLING SALESMAN

The travelling salesman problem is considered as equal to robotics problem. In a robot, whatever may be the shape of its links, there are some state variable points, specifically, in its work area that cannot be reached. The collection of all the state variables is state vector and the robot cannot reach certain state vectors. We can visualize them as all the reachable states that are figured as a tree. The tree is made up of nodes that represent the reachable states of a system. Then the system can be considered as an example of travelling salesman. The travelling salesman has to travel all the cities, at the maximum, using shortest travel paths. Figure 12.3 illustrates possible routes by the salesman.

There can be inconsistencies. Logically, a salesman cannot go to all the required cities but can only travel along some routes. Other routes that are not taken by this salesman will be taken care of by other salesman.

This way of solving a problem is known as 'forward reasoning' since we ourselves have to go forward to find a solution. The minimum distances are ABCD=ACDB=18.

With this as an example, we analyse and discuss the happenings in the search techniques listed in Table 12.2.

At this point it is worth reviewing what has been covered so far. Knowledge or data can be represented in many ways. Not only data are important but also the relation between data and others also important. Moreover, manipulation methods are also needed. Problems may be solved by forward and backward reasoning.

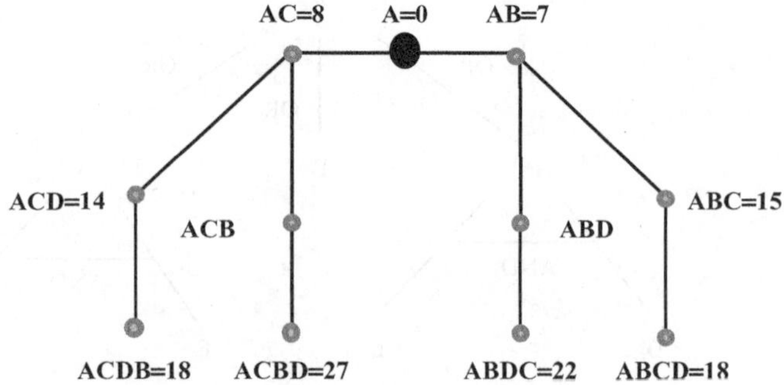

Figure 12.3 Travelling Salesman Problem

Table 12.2 Search Techniques and Explanation

Search Techniques	Explanation
Depth-first search	Figure 12.4 illustrates this way of search. This search technique does an in-depth search until it finds a terminal node. If this node does not have a solution, then it 'back tracks'. This back tracking is performed until it finds the first available branch point and goes down the next branch. In the example illustrated the system will eventually find the solution node. In some system it is not possible to determine a solution. Then we can ask the network to go to next breadth-first search so that at least one solution is found
Breadth-first search	Figure 12.5 describes this way of search. The system evaluates all nodes at the same level in the tree before going to the next level. The system is more conservative compared with depth-first search; it cannot be easily trapped. If the network has broad branches with less depth, the depth-first search is not advantageous
Hill climbing	Hill climbing method provides a variation on the depth-first searches. The hill climbing method attempts to make the best choice among the possible branches from the current node. We may be making the best local choices, or we may be missing an overall solution that had one bad branch
Best-first search	This has a variation on hill climbing. In this case, rather than choosing the best next branch choose from a node regardless of its position in the system. This generally provides an optimum solution, but we can never guarantee it
Branch and bound	This is similar to the best depth-first and best breadth-search next. The system evaluates all the partial paths and expanding the same in the next level. The system expands to understand a new most promising path. In this way it is always investigating an optimum path to a deepest level necessary
Constraints	In some cases it is not needed to consider all the possible options. Very often in the real world, constraints are placed on information by its context. Applying these constraints we are travelling through a tree that may lead to a significant reduction in the possible number of nodes to be searched for

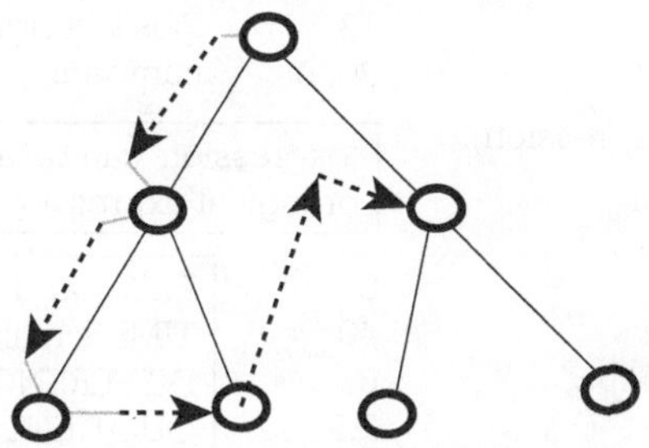

Figure 12.4 Depth-first Search

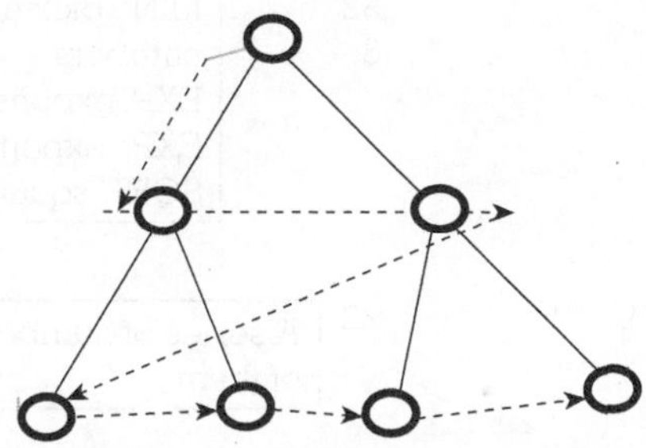

Figure 12.5 Breadth-first Search

12.7 LISP

LISP is usually adopted in AI research. LISP has some interesting features that are not available in other languages. One of many features is that both LISP data and LISP program are created out of 'lists'. Hence, LISP can modify a general statement into smaller statements.

We shall introduce several statements which can be used in generating a LISP program. LISP program consists of lists which are constructed from elements named atoms.

12.7.1 LISP Statements

Following are the LISP statements and details that usually occur in the LISP program:

LISP statements	**Answers/Details**
(A, B, C, D, E) (Item1 Item2) (Item1 A)(Item2 B)	LISP program consists of lists constructed from elements called 'atoms'. Some examples are given in the right column.
A B Item1 NIL	These are some of the atoms

Expression	Result	Explanation
(PLUS 8 5)	13	Plus is a sign which can add two numbers
(PLUS 3000 5000)	8000	
(expression, expression, expression).		'expression' can be a list of mathematical or logical expression
(PLUS 4 4)	8	PLUS: add TIMES: multiplication DIFFERENCE: subtraction QUOTIENT: division ADD: addition SUB: subtraction MAX: pick up the maximum of numbers MIN: pick up the minimum of numbers EXP: exponential EXP: exponential of a number SQRT: square root of a number
(TIMES 5 8)	40	
(DIFFERENCE 7 1)	6	
(QUOTIENT 5 2)	2.5	
(ADD1 9)	10	
(SUB1 8)	7	
(MAX 1 2 3)	3	
(MIN 1 2 3)	1	
(EXP 2 5)	32	
(SQRT 9)	3	
(PLUS (PLUS 4 5)(PLUS 6 7))	22	A series of numbers are given; add all of them
(CAR (A B C))	A	In the first example, the argument need not be actual numbers, they may be of expressions. Two functions which are used for this purpose are CAR and CDR. CAR returns the first element in a list and CDR returns every things else
(CAR ((A B)((C D))	(A B)	
(CDR (A B C))	(B C)	
(CDR ((A B)(C D)))	((C D))	
(APPEND (A B)(C D))	(A B C D)	If we detach expressions apart, there must be some way to put them together. APPEND, LIST and CONS are the three functions which can be used to point them together
(LIST (A B)(C D))	((A B) (C D))	
(CONS (A B) (C D))	((A B)C D)	
(CONS(CAR (A B C D))(CDR (A B C D)))	(A B C D)	CONS can 'redo' what CAR and CDR undo
(SETQ VW (A TYPE OF CAR))		SETQ is used to 'name' an expression (subroutine) which assigns a value, 'A TYPE OF CAR' to the arguments VW. From now on wherever LISP sees VW, it will return 'A TYPE OF CAR'
(CAR VW)	A	
(CDR VW)	TYPE OF CAR	

(DEFUN <function name><parameter 1>, <parameter 2>, ... ,<parameter n>)<functiondescription>)

DEFUN helps to define a new function. The command for this is DEFUN (DEFine FUNction). DEFUN uses the format indicated in this table. The example indicates to convert feet into inches

(DEFUN FT-TO-IN (FT IN)(PLUS(TIMES FT 12) IN))

(FT-TO-IN (4 1))	49
(LESSP 5 4)	NIL
(LESSP 4 5)	T
(GREATERP 2 3)	NIL
(GREATERP 3 2)	T
(ANDP 4 5)	T
(ORP 4 5)	T
(NOTP 2 3)	NIL
(EQUALP 3 3)	T
(ZEROP 0)	T

LESS, GREATER, AND, NOT, OR, ZERO, EQUAL are logic expressions; values are NIL or T

(COND ((test1)(result1) .... (testn)(resultn)))

LISP evaluates each clause until it finds 'nonNIL' result (not necessary a T). When it finds a nonNIL clause, it evaluates that list till it reaches the last expression and returns that value

12.8 EXAMPLES

EXAMPLE 12.8.1

Prove that
(CAR(CDR(CDR(CDR (A B C D E))))) is D

Solution:
(CDR(ABCDE)) = BCDE
(CDR(CDR(ABCDE))) = CDE
(CDR(CDR(CDR(ABCDE)))) = DE
(CAR(CDR(CDR(CDR(ABCDE))))) = D

EXAMPLE 12.8.2

Prove that
(SETQ STUFF (ALL THINGS ARE SILLY))
STUFF
(CAR STUFF)
(CDR STUFF)
(CAR (CDR STUFF)) = ?

Solution:
(SETQSTUFF (ALL THINGS ARE SILLY))
= STUFF (ALL THINGS ARE SILLY)
STUFF = ALL THINGS ARE SILLY
(CAR STUFF) = ALL
(CDR STUFF) = LL
(CAR (CDR STUFF)) = L

REVIEW QUESTIONS

1. How do you compare LISP with PROLOG? Which is more suitable to AI and why? Is there any book which says other way is better?
2. Expand 'LISP' and 'PROLOG'
3. Define and explain AI.
4. What are AI research goals? Explain.
5. Explain four ways of knowledge representation?
6. Explain six further ways of knowledge representation?
7. Explain search techniques.
8. Travelling salesman is one of the important problems in AI. Why?
9. List and explain six ways that describe search techniques.
10. What do you understand from the listing of LISP statements?
11. Solve the following LISP statements:

    (PLUS −1000 2000)　　(TIMES 5 −4)
    (SQRT 9.3)　　(CAR ((C D)((A B))
    (CDR (B C D))　　(LIST (A B) (C D))

ROBOT DYNAMICS

13

The hand is an exquisite machine. The same basic structure can be trained for use by a weight lifter or to perform extraordinarily delicate tasks, as evidenced by the micro-vascular surgeon who is able to sew minute arteries and nerves together

David Thompson

This chapter covers the following key topics:

Introduction – Lagrangian Design – Generalized Coordinates – Generalized Forces – Lagrangian Equation – Example – n-link Robot Manipulator – Slender Rod as a Robot Link – Example

---

13.1 INTRODUCTION

The general task of an industrial robot is a set of movements with a motion to start, accelerate, move with constant velocity and finally decelerate. This movement perfectly fits to a pick and place work cycle. This manipulator is considered as a multiple input and multiple output dynamical system with its actuator torques and forces as the inputs and its continuous joint positions and joint velocities as the outputs.

The dynamics of n-link manipulator can be represented by n-coupled differential equations. The *Newton–Euler dynamics* and the *Lagrangian dynamics* are often used in the area of robotics to derive its dynamics of motion. The Newton–Euler dynamics uses Newton's second law of motion to obtain robot system differential equations considering all forces and moments acting on arm and links. This includes the constraint forces and constraint moments between adjacent links.

The Lagrangian dynamics uses the Lagrangian function which is the difference between the total kinetic energy and total potential energy stored in the system. This function is to obtain robot differential equations without considering work forces and constraint forces. The Newton–Euler dynamics considers the constraints in robot motion, but not in the Lagrangian dynamics.

Table 13.1 Questions and Explanation

No.	Questions	Explanation
1	Who invented Lagrangian concepts and in which year?	A French scientist, J. L. Lagrange in 1780
2	What are the two concepts that help us to develop robot dynamics?	(i) Lagrangian dynamics and (ii) Newton–Euler dynamics
3	What is the main concept of Lagrangian dynamics?	Minimizing a Lagrangian function
4	What is Lagrangian function?	It is the difference between total kinetic energy and the total potential energy
5	Why are Cartesian coordinates not preferred in developing robot dynamics?	Cartesian coordinates of particles (include robots) are not independent and related to some constraint equations, which make the derivation of dynamics equation complex
6	What are generalized coordinates?	Generalized coordinates are independent among themselves and they are the same as degrees of freedom (DoF)
7	What are generalized forces?	Suppose that we use n generalized coordinates (q_1, q_2, ... q_n). Any small increment in Δq_i of the ith coordinate will not affect the other coordinates. The work done is ($w_i = \Delta q_i Q_i$) where Q_i is the force in the direction of motion
8	What is inertia matrix?	The inertia matrix $D(q)$ is symmetric (i. e. $D(q) = D^T(q)$) and is also positive definite (i.e. $D(q) > 0$)
9	What are gravity loading vector and velocity coupling vector?	$[G(q)]$ and $[H(q, q')q']$ are gravity loading vector and velocity coupling vector, respectively
10	What is inertia tensor?	The inertia tensor of a rigid body is a (3×3) matrix in (x, y, z) coordinates
11	What is rigid body coordinate frame?	Let D and v be the density and volume of a rigid body, respectively. Then the rigid body coordinate frame is defined as Dv
12	What are products of inertia?	$I_{xy'}$ $I_{yx'}$ $I_{zx'}$ $I_{xz'}$ $I_{yz'}$ $I_{zy'}$ $I_{xx'}$ I_{yy} and I_{zz} are known as the products of inertia. These products come from (x, y, z) moments of inertia
13	What is moment of inertia?	The moment of inertia of the rigid body with respect to (x, y, z) body coordinate frame is a (3×3) matrix of two combinations of coordinate frame. Sometimes the moment of inertia is referred to as inertia tensor
14	What is principle axes of rigid body?	The (x, y, z) frame of body coordinate axes is called the principle axes of rigid body. The advantage of using the principle axes of rigid body is that the inertia tensor is time invariant even in the time-varying situation

The mathematical modelling by Lagrangian dynamics is, in a way, simpler and easier than the other one. Table 13.1 illustrates some of nomenclatures of important features of Lagrangian concepts and on how the robot's dynamic equations are obtained.

Definitions of Variables

r: length of link; θ: angle of link; p_x, p_y: torques resolved in *x* and *y* directions; (x, y, z): three-dimensional axes; (x_j, y_j, z_j): describes the *j*th particle; $(q_1, q_2, ..., q_n)$: *n*-vector in generalized coordinate system; (f_{xj}, f_{yj}, f_{zj}): are forces in *j* direction of (x, y, z); *p*: number of particles, each varying in *n*-DoF; q_i and Δq_i: selected coordinate and its increment in the direction of motion; Q_i: torque/force that creates this increment; $(Q_i \Delta q_i)$: the work done on the *i*th particle to move the distance Δq_i; *L*: Lagrangian function; *K* and *P*: the total kinetic energy and the total potential energy; $\iiint$: volume integrals; *g*: a (3×1) gravity vector; p_{oi}: position vector of centre of gravity.

13.2 LAGRANGIAN DESIGN

The Cartesian coordinate frame is not convenient to solve the set of particles motion and orientation. This is often restricted by a set of coordinate constraints that makes this sort of clumsiness. In order to avoid such an inconvenience, let us discuss the design by Lagrangian mechanics in *generalized coordinates.*

13.2.1 Generalized Coordinates

An arrangement of a three-dimensional space (x, y, z) that describes a system of particles can be considered. This space describes the positions and the orientations of particles to circumvent a complexity of constraint equations in Cartesian coordinates. Let us first investigate the power of the generalized coordinates.

A set of generalized coordinates gives a simplified approach and results in a set of coordinates that are independent among themselves. The set of generalized coordinates is equal to the number of DoF of a system. Let us review the system of planar robot of Figure 13.1 that involves the set of generalized coordinates. The $\theta - r$ coordinates of Figure 13.1 indicate

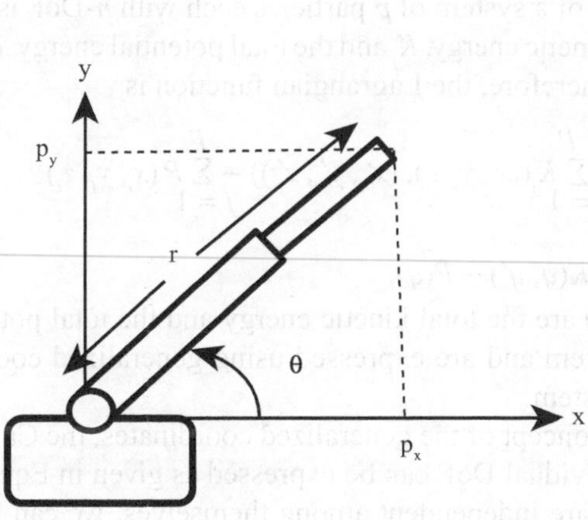

Figure 13.1 θ–*r* Robot

$$p_x = r \cos \theta \quad \text{and} \quad p_y = r \sin \theta, \tag{13.1}$$

that is, the coordinates of p_x and p_y are the functions of the variables θ and r, the angle in degrees and the radius in millimetre (or in metre), respectively. We may choose these variables as independent and hence $\theta - r$ are the generalized coordinates.

In fact, if the variables θ and r are known, then the position of tip of the arm is known. Let us consider a set of particles p with motions of n-DoF each. When (x_j, y_j, z_j) describes the jth particle in generalized coordinates of the system chosen as $(q_1, q_2, ..., q_n)$

$$\begin{aligned} x_j &= f_{xj}(q_1, q_2, ..., q_n), \\ y_j &= f_{yj}(q_1, q_2, ..., q_n), \\ z_j &= f_{zj}(q_1, q_2, ..., q_n). \end{aligned} \tag{13.2}$$

Similarly, for a robot with n joints, the n-joint variables $(q_1, q_2, ..., q_n)$ are chosen as the generalized coordinates. Each joint angle can be altered independently since the number of DoF of a robot is the same as the number of joints and, hence joint angles.

13.2.2 Generalized Forces

We use n generalized coordinates $(q_1, q_2, ..., q_i, ..., q_n)$ to describe the position and orientation of a system of p particles each one with n-DoF. Any small increment Δq_i to the coordinates q_i will not affect other coordinates such as $q_j, j \neq i$, since all generalized coordinates are independent. A force Q_i moves the coordinate q_i to a distance Δq_i; the work done is $\Delta w_i = Q_i.\Delta q_i$. If the generalized coordinate q_i is an angle, then Q_i is a torque required to drive the system to rotate through an angle Δq_i and the work done is the same as $\Delta w_i = Q_i.\Delta q_i$. In dynamics, Q_i, which may be a force or a torque as discussed in the above is called *generalized force* for the generalized coordinate q_i. The concept of the generalized force will be used in the following section on Lagrangian equation.

13.2.3 Lagrangian Equation

Lagrangian function, L, of a system of p particles each with n-DoF is defined as the difference between the total kinetic energy, K_j and the total potential energy, P_j, where $j = 1, 2, ..., p$, stored in the system,. Therefore, the Lagrangian function is

$$L = \sum_{j=1}^{p} K_j((x_i, y_i, z_i), (x'_i, y'_i, z'_i)) - \sum_{j=1}^{p} P_j(x_i, y_i, z_i) \tag{13.3}$$

or
$$L = K(q, q') - P(q), \tag{13.4}$$

where $K(q, q')$ and $P(q)$ are the total kinetic energy and the total potential energy, respectively, stored in the system and are expressed using generalized coordinates. The system can also be a robotic system.

As discussed in the concept of the generalized coordinates, the Cartesian coordinates of the jth particle with individual DoF can be expressed as given in Equation (13.2).

Since $(q_1, q_2, ..., q_n)$ are independent among themselves, we can let q_1 be incremented by Δq_1 and other coordinates $(q_2, ..., q_n)$ are maintained constants. Then, Equation (13.2) is simplified as

$$\Delta x_j = (\partial x_j/\partial q_1)\,\Delta q_1,$$
$$\Delta y_j = (\partial y_j/\partial q_1)\,\Delta q_1,$$
$$\Delta z_j = (\partial z_j/\partial q_1)\,\Delta q_1. \tag{13.5}$$

The work done is

$$\Delta w = \left[\sum_{j=1}^{p}\left\{F_{xj}\left(\partial x_j/\partial q_1\right)+F_{yj}\left(\partial y_j/\partial q_1\right)+F_{zj}\left(\partial z_j/\partial q_1\right)\right\}\right]\Delta q_1 = Q_1\Delta q_1, \tag{13.6}$$

where Q_1 is the generalized force corresponding to q_1. The set (F_{xj}, F_{yj}, F_{zj}) is generalized force that gives the minimal displacement Δq_1.

Using Newton's second law, we have three equations as follows:

$$F_{xj} = m_j\,x_j'',\ F_{yj} = m_j y_j'' \text{ and } F_{zj} = m_j z_j''. \tag{13.7}$$

Hence, the term Q_1 on the right-hand side of Equation (13.6) now becomes

$$Q_1 = \sum_{j=1}^{p} m_j\left[x_j''(\partial x_j/\partial q_1) + y_j''(\partial y_j/\partial q_1) + z_j''(\partial z_j/\partial q_1)\right]. \tag{13.8}$$

Using the following relationships:

$$x_j''[\partial x_j/\partial q_1] = ½\,[\mathrm{d}/\mathrm{d}t]\,\{(\partial(x_j'^2)/\partial q_1'\} - ½\,\{\partial(x_j'^2)/\partial q_1\},$$

$$y_j''\,[\partial y_j/\partial q_1] = ½\,[\mathrm{d}/\mathrm{d}t]\,\{(\partial(y_j'^2)/\partial q_1'\} - ½\,\{\partial(y_j'^2)/\partial q_1\},$$

$$z_j''\,[\partial z_j/\partial q_1] = ½\,[\mathrm{d}/\mathrm{d}t]\,\{(\partial(z_j'^2)/\partial q_1'\} - ½\,\{\partial(z_j'^2)/\partial q_1\}, \tag{13.9a}$$

and on simplification, Equation (13.9a) reduces to

$$\left[Q_1 = \frac{\mathrm{d}}{\mathrm{d}t}\left[\frac{\partial}{\partial q_1'}\left\{\sum_{j=1}^{p}\frac{m_j\left(x_j'^2+y_j'^2+z_j'^2\right)}{2}\right\}\right] - \frac{\partial}{\partial q_1}\left\{\sum_{j=1}^{p}\frac{m_j\left(x_j'^2+y_j'^2+z_j'^2\right)}{2}\right\}\right]. \tag{13.9b}$$

Hence, the total kinetic energy, K, of the system is

$$K = \sum_{j=1}^{p}(½)\,m_j(x_j'^2+y_j'^2+z_j'^2). \tag{13.10}$$

Then, after substituting Equation (13.9b) into Equation (13.10) we get

$$Q_1 = \{(\mathrm{d}/\mathrm{d}t)\,\{\partial K/\partial q_1'\}\} - \{\partial K/\partial q_1\}. \tag{13.11}$$

The generalized force Q_1 has two components:

(i) An external force/torque, τ_1

(ii) A conservative force of value $\{-\partial P/\partial q_1\}$.

The element, P, is the total potential energy of the system.

Equation (13.11) can then be written as follows:

$$(\mathrm{d}/\mathrm{d}t)\,\{\partial K/\partial q_1'\} - \{\partial K/\partial q_1\} = \tau_1 - \{\partial P/\partial q_1\} \tag{13.12}$$

or
$$(\mathrm{d}/\mathrm{d}t)\,\{\partial K/\partial q_1'\} - \{\partial\,(K-P)/\partial q_1\} = \tau_1. \tag{13.13}$$

Considering the Lagrangian function in

$$\frac{\partial L}{\partial q_1'} = \frac{\partial (K(q,q') - P(q))}{\partial q_1'} = \frac{\partial K(q,q')}{\partial q_1'}. \tag{13.14}$$

Equation (13.13) can then be written as

$$(\mathrm{d}/\mathrm{d}t)\{\partial L/\partial q_1'\} - \{\partial L/\partial q_1\} = \tau_1. \tag{13.15}$$

In a similar way, let q_i be modified by Δq_i and other coordinates kept constants, we obtain the general Lagrangian equation as follows:

$$(\mathrm{d}/\mathrm{d}t)\{\partial L/\partial q_i'\} - \{\partial L/\partial q_i\} = \tau_i,\ i = 1, 2, \ldots, n. \tag{13.16}$$

Equation (13.16) is generally used to obtain the differential equations of motion for non-conservative systems (the systems having forcing functions).

If the system is conservative, the external force/torque $\tau_i = 0$ and the corresponding Lagrangian equation is

$$(\mathrm{d}/\mathrm{d}t)\{\partial L/\partial q_i'\} - \{\partial L/\partial q_i\} = 0,\ i = 1, 2, \ldots, n. \tag{13.17}$$

The Lagrangian equations derived above are applicable not only to rigid body problems but also to robots.

We shall investigate this approach to inverted pendulum.

EXAMPLE

Consider an inverted pendulum as shown in Figure 13.2. This pendulum works with a weightless rod of length l and the point mass m at the end of rod.

(a) Determine potential energy and kinetic energy of the inverted pendulum.

(a) Determine Lagrangian function and Lagrangian equation.

(b) Derive the dynamical equation when θ is very small.

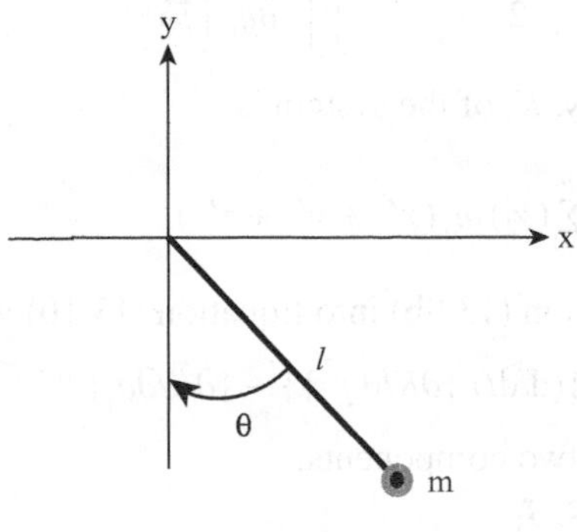

Figure 13.2 An Inverted Pendulum

Solution:

(a) Cartesian coordinates of the mass = (x, y)

Mass of the pendulum = m

Generalized coordinates of the system = (l, θ)

The coordinates (x, y) can be expressed in terms of (l, θ) as follows:

$$x = l \sin \theta,$$
$$y = -l \cos \theta.$$

The first-order derivatives of (x, y) with respect to time are as follows:

$$x' = l \cos \theta\theta'$$
$$y' = l \sin \theta\theta'$$

Then, the total potential energy and total kinetic energy of the mass m are as follows:

Total potential energy: $P = mgy = -mgl \cos \theta.$

Total kinetic energy:

$$K = ½\,(mx'^2 + my'^2) = (m/2)\,(l^2\theta'^2 \cos^2\theta + l^2\theta'^2 \sin^2\theta)$$

or $K = (m/2)\, l^2\theta'^2.$

(b) The Lagrangian function is

$$L = K - P = ½\, ml^2\theta'^2 + mgl \cos \theta.$$

The Lagrangian equation is

$$(\partial L/\partial\theta) = -\, mgl \sin \theta,$$
$$(\partial L/\partial\theta') = ml^2\theta'$$

and $(d/dt)\{\partial L/\partial\theta'\} = ml^2\, \theta''.$

(c) The dynamical equation

This results in Lagrangian equation as given in Equation (13.9a). The differential equation of motion of inverted pendulum is hence (with no external force/torque)

$$ml^2\, \theta'' + mgl \sin \theta = 0$$

or $\theta'' + (g/l) \sin \theta = 0.$

When θ is very small, $\sin \theta = \theta$, then the system equation becomes linear and results in

$$ml^2\theta'' + mgl\theta = 0$$

or $\theta'' + (g/l)\, \theta = 0.$

13.3 N-LINK ROBOT MANIPULATOR

Using the Lagrangian equations, we can individually obtain a set of differential equations for each link of an n-link rigid robot manipulator. In this section, we develop an algorithm which can help the reader to obtain the vector differential equation for any n-link rigid robot manipulator. Let us start with the concept of *inertia tensor* of a rigid body.

An inertia tensor of a rigid body is a (3×3) matrix. This is used to describe the mass distribution of rigid body with respect to selected coordinate frame.

Let D and v, respectively, be the mass density of the rigid body and the volume occupied by the *body coordinate frame* as shown in Figure 13.3.

In Figure 13.3:

(i) A vector ω comes out of paper, parallel to x-axis,
(ii) The elemental volume is dv,
(iii) D is the density; (D dv) is the elemental mass,
(iv) v is the direction vector of whole mass of the body,
(v) The length of arm is r,
(vi) The gravity force is G.

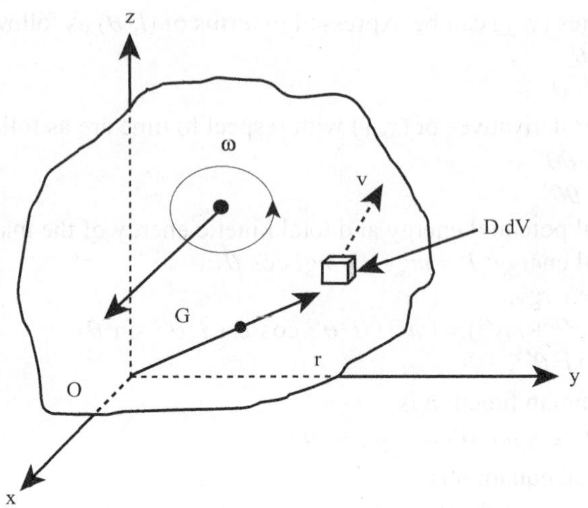

Figure 13.3 Rigid Body, I, Attached to Coordinate Frame

The inertia tensor of the rigid body with respect to body coordinate frame is

$$I = \begin{bmatrix} I_{xx} & I_{xy} & I_{xz} \\ I_{yx} & I_{yy} & I_{yz} \\ I_{zx} & I_{zy} & I_{zz} \end{bmatrix}. \quad (13.18)$$

The elements are called the *moments of inertia* about (*x*, *y*, *z*) axes, respectively.

Then $I_{xy} = I_{yx} = -\iiint xyD \, \mathrm{d}v$

$I_{zx} = I_{xz} = -\iiint zxD \, \mathrm{d}v$

$I_{yz} = I_{zy} = -\iiint yzD \, \mathrm{d}v$ (13.19)

are known as the *products of inertia*.

The angular velocity: $\omega = [\omega_x \omega_y \omega_z]^T$.

The elemental mass: (D dv).

In the frame *xyz*, the length of link (or arm): $r = [x, y, z]^T$.

The elemental part of angular momentum: $\mathrm{d}H = (r \times v) D \, \mathrm{d}v$, where ($D$ dv) is the elemental mass, v is the linear velocity (or linear speed) vector of the mass (D dv) and $v = \omega \times r$, ω is the angular velocity (or angular speed) vector.

The angular momentum of the whole rigid body is

$$H = \iiint [r \times (\omega \times r) D \, \mathrm{d}v]$$

$$= \iiint [\{r^T r\} \omega - \{r^T \omega\} r] D \, \mathrm{d}v$$

$$= I \omega, \quad (13.20)$$

where I is the inertia tensor (Equation 13.18) and ω is the angular (rotational) velocity.

The angular (rotational) kinetic energy of whole rigid body

$$K_{rot} = ½\,(\omega^T I\,\omega). \tag{13.21}$$

The inertia tensor I in Equation (13.18) is obtained with respect to origin of the body coordinate frame (x, y, z). If the inertia tensor is obtained with respect to G, the centre of gravity of rigid body, the rotational kinetic energy, K_{rot}, of rigid body can then be expressed as follows:

$$K_{rot} = ½\,(mv_g^{\,T} v_g) + ½\,(\omega^T I_g \omega), \tag{13.22}$$

where m is the total mass of rigid body, v_g is the velocity of centre of the mass and I_g is the inertia tensor with respect to the centre of mass.

If the rigid body undertakes not only the rotation but also the translation, then the translational kinetic energy, K_{tran}, of the rigid body is

$$K_{tran} = ½\left(mv_m^T v_m\right), \tag{13.23}$$

where v_m is the velocity of translation of whole rigid body.

The total kinetic energy is

$$\begin{aligned} K = K_{tran} + K_{rot} &= ½\,\{(mv_m^T v_m)\} + ½\,\{(mv_g^{\,T} v_g) + (\omega^T I_g \omega)\} \\ &= ½\,\{mv^T v + \omega^T I_g \omega\}, \end{aligned} \tag{13.24}$$

where $v = [v_m v_g]^T$.

13.4 SLENDER ROD AS ROBOT LINK

It is always possible to choose a body coordinate frame in the sense that all off-diagonal elements of the inertia tensor, I, are zero. The (x, y, z) frame of body coordinate is called the *principle axes of rigid body.* The advantage of using the principle axes of rigid body is that the inertia tensor is time invariant even in the time-varying situation. The slender rod is generally considered as an approximation to the robot link.

By slender rod, we mean that the rod has no breadth but has length, l, as shown in Figure 13.4. In addition, there are two others, the cylindrical rod and the rectangular prism, but these two are not discussed in this book.

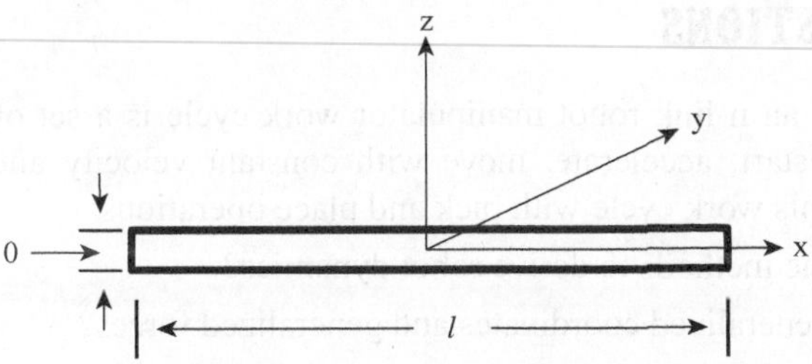

Figure 13.4 A Slender Rod

The inertia tensor, $\hat{I}$, of a rigid slender rod is

$$\hat{I} = \begin{bmatrix} 0 & 0 & 0 \\ 0 & (1/12)\,ml^2 & 0 \\ 0 & 0 & (1/12)\,ml^2 \end{bmatrix}. \tag{13.25}$$

The matrix, $\hat{I}$, will be used in finding kinetic energy of the *i*th link of a rigid slender rod of a robot link.

With the above background of inertia tensor and kinetic energy, we are now ready to consider how to obtain the vector differential equation of any n-link manipulator.

Let $\dot{\upsilon}_i$ be the linear velocity vector of link, *i*, relative to the centre of mass, m_i, and $\dot{\omega}_i$ be the angular velocity vector of link, *i*, relative to the centre of mass, m_i.

Then, the total kinetic energy is

$$K = \sum_{i=1}^{n} K_i = 1/2\left\{\sum_{i=1}^{n} m_i \upsilon_i^T \upsilon_i + \sum_{i=1}^{n} \omega_i^T I \omega_i\right\}$$

$$= ½\,(q'^{\mathrm{T}} D q'), \tag{13.26}$$

where q' is the vector $[\dot{\upsilon}\dot{\omega}]^{\mathrm{T}}$.

Potential Energy and Lagrangian Function

In getting the general expression of the total potential energy, let g be the (3×1) gravity vector with respect to the robot base frame. The potential energy stored in the *i*th link is then given by $(m_i g^{\mathrm{T}} p_{\mathrm{oi}})$, where m_i is the centre of the mass, g is the gravitational force and p_{oi} is the position vector of the centre of gravity of the link *i*.

The total potential energy stored in an n-link robot is then computed by

$$P = \sum_{i=1}^{n} (m_i g^{\mathrm{T}} p_{\mathrm{oi}}) \tag{13.27}$$

and the Lagrangian function L can then be

$$L = K - P$$

$$= 1/2(q'^T Dq) - \sum_{i=1}^{n} (m_i g^{\mathrm{T}} p_{\mathrm{oi}}). \tag{13.28}$$

The Lagrangian function L can be further used in developing the equations of motion of robot.

REVIEW QUESTIONS

1. In general, an n-link robot manipulator work cycle is a set of movements with a motion to start, accelerate, move with constant velocity and finally decelerate. Compare this work cycle with pick and place operations.
2. What are the methods to derive robot dynamics?
3. What are generalized coordinates and generalized forces?
4. What are velocity coupling vector and gravity loading vector? Where are they used?

5. Why are Cartesian coordinates not easier than other methods to compute robot dynamics? What are other methods to compute robot dynamics?
6. How are DoF and the generalized coordinates connected?
7. List French scientist's invention, his formulation, Lagrangian function, Lagrangian equation and Lagrangian dynamics?
8. List atleast five important aspects? How do you make sure their importances?
9. Draw a diagram of rigid body attached to coordinate frame of robot link and describe them?
10. How do you make sure that the slender rod is sufficient in explaining some of the aspects of the robot basics? When is it not acceptable to the basic rules?

FLC OF ROBOT JOINTS 14

It is the language that decides how well the robot performs

Peter Darey

This chapter covers the following key topics:

Introduction – Advantages and Drawbacks of FLC – Probability and Possibility – Possibility Distribution Functions – Description of a Fuzzy Set – Fuzzy Set Operations – Fuzzy Relation – Designing FL Controllers – A General Scheme of a FLC – Steps in Designing FLC – The Inverted Pendulum – Fuzzification of θ, ω and V – Development of FAM Table and Fuzzy Rules

---

14.1 INTRODUCTION

Let us refer to the following statements. These statements are pertaining to parallel parking of my car in a parking lot. I am using fuzzy logic (FL) variables without my knowledge. The variables are known as linguistic variables and I have to use them appropriately.

> 'There are *almost* hundred cars in the parking lot. I go round to find an empty place. I line up my car next to the one in front of the empty space. Then, I angle the car back in to space turning the steering wheel *slightly* to adjust the angle as I get *closer* to the boundary. Now, I turn the wheel to straight the car *a bit* to go *nearer* to the boundary. Alas! I have jammed on the boundary. Now, I go forward *slowly* steering towards the curb until the rear tires straightened out. Fine, I am not *too far* from the boundary. I drive back and forth again using *smaller* angles. I straight the car forward *a little*. It is now *too close* to the car ahead. I back up *a few* inches. Now, I have successfully completed the process of parallel parking my car'.

The words in italic are the linguistic variables which indicate the 'descriptive' factors. Students can describe such a paragraph on their own from the incidents that had given an experience to them in using the linguistic variables.

Now, I can try to use the crisp logic to see the difference. In doing so, some requires counting and other requires some other methods of measurements.

These linguistic words are devised by Lofty Zadeh in 1960. For about 10 years, there was a silence in the application of FL control (FLC). Its first application was for the purpose of controlling the speed of a steam engine. Initially, people were reluctant to accept the applications of FLC. Then there were various applications such as in SONY TV, MATSUSHITA vacuum cleaner, NISSAN cars, SHARP air conditioners, MITSUBHISI lifts, OTIS elevators and in many others.

FL has been considered as an important artificial intelligence (AI) to model vagueness and to model those functions not clearly defined. Modelling the possibilities is a method of application of FL. In fact everything around us is considered as vague. Understanding a lecture by a FL expert in a class, mixing of certain colours, classifying persons by age and characteristics of a system are all vague and uncertain. We define all those we do not know how this can come in as FL.

The possibilities are marked in y-axis. Maximum possibility is unity or maximum $\mu(x) = 1$

14.2 ADVANTAGES AND DRAWBACKS OF FLC

There are several advantages and a few drawbacks of FLC. Several advantages are as follows:

1. Linguistic variables are used in FLC; this is similar to the way human beings think.
2. No mathematical model is required; only the logical relations between inputs and outputs are required.
3. The FLC gives a reasonable accuracy ensuring the stability of systems.
4. Rapid prototyping is possible. Control engineers need not know everything about the behaviour of the system for designing FLC.
5. For FLC, several software platforms are available and applications are possible.
6. FLC is cheaper and easier to design compared with conventional advanced controllers.
7. FLC has increased robustness.
8. FLC helps to incorporate our knowledge and experience in controlling the systems.
9. A few FLC rules are sufficient in many control applications. Few rules means faster processing and quicker control.

Following are a few drawbacks of FLC:

1. It is hard to model a system with FLC.
2. The systems designed with FLC requires several simulations and fine tunings before it is finally implemented.
3. Cultural bias: people do still think that crisp logic is better in designing linear mathematical model which has better behaviour than the existing one.

Some times FLC combined with other advanced AI control techniques, such as neural networks and/or genetic algorithms, are used for a greater advantage.

14.3 PROBABILITY AND POSSIBILITY

Probability and possibility are two terms we normally come across in our daily life. In most cases we mean them as 'something uncertain'. However, we can realize a lot difference if we carefully analyse the meanings of these terms. In fact, the concept of probability leads us to the statistical theory and the possibility leads to the FL theory.

Probability indicates 'the chance' of an event *to take* place; possibility indicates 'the precision' if the event has *already taken* place.

Example: Let us consider two glasses of liquids. The liquid in the first glass is said to be 95% *probability* of being 'healthy and good'. The liquid in the second glass is said to be 95% *possibility* of being 'healthy and good'.

The first statement indicates clearly that there is a 5% chance of 'liquid not healthy and good'; that is, the first glass may contain unhealthy liquid – even poison; something *not recommended for drinking.* Here, the liquid belongs to the set of 'healthy and good' can be *true* or sometimes *not true.*

It is a chance of event taking place.

The second statement indicates that 'healthy and good' is not precise. That is, this is not exact; there can be error by 5%. This second glass certainly contains a healthy and good liquid but the quality of content is not precisely 100% healthy and good. In other words, the quality of the content of the second glass being 'healthy and good' has a value 0.95. The event 'healthy and good' has been already established but it is not precise.

It is the precision of event already taken place.

Another example: let us consider 'tall' is of height 2 m.

Then, a 90% probability of a man being 'tall' can indicate that the man can be a dwarf.

A 90% possibility of a man being 'tall' indicates that he can be of 1.8 m in height. In no way he is dwarf.

In the examples:

Probability of 90% means a membership value = 0.90 (maximum is 1.00).
Possibility of 90% means a membership value = 0.90 (maximum is 1.00).

14.4 POSSIBILITY DISTRIBUTION FUNCTIONS

In this section, we compare 'linear form' of membership functions with 'the actual form' of membership function. In many cases, the linear form is advantageous right from the linear relation between input and output. Writing mathematical relation between input (the number of happenings) and output (the possibility) is easier in the case of linear form. In most of the cases we follow the linear form.

Consider an example: Rajan eats x eggs for his breakfast, where

$$x \,\varepsilon\, U = \{1, 2, 3, 4, 5, 6, 9, 12, 20, 30, 50, 60\}, \qquad (14.1)$$

where U is known as universe of discourse (UoD).

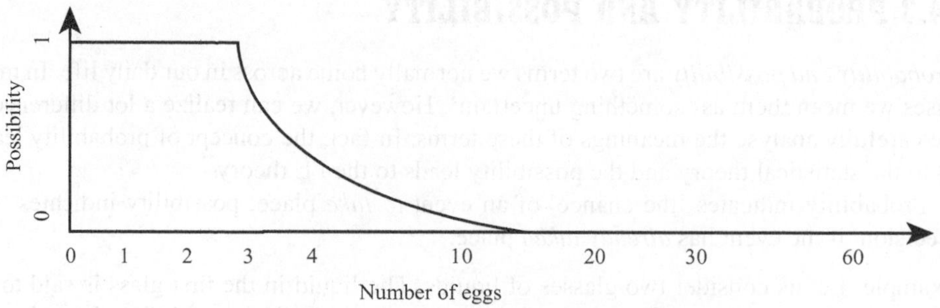

Figure 14.1 Membership Function

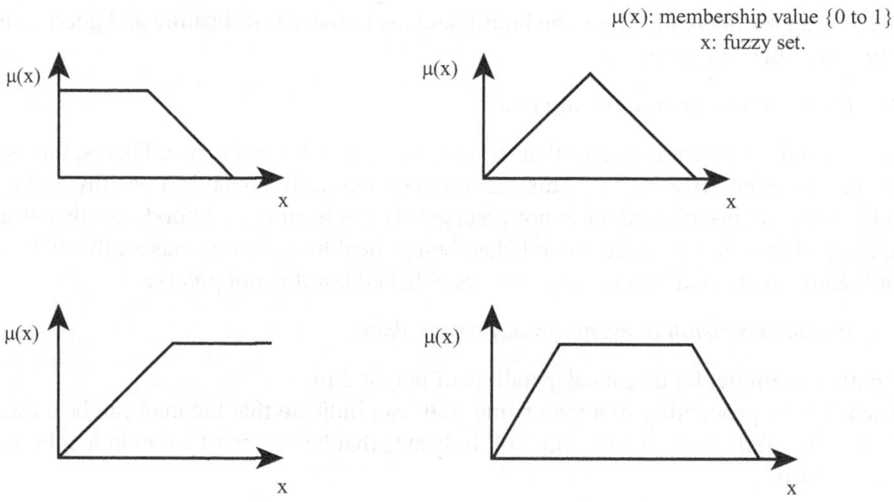

Figure 14.2 A Few Possible Membership Functions

The membership function can be represented in a graphical way as shown in Figure 14.1.

That is, Rajan can eat a maximum of three eggs for morning breakfast. After which he becomes tired of eating more eggs.

The relation between number of eggs and possibility of eating are either both or any one is a linear function.

Figure 14.2 describes some of the possible membership functions which we normally come across in every effort. Students can find a suitable task from the indicated linear membership functions.

14.4.1 Description of a Fuzzy Set

Figure 14.3 illustrates one of the forms (singleton) of linear membership functions. The fuzzy variable x can be of feet. We can have several sections of fuzzy variables:

$$\mu(x) = \begin{cases} 0 & x \le 1 \text{ foot} \\ (x-1) & 1 \le x \le 2 \text{ feet} \\ (3-x) & 2 \le x \le 3 \text{ feet} \\ 0 & x \ge 3 \text{ feet} \end{cases} \quad (14.2)$$

Fuzzy variable x can thus be divided into several sections to meet all portions of the x-axis.

A fuzzy set A is shown in Figure 14.4; set A can be divided into several sections within its x-axis such that

$$A = \{0.0/2.0 + 0.6/2.2 + 1.0/3.0 + 0.4/3.8 + 0.0/4.0\}. \quad (14.3)$$

This A can be used, with the above form, whenever we have to indicate any FL linguistic variable.

Figure 14.5 illustrates a combined form of fuzzy sets. The relation between height and membership values are depicted. A vertical line can be seen representing a connection between 'short' and 'medium' at a height of 4.25 feet. We can conclude that this line passes through points of heights of 25% medium and 75% short. Such decisions are to be made at several places in solving FL problems.

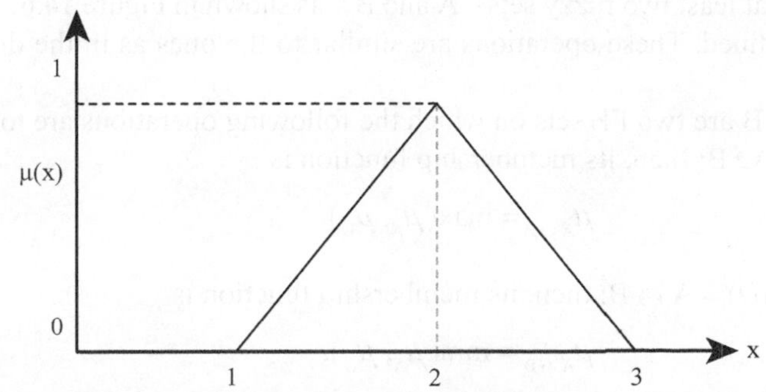

Figure 14.3 A Description of a Fuzzy Set

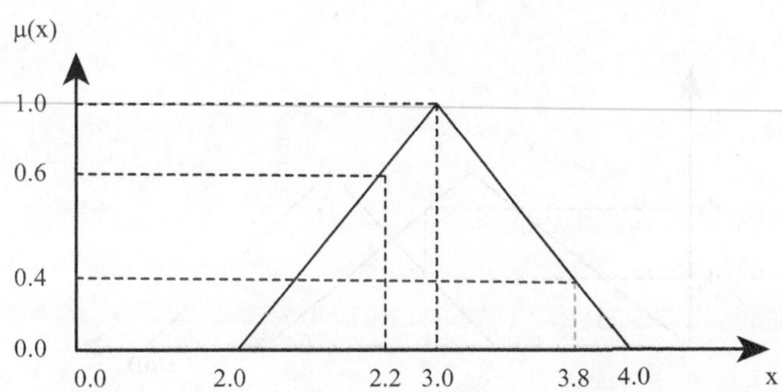

Figure 14.4 Another Description of a Fuzzy Set

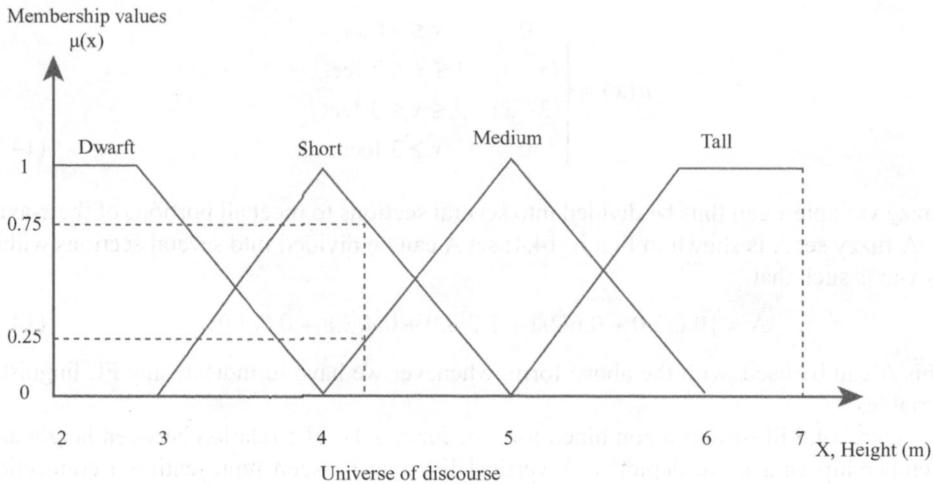

Figure 14.5 A Combined Form of Fuzzy Sets

14.5 FUZZY SET OPERATIONS

There should be at least two fuzzy sets – A and B – as shown in Figure 14.6. Then, the operations can be defined. These operations are similar to the ones as in the digital examples described earlier.

Set A and set B are two FL sets on which the following operations are to be done:
Union (OR) = A ∪ B; then, its membership function is

$$\mu_{A \cup B} = \max(\mu_A, \mu_B). \tag{14.4}$$

Intersection (AND) = A ∩ B; then, its membership function is

$$\mu_{A \cap B} = \min(\mu_A, \mu_B). \tag{14.5}$$

Complement (NOT) of A = A′; then its membership function is

$$\mu_{A'} = (1 - \mu_A). \tag{14.6}$$

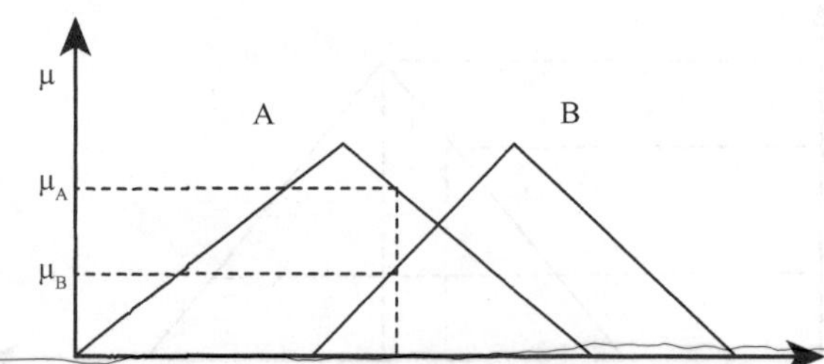

Figure 14.6 Two Coupled Fuzzy Sets

The students can imagine that there are three fuzzy sets – set A, set B and set C; then the following properties hold good; these are similar to the crisp sets.

Let X be the universal set and ∅ be the null set.

Commutativity: $A \cup B = B \cup A$

$$A \cap B = B \cap A \tag{14.7}$$

Associativity: $A \cup (B \cup C) = (A \cup B) \cup C$

$$A \cap (B \cap C) = (A \cap B) \cap C \tag{14.8}$$

Distributivity: $A \cup (B \cap C) = (A \cup B) \cap (A \cup C)$

$$A \cap (B \cup C) = (A \cap B) \cup (A \cap C) \tag{14.9}$$

Idempotency: $A \cup A = A;\ A \cap A = A$ (14.10)

Identity: $A \cup \varnothing = A;\ A \cap X = A$

$$A \cup X = X;\ A \cap \varnothing = \varnothing \tag{14.11}$$

Transitivity: If $A \subseteq B \subseteq C$, then $A \subseteq C$ (14.12)

Involution: $(A')' = A$ (14.13)

Excluded middle laws:

(a) Law of excluded middle: $A \cup A' = X$

(b) Law of contradiction: $A \cap A' = \varnothing$ (14.14)

De Morgan's laws: $(A \cup B)' = A' \cap B'$

$$(A \cap B)' = A' \cup B'. \tag{14.15}$$

NOTE: Only the excluded middle laws are the only set operations that are different from crisp logic set operations.

De Morgan's law states that the complement of union of two sets is the same as the intersection of complement of individual sets (intersection of complement sets). In a similar way, the complement of intersection of two sets is the union of complement sets.

14.6 FUZZY RELATION

A fuzzy relation is defined as the mapping of elements of one fuzzy set to the other fuzzy set – a sort of connection between fuzzy sets. A set can be a UoD. Hence we can describe, in general, the relation between two universes, say universe X and universe Y.

The Cartesian product X × Y gives the relation. That is, the relation between X and Y is denoted by R = X × Y.

The strength of the relation, R, is measured by membership function within the interval [0, 1].

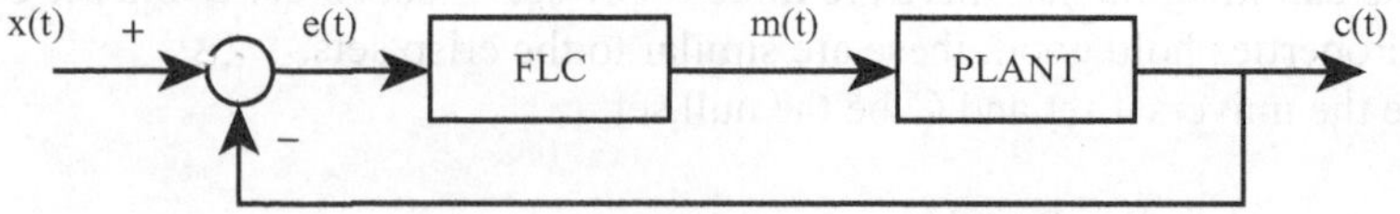

Figure 14.7 A FLC System

Let x ε X and y ε Y, then (x, y) is an ordered pair from these two universes. The strength of relation X × Y is, then, represented by the membership function $\mu_R(x, y)$ where

$$\mu_R(x, y) = \min \{\mu_X(x), \mu_Y(y)\}. \tag{14.16}$$

We use this equation repeatedly when we solve a FLC system.

Let us consider a FLC system as shown in Figure 14.7. The input to the controller e(t) and its output m(t) have relations so that the plant output c(t) follows the reference input x(t).

Let set E, set M and set C denote three fuzzy sets on the universe of error E, universe of controller output M and the universe of plant output C, respectively.

Let the membership functions be

$$\begin{aligned} E &= \{(0.3/e1) + (0.8/e2) + (0.7/e3)\} \\ M &= \{(0.2/m1) + (1/m2) + (0.5/m3)\} \text{ and} \\ C &= \{(0.1/c1) + (0.8/c2) + (0.2/c3)\}. \end{aligned} \tag{14.17}$$

Then, the relation R between E and M is denoted by

$$\mu_R(x, y) = \min \{(\mu_X(x), \mu_Y(y))\}, \tag{14.18}$$

$$\begin{aligned} \mu_{E\times M} &= E \times M, \\ \mu_{M\times C} &= M \times C, \\ \mu_{C\times E} &= C \times E, \end{aligned} \tag{14.19}$$

which are relations among E, M and C.

14.7 DESIGNING FUZZY LOGIC CONTROLLER

Human beings make decisions based on rules. Human knowledge is acquired, created and applied in this way.

Example:

IF (weather is FINE), THEN (we GO OUT)
IF ((weather is FINE) AND (today is HOLIDAY)), THEN (we GO OUT)
IF ((weather is FINE) AND (today is SATURDAY) OR
(today is SUNDAY)), THEN (we GO OUT). (14.20)

That is,

$$\text{IF } \{(A \cap (B \cup C))\} \text{ THEN } \{D\} \tag{14.21}$$

in which A, B, C and D are linguistic variables.

Similar rules are formed and applied in a FLC system. Since the control system is to be controlled in real time, there should be repeated use of such rules. Faster execution of rules results in faster response of control system. Hence, the rules are to be as simple as possible.

14.7.1 A General Scheme of a FLC

Fuzzification and defuzzification stages have four modules as shown in Figure 14.8, which are illustrated as follows:

1. A fuzzy rule base: this keeps the essence of rules and rule base
2. A fuzzy inference engine: this temporarily keeps the pertinent rules
3. A fuzzification module: this fuzzifies the fuzzy variables
4. A defuzzification module: this makes crisp logic which are already in fuzzy stage

In addition, we have one controlled process. This involves making of a FLC system with its input and its output as all the variables that include the required FLC variables.

With regard to items 3 and 4, FLC operates based on how many steps are given to the stage fuzzification. More steps may create delay time. The steps reflects on the uncertainties in the measurements. In addition, more time is required for item 4 to defuzzify. Altogether, more time is spent on items 3 and 4 if the rules are complex. Hence, the human operator comes in deciding fuzzy quantization and the UoD. In addition, he should decide what types of quantization are required – whether a triangular form or any other form.

Students should observe that there is no mathematical model of the plant used anywhere. This block diagram of Figure 14.8 can also be applied to multinput and multioutput control systems.

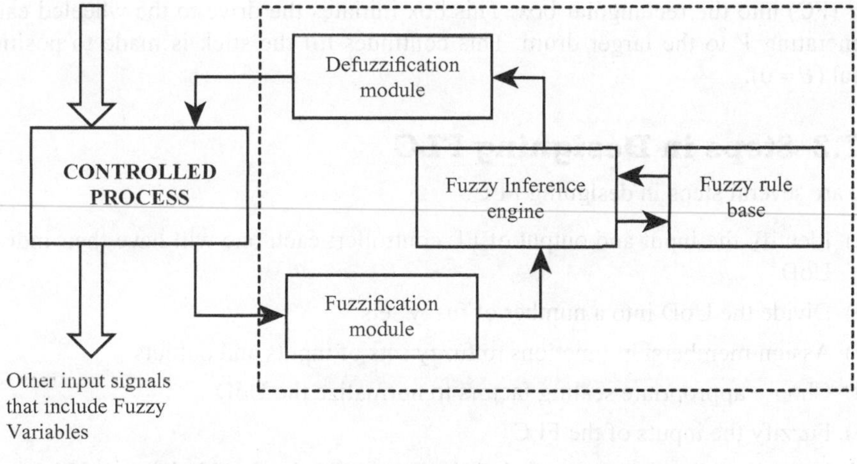

Figure 14.8 A General Scheme of FL Controller

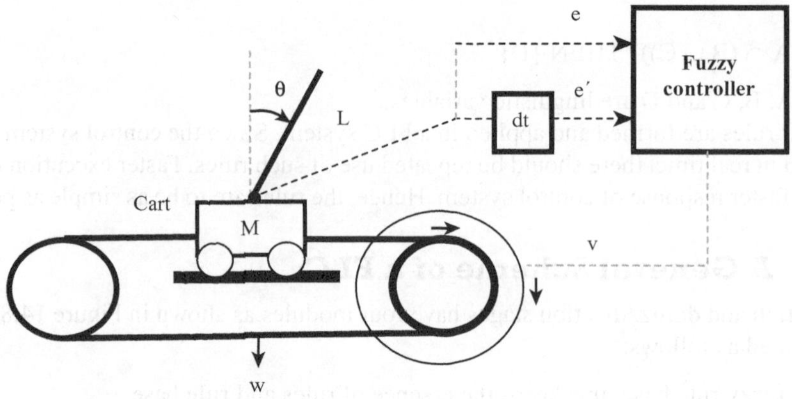

Figure 14.9 Inverted Pendulum

Figure 14.9 illustrates following aspects regarding inverted pendulum:

(a) M is the cart which can move left/right
(b) *L* is the weightless stick which can assist the whole system in motion
(c) θ is the angular position in radians; at this instant θ angle is measured as negative
(d) *W* is the weight of the belt
(e) Fuzzy controller block is basically an electronic box
(f) d*t* is a derivative symbol
(g) *e* and *e′* are error and derivative of error, respectively
(h) *V* is the drive signal to the larger drum

M takes care of the whole system dynamics. *L* creates and assists the system in motion. This is because the stick not being vertical (θ not zero) and makes the cart to move right and left. The entire system is operated by electronic circuit embedded as fuzzy controller (FC) into the rectangular box. This box initiates the drive to the wheeled cart, M, by generating *V* to the larger drum. This continues till the stick is made to position as vertical ($\theta \approx 0$).

14.7.2 Steps in Designing FLC

There are several steps in designing FLC:

(a) Identify the input and output of FL controller; each one will have their individual UoD
(b) Divide the UoD into a number of fuzzy sets
(c) Assign membership functions to fuzzy sets of inputs and outputs
(d) Choose appropriate scaling factors to normalize the UoD
(e) Fuzzify the inputs of the FLC
(f) Use appropriate reasoning (rule firing) and infer the fuzzified form of FLC output

(g) Apply any suitable defuzzification method to the output in order to get the crisp values of FLC output of the controller
(h) This crisp value is applied to the plant
(i) From the plant output, obtain new inputs of FLC
(j) Repeat step (e) to step (j) till the desired plant response is obtained. If not, go back to step (e).

With regard to step (g), there are three defuzzification methods that are commonly employed: centre of area method (centre of gravity method, centroid method), centre of maximum method and mean of maxima method which are very commonly used in the literature. Of these the centre of area method is well known and used frequently. The other defuzzification methods are left to interested students to investigate.

In the centre of area method, the defuzzified value, $d_{CA}(C)$ is calculated by using the following formula:

$$d_{CA}(C) = \{\Sigma\, C(z_k)\, z_k / \Sigma\, C(z_k)\}, \tag{14.22}$$

where the summation is done from $k = 1$ to $k = n$.

14.8 THE INVERTED PENDULUM

The relevant block diagram used in the example of inverted pendulum is shown in Figure 14.10. The pendulum system has all the subsystems of Figure 14.9.

One has to decide the UoDs for inputs and outputs and also the fuzzy variables. This requires a long and dedicated experience and knowledge of the inverted pendulum. We shall decide the following UoDs:

(a) The position (θ radians): $\{-2, 2\}$
(b) The velocity (ω radians/second): $\{-5, 5\}$
(c) The voltage to cart-motors (V volts): $\{-16, 16\}$

In the following paragraphs, (*) indicates the measured signals of θ and ω. These signals are fuzzified and applied.

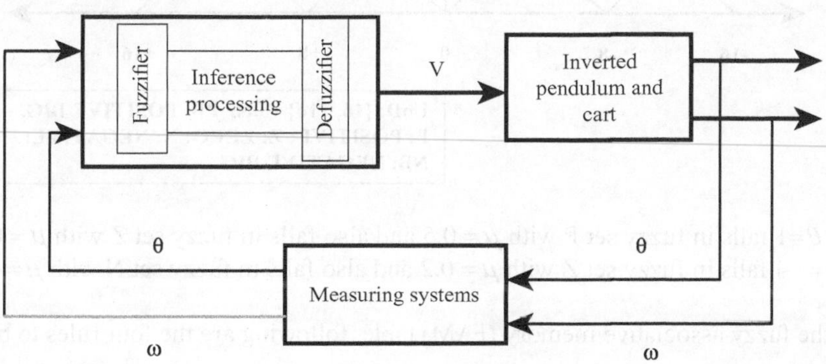

Figure 14.10 Simplified Inverted Pendulum System

Fuzzification of θ, ω and V:

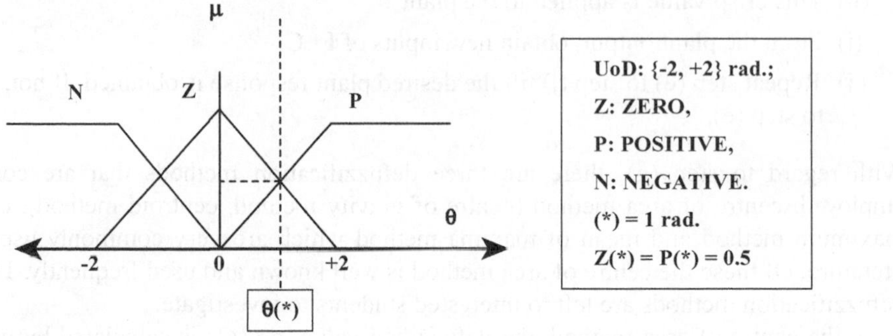

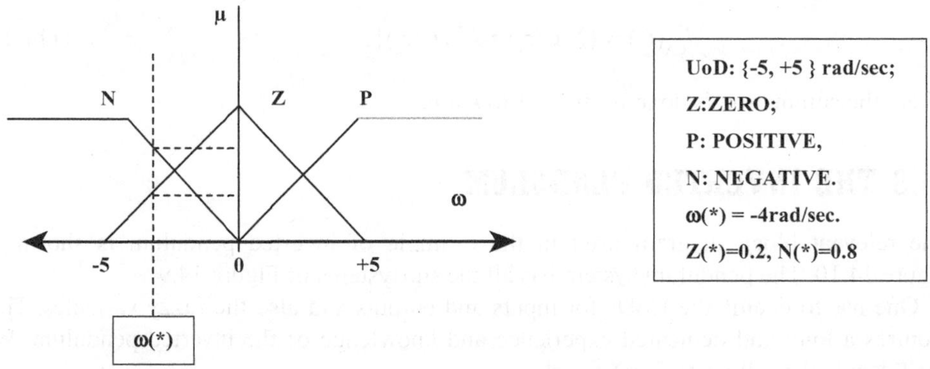

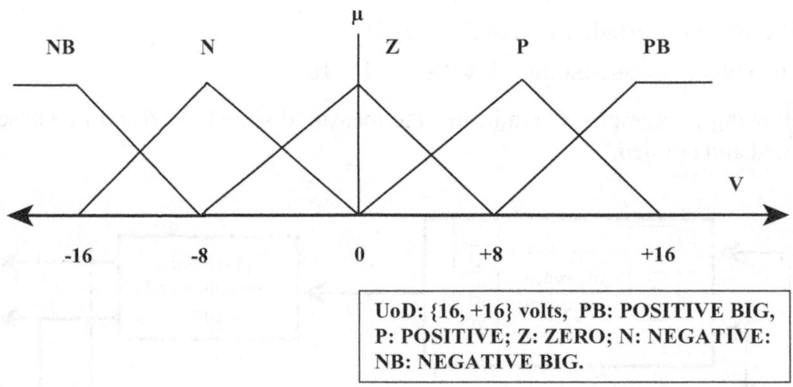

When θ=1 falls in fuzzy set P with μ = 0.5 and also falls in fuzzy set Z with μ = 0.5 and ω = –4 falls in fuzzy set Z with μ = 0.2 and also falls in fuzzy set N with μ = 0.8.

From the fuzzy associative memory (FAM) table, following are the four rules to be fired:

IF {(θ = *P*) <u>and</u> (ω = *Z*)}, THEN (*V* = *P*) with μ = <u>min</u> (0.5, 0.2) then *P*(0.2)
IF {(θ = *P*) <u>and</u> (ω = *N*)}, THEN (*V* = *Z*) with μ = <u>min</u> (0.5, 0.8) then *Z*(0.5)

IF {($\theta = Z$) <u>and</u> ($\omega = Z$)}, THEN ($V = Z$) with $\mu = \underline{\min}$ (0.5, 0.2) then $Z(0.2)$
IF {($\theta = Z$) <u>and</u> ($\omega = N$)}, THEN ($V = N$) with $\mu = \underline{\min}$ (0.5, 0.8) then $N(0.5)$

Development of FAM table and fuzzy rules:

FAM table

θ \ ω	**P**	**Z**	**N**
P	PB	P	Z
Z	P	Z	N
N	Z	N	NB

Fuzzy rules

IF {θ is P} and {ω is P} THEN {V is PB}
.
.
.
.
IF {θ is N} and {ω is N} THEN {V is NB}

Students to fill up

This is the way of solving FLC problems – not the only way. There are several ways that include charting a table (similar to FAM) and solving for the result. The result, though known, never reaches in one iteration but may take several steps. In this problem we come to know that reducing θ to 0 is one of the answers. We can also try to solve a two-dimensional problem in which we are given an area in which the stick has to have its θ convergent to 0.

REVIEW QUESTIONS

1. Write a paragraph similar to the one indicated in INTRODUCTION of this chapter.
2. What is FLC? Does it belong to AI? What are the characteristics of FLC?
3. What are possibilities and probabilities? What is a possibility distribution function?
4. Which is suitable – 'A description of a fuzzy set' or 'Another description of fuzzy set'? Why?
5. State De Morgan's law.
6. What is a fuzzy relation? Where can this be useful?
7. Discuss four modules of FLC?
8. Draw a sketch of inverted pendulum? What principle does it teach?
9. Explain why θ and ω are to be simultaneously made zero?
10. What is FAM? Explain briefly

MEDICAL APPLICATIONS OF ROBOTS

15

What is is and what isn't ain't

Werner Erhard

The chapter covers the following topics:

Introduction – Classification of Medical Robots – Slow Growth – Cultural Gap – Technology – Rehabilitation Robots – Guide Robot – Guide Cane – Prosthetic Limb – Prosthetic Arms – Exoskeleton – Hospital Service Robot – Hospital Transport Robot – Hospital Pharmacy Robot – Cell Culture Robot – Clinical Robot – The PAM Robot – Robot for Hip Surgery – Tumour Extracting Robot

15.1 INTRODUCTION

Unconventional application of robotics, the medical robotics, designed for guiding humans are known as intelligent robotics. This is based on replacing artificial intelligence, the power of decision making by human intelligence. Several of them are finding applications in medical fields, thus these robots are named as *medical robots*. The concept of these robots culminated as early as 1921 when Karel Capek's play *Rossum's Universal Robots* went on stage in Prague. This stage play made him to develop the famous three laws of robots that are still being considered as a standard in designing robots.

Medical robotics is a newly developed concept with tremendous potential for research and commercialization. Medical robots are intended for assisting doctors, surgeons, nurses, patients and disabled in several ways. Their applications vary from artificial limb which is activated by bio-signals and which can perform with an appreciable level of intelligence to micro-surgery in the brain, a task that requires a very high level of human expertise. Some applications can reduce man power that is otherwise used in a more intelligent way.

15.2 CLASSIFICATION OF MEDICAL ROBOTS

In contrast to industrial robots, the medical robots are designed to interact with patients often in intimate procedures. Some of medical robots are to be in invasive applications on patients such as in brain surgery for tumour removal. Some of the others are to be powerful enough in non-invasive applications such as lifting and holding patients from bed as an assistance to nurses. Some are to repeat their functions for performing physiotherapy exercises for knee-operated patients. The application sphere of medical robots is unbounded and limited only to human intelligence.

A series of discussions between physicians, surgeons and engineers of various disciplines organized by Fulmer Research, UK, in 1988, as a part of study sponsored by UK Department of Trade and Industry resulted in identifying 400 highly demanded and feasible applications on medical robots ranging from intelligent artificial limb to micro-surgery in the brain.

Application of medical robots is classified into three groups having the following functional characteristics:

1. *Rehabilitation functions*: Characteristics exhibited by mechanisms for restoring artificial limb actions, restoring mobility to disabled such as exoskeleton, assisting visually impaired for their navigation such as intelligent wheel chairs.
2. *Hospital service functions*: Handling of materials such as reports, dispatching medicines from hospital pharmacy, delivering food to patients, handling and taking away laundry and waste.
3. *Clinical functions*: Involving direct interaction with patients such as diagnosis, therapy, surgery and patient handling.

Universities, hospitals and other research centres of various countries have designed a few medical robots in each of the above groups and tested in real time on patients in the hospitals and health care centres. Some of them are successfully performing their intended services with zero defect in selected organizations around the world.

15.3 SLOW GROWTH

Current applications of medical robots are only a few compared with necessity and demand. The advancement of medical robot technology is far behind that of industrial robot. Specifications of several types of medical robots are very intricate and stringent, especially in clinical functions than those of sophisticated counterparts in industries.

15.3.1 Cultural Gap

Several aspects contribute towards the slow growth of medical robot development and application. The foremost among them are the well-known cultural gap between engineers and medical experts. Medical experts have nothing to gain from conducting clinical trials with newly developed medical robot equipment but a lot to lose in terms of reputation if the trails fail. Engineers, on the other hand, have to perform rigorous tests, especially in real time, on the developed medical robot equipment for the approval of medical experts. Failure analysis and redesign are inevitable tasks of engineers. A strong partnership among engineers and medical experts in medical robot projects right from the start is, hence, vital, thus creating a mutual confidence in the development of any new medical robot system.

Another break in the development of medical robot is due to psychological aversion of patients. A patient expects that he/she should be handled by a human expert who is always sympathetic and caring. There can be no suggestion that the nurses and doctors can be replaced by robots. The robot is only an intelligence tool which assists these professionals to do their jobs more efficiently.

15.3.2 Technology

Requirement of sophisticated high technology plays an important role in obstructing the growth of medical robots. Human doctors and nurses around a surgical table perform highly coordinated tasks and are ready to react quickly to any abnormality that occurs in patient. This is a striking example of sensor integration which coordinates various senses of human being all at a time, and a remarkable illustration to real-time interactive feedback among various human experts. A medical robot which assists these experts in surgical process is expected to have these capabilities. Technology of today is yet to reach this level of sensor integration, real-time interactive feedback and precision. Functions such as handling a severely affected patient or removing a tissue from an intricate part of the body require very precise procedures effectively done by human experts. In addition, a medical robot, particularly in rehabilitation function, should have to be compassionate. Compassion does not exist without emotion. The concept of controlling a machine by human emotion is yet to be conceived.

15.4 REHABILITATION ROBOTS

Industrial robots can function only when a human operator is adapted to its capabilities. A sophisticated rehabilitation medical robot reverses this pattern. In general, these robots should be adaptable to the patients and they are to be as friendly as possible.

The research efforts on rehabilitation robots have been substantial and several robots of this kind are now successfully working around the world. This is due to the fact that rehabilitation robots have more in common with service robots and their design is simpler. The general design requirements of many varieties of rehabilitation robots in comparison with conventional industrial types are low cost, lower pay load, lower accuracy, larger work space, lower duty cycle and low speed.

15.5 GUIDE ROBOT

This is basically a four wheeled robot which takes the blind anywhere as planned. This robot has a long handle and a set of sensors at the robot end (Figure 15.1).

The blind has to just hold the handle in his/her hand. The sensors at the robot end are capable of identifying obstacles. It then gives an indication to the blind person so that he/she avoids this obstacle.

Figure 15.1 Guide Robot (*UniMAP UG PROJECT*)

15.6 GUIDE CANE

This invention received a great attention of many totally blind people who really wanted some way of finding obstacles in their way. The guide cane consists of long handle (Figure 15.2) the way it turns its wheel-based head and shows the correct way to the blind.

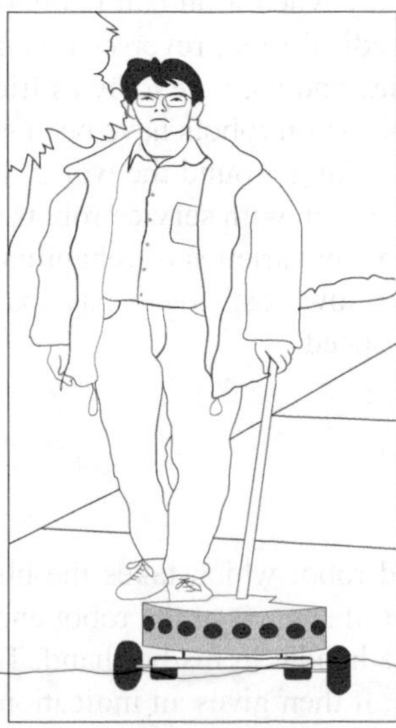

Figure 15.2 Guide Cane

This device has to be pushed through the long handle. Upon detection of any obstacle in its way the head turns the way in correct direction. The blind user recognizes a change in the direction of motion and follows the way shown by the handle. This guide cane received Invention Award from the Christopher Columbus Foundation.

15.7 PROSTHETIC LIMB

Amputees normally exhibit phantom limb phenomenon. They feel as though that the limb is still attached, which is due to the nerve being still intact between brain and amputee's limb. This phenomenon is exploited in developing prosthetic arms and legs.

15.8 PROSTHETIC ARMS

A couple of prosthetic arms were demonstrated in the University of California, USA. The arms were attached with electronic control device (Figure 15.3). The brain sends messages to the lost arms and these bio-impulse signals simulate an in-built micro-computer circuit which activates the prosthesis. The amputees can thus rotate their arms to perform the desired tasks such as grasping, rotating and lifting. An Oklahoma company manufactured these arms. A temperature sensor in the finger tips sends the data to an electrode touching the nerve ending of patient's residual arm to forward the information to the brain. This allows an amputee to feel the difference between hot and cold.

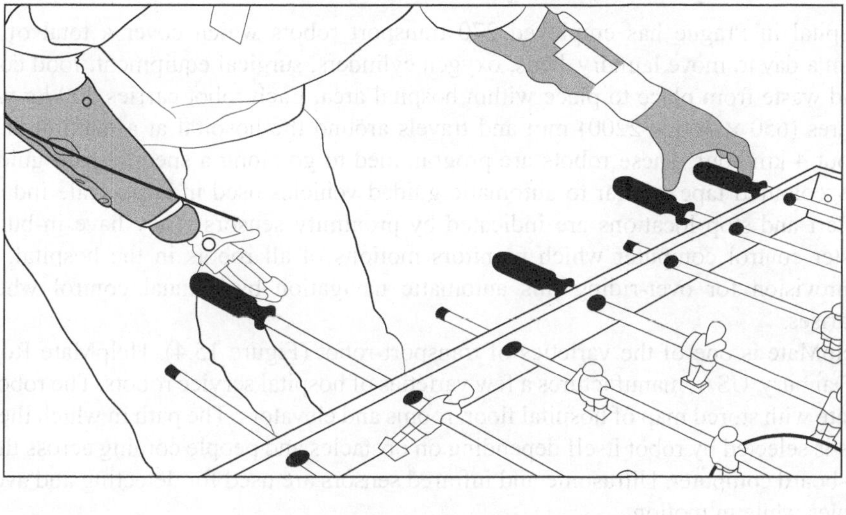

Figure 15.3 Prosthetic Arms

15.9 EXOSKELETON

Limb paralyses are very common in road accidents. One of several approaches is aimed at developing a robotic or otherwise exoskeleton training. At the Cleveland Veterans Medical Centre, Ohio, at least 13 people were using an experimental system which enabled them to walk, step sideways and even climb stairs. Figure 15.3 illustrates an exoskeleton for limb amputee.

AID 1, a rehabilitation robot, developed by a Japanese company is used for gait training. This robot has a provision to move 360° horizontally and 30° vertically up and down. The arm holds the trunk of a patient and reduces patient's weight, thus creating an ability of walking with a small amount of muscle power. More than 800 cases have been gait trained for the last 12 years which enhanced walking ability for over 80% of these cases.

The MULOS project funded by European Union Telematics Program involves a 5-degree of freedom (DoF) of an articulated robot for upper limb motion. Higher DoF offers higher flexibility in robot arm articulation.

15.10 HOSPITAL SERVICE ROBOT

Handling and transporting laundry, reports, pathology specimen, food and pharmaceuticals are the most important functions of hospital service robots. Even an average capacity hospital does require these services to reduce expensive man power. These service robots are required to undertake the handling and transporting of these materials in selected time schedules, from one location in the hospital to another location under programmed control directives.

15.10.1 Hospital Transport Robot

A hospital in Prague has employed 270 transport robots which cover a total of about 560 km a day to move laundry, beds, oxygen cylinders, surgical equipment, food containers and waste from place to place within hospital area. Each robot carries 500 kg weight, measures (650 × 360 × 2200) mm and travels around the hospital at a maximum speed of about 4 km/hour. These robots are programmed to go along a specific path guided by yellow coloured tape similar to automatic guided vehicles used in large-scale industries. The start and stop locations are indicated by proximity sensors. They have in-built and a master control computer which monitors motions of all robots in the hospital. There is a provision for over-riding this automatic navigation by manual control whenever need arises.

HelpMate is one of the varieties of transport robot (Figure 15.4). HelpMate Robotics Inc., Danbury, USA, manufactures a few varieties of hospital service robots. The robots can navigate with stored map of hospital floor, rooms and elevators. The path in which the robot moves is selected by robot itself depending on obstacles and people coming across through its on-board computer. Ultrasonic and infrared sensors are used for detecting and avoiding obstacles while in motion.

One variety of HelpMate being used in Danbury University Hospital is priced at US$ 60,000, weighing 400 lb, is less than 5 feet tall and moves with an average speed of 3 times slower than a man. Another variety can be rented for US$ 1.00–5.00 per working hour. Mercedes, Rosie and Hilda are some varieties of the hospital service robot type from HelpMate used in more than 100 hospitals in the US and about 50 hospitals in the UK. They transport food and drugs to an average of 800–1000 patients per day, thereby replacing about 30% service staff.

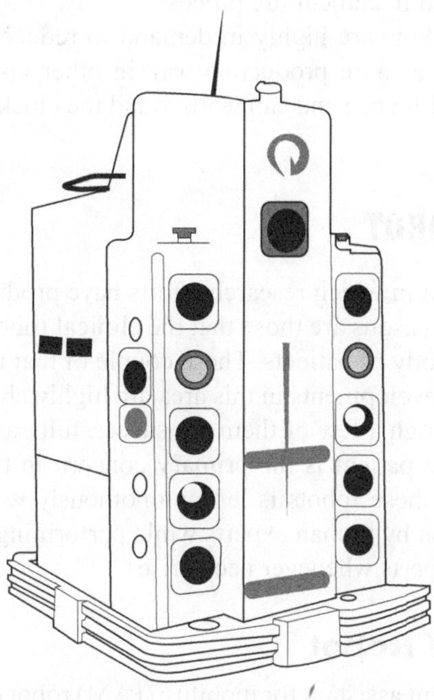

Figure 15.4 The HelpMate

15.10.2 Hospital Pharmacy Robot

Replacing hospital pharmacist with a service robot is another special application of medical robots. Robo-Rx is a 5-DoF, 3-axis gantry type robot used in Laxington's Baptist Hospital.

These robots place the drugs in an orderly manner. In addition, following a medical prescription entered in a computer as a coded data, the robot goes to the correct location where the required medicines are stored, identifies that particular medicine, checks the expiry date through barcode on the container, retrieves and delivers it to a tray. The robot thus delivers medicines to 7000 patients over 24 hours a day, or takes more than 2.5 million orders in a year. It replaces at least two pharmacists and four assistants per shift in the duty of three shifts per day. P2-D2 and 5/20 are other pharmacy robots. Over 150 US hospital pharmacies have employed such a family of hospital service robots of one kind or the other.

15.10.3 Cell Culture Robot

Cellmate, a British designed hospital service robot for processing living cells cultured in sterile vessels was one among seven winners of British Design Awards for developing biotechnology equipment. This robotic machine, designed by Technology Partnership Ltd., Melbourne, imitates the processing skill of a laboratory technician. *Cellmate* is used for the manufacture of viral vaccines and *therapeutic* hormones and for the biological process development.

This robot is regarded as the first automated system in the world that is capable of undertaking all the steps involved in cell culture process.

The hospital service robots are highly in demand to reduce human labour which can be trained and utilized in a more productive way in other tasks. The service robots of today have reasonable intelligence and can work round the clock with acceptable accuracy, repeatability and cleanliness.

15.11 CLINICAL ROBOT

Clinical robotics is the area in which research efforts have produced only a few successful medical robots. The main reasons are those that the clinical robots are not invasive and not directly interact with the body of patients. These couple of that the clinical robots are used in surgery. Research and developments in this area are highly challenging, the outcome are not that excellent even though a few of them are successfully assisting surgeons in operation theatres. Safety of the patient is the primary concern in the applications of clinical robots and hence none of these robots is left autonomously working. Clinical robots are under the direct supervision by human experts while performing their tasks and their tasks will be modified by the experts whenever need arises.

15.11.1 The PAM Robot

Figure 15.5 shows the patient assistant for mobility (PAM) robot designed by British robotic solution for handling the bedridden patients. This is the nurse assistance robot which avoids nurses lifting patients physically, thereby preventing their back injuries. PAM is based on the following three subsystems:

1. A patient lifting mechanism to lift and hold the patient in a comfortable posture
2. An intelligent mobile base to transport the patient from one room to another around the hospital
3. A human computer interface to ensure gentle handling of patients

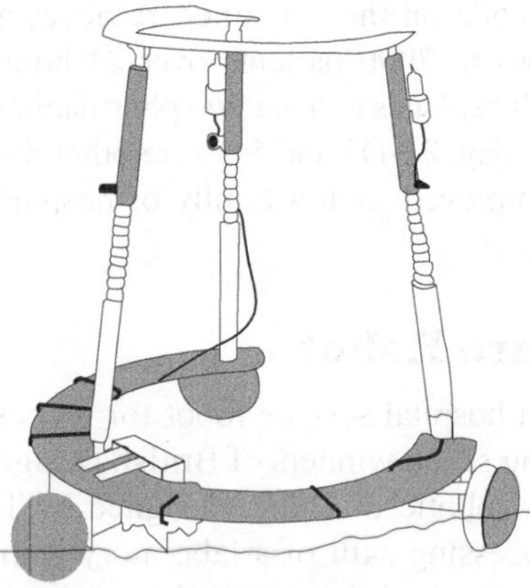

Figure 15.5 A Patient Assistant for Mobility

The computer and the servomechanism maintain stability. Several sensors are mounted on the outer surface of the PAM. The sensor feedback ensures avoidance of obstacles on its path and guarantees a collision-free motion. A nurse accompanies PAM and monitors the robot as it moves with the patient. The PAM is designed in such a way that the nurse can exert a different computer control to apply only human control whenever need arises or in emergency. Also, the presence of nurse with the PAM makes the patient comfortable and confident as the nurse is always with the patient and does not leave the patient under the complete care of a machine.

15.11.2 Robot For Hip Surgery

Hip replacement is a blessed application of clinical robots for people who have involved in accidents and damaged hip joints. A total hip replacement surgery involves implanting a metal prosthesis in the patient's thigh bone (femur; Figure 15.6). Each year about 25,000 Americans undergo this arthoplasty surgery, a surgery to replace a damaged or diseased hip.

In this surgical process, the surgeons first create a cavity in the femur to accommodate the prosthesis. An accuracy of close fit is crucial to implant a cementless fitting. Rough surface and gaps inside the cavity are detrimental to patients. Cementless implant systems rely on 'natural' fixation from bone growth into the metal surface of the implant. Clinical research has shown that for the in-growth to occur, the cavity of femur must be very appropriate such that any gap between bone and porous surface less than 0.25 μm (half the thickness of hair) is detrimental to patient.

In traditional method, the surgeons use several instruments such as reamers and chisels to create a cavity. Even in the case of most experienced surgeons and instruments of

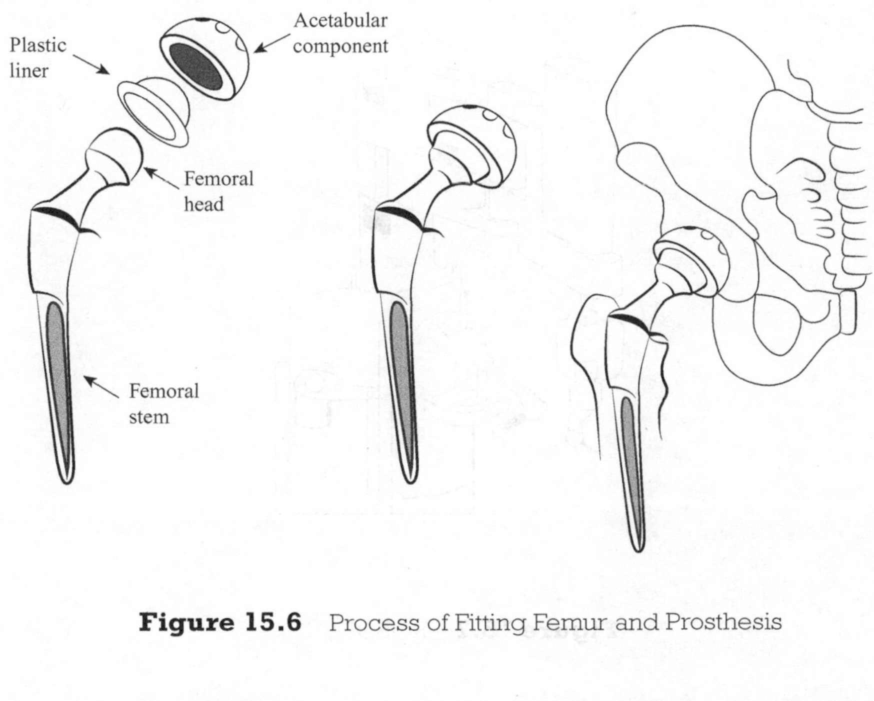

Figure 15.6 Process of Fitting Femur and Prosthesis

high level precision, the method of preparation of bone cavity has been much less precise. Imprecision can always result in improper in-growth and early loosening of implant.

ROBODOC, a surgical assistance system (Figure 15.7) is a solution to precisely create the cavity for the implant. The ROBODOC uses computer tomographic imaging (CT scan) to provide surgeons with a three-dimensional view of femur upon which the surgeons can precisely select the implant size and placement prior to surgery. The work plan creates data on a software, ORTHODOC, which prepares a set of robot cutting instructions. These instructions are transferred to robot in the surgical room. The femur is attached to the robot base and its tool, a high-speed cutting burr, is precisely guided by robot when it follows the cutting instructions prepared by ORTHODOC. After the preparation of cavity, the surgery is proceeded in the usual way. By 2013, over 1500 patients worldwide underwent total hip replacement with ROBODOC system. ROBODOC is currently in clinical use in several parts of Europe.

HipNav is another software which guides surgeons in total hip replacement, developed recently by Carnegie Mellon University, Pittsburgh, in collaboration with Harvard Medical School, Boston and Shadyside Hospital, Pittsburgh. This includes following three components:

1. A preoperative planner
2. A range of motion simulator
3. An intra-operative tracking and guidance system

This system allows a surgeon to optimally determine and accurately achieve the desired implant placement during surgery.

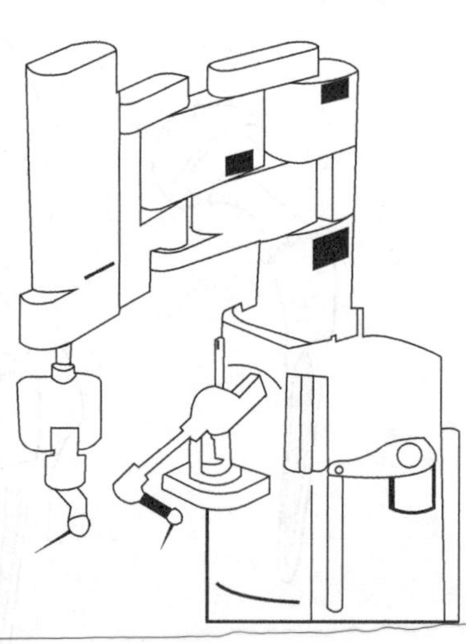

Figure 15.7 ROBODOC

15.11.3 Tumour Extracting Robot

Stereotactic (or image guided) surgery, particularly in brain, is a technique for guiding a probe or other instrument to a predetermined point in the nervous system without directly viewing the surgical sight but through images. Images provided by X-ray, CT or magnetic resonance imaging stand as reference which are correlated against some physical land mark before performing the surgery. Free hand and instrument-based stereotactic surgeries have been used for almost 80 years. Recent robotic application in brain surgery enjoys a significant advantage over instrument surgery in terms of repeatability in performing a number of surgeries in a short time duration. A PUMA 260 Robot arm system manufactured by Unimation Inc., USA has demonstrated extraction of brain tissue for biopsy.

This process of robotic extraction of tumour starts with obtaining CT images of skull. The surgeon identifies the location of tumour and enters the data in a robot computer. The robot selects the point of entry on the skull. Once this has been examined and unanimously approved by surgeons, the robot creates a hole in the skull, 2–3 mm across. The robot then drills through the skull and steers its surgical instrument through the brain towards the tumour at a controlled speed as decided by the surgeons. The surgeons and other staff in the operation theatre closely monitor the robot motion through X-ray or other visual means. Every provision is available in the process to modify the tool trajectory and speed whenever the robot is about to encounter with a no-go regions such as blood vessels or critical and restricted areas of brain. Such robotic surgery has been successfully performed on several patients, in more than 50 hospitals in the US.

REVIEW QUESTIONS

1. What are the unconventional applications of robots? Briefly explain them.
2. What are the applications of robotics in hospitals and in health care centers?
3. Explain the word ‘cultural gap’ in the development of medical robotics.
4. Explain the word ‘compassion’ in medical robotics.
5. What are ‘eight design requirements’ of medical robotics? However the design requirements are listed as six in the book.
6. What is the main difference between guide robot and guide cane ?
7. What are the main robotic applications in (a) hospital transport, (b) hospital pharmacy, (c) cell culture and (d) hip surgery ?
8. What are the main difficulties in implementing tumour extraction?
9. What is the PAM robot? Briefly explain three subsystems of PAM.
10. Do you think implementing medical robots in hospitals will improve the facilities of hospitals? Explain your understanding.

HELPING THE VISUALLY IMPAIRED FOR THEIR AUTONOMOUS NAVIGATION

16

The important thing is to learn a lesson every time you lose

John McEnroe

The chapter covers the following topics:

Introduction – The Blind – The Blind Population – Blind Navigational Aids – Classification of ETA – Electronic Sensor-Based Navigational Aids – Dobelle Eye – NAVI – NAVI-1 and NAVI-2 – Image to Sound Conversion – SVETA – Computation of Object Distance – SVETA-1 – Stereo Image to Stereo Sound Conversion – Use of Musical Board – SVETA-2 – Summary.

16.1 INTRODUCTION

The requirements of blind people of today are vast. They wish to enjoy all aspects of life in spite of the fact that they are blind. Moving through obstacles, stationary or dynamic, collision free, without human assistance is their most important requirement. This is known as autonomous navigation. Other requirements associated with autonomous navigation are identifying objects, knowing the object size, computing object distances and recognizing the directions of its movements. This research is about designing and developing a computer-based equipment for autonomous navigation of blind individuals.

The main technique used in this work is of vision substitution for the blind using a set of digital video cameras. The images captured by the cameras are processed by a portable processing equipment to produce specialized stereo sounds. The sounds provide information on understanding the presence of the object nearby, its size and its distance so that the blind can either avoid the object and comfortably walk collision free or reach the object, touch and grasp it. The image processing techniques and image-to-sound conversion procedures are briefly described below.

16.2 THE BLIND

God has bestowed us vision, the most beneficial among the five senses. Human navigation in an obstacle cluttered environment becomes easier with a reasonably accurate vision. Human navigation consists of two distinct components:

1. Sensing of the immediate environment for a collision-free travel
2. Navigating to remote destinations beyond the immediately perceptible environment.

The first requires information on the presence of obstacles and hazards around the human. This component is inevitable for the process of collision-free navigation. The second component is the way of learning the directions at every instant to reach the destination. Much of the information that humans get from the outside world is obtained through sight. With defects in eyesight, visually impaired people have many difficulties in their daily life. A total loss of eyesight is one of the most serious misfortunes that can befall a person.

Persons with severe visual impairment experience significant problems in achieving independent mobility since the appropriate information about the environment is not available to them. The term blindness or visual impairment refers to the malfunctioning of eye or absence of sight. World Health Organization (WHO), in 1998, has defined visually impaired as people with no sight, or those, whose vision cannot be corrected or treated. Most visual impairment is caused by diseases and malnutrition. The most common causes of blindness are cataracts, glaucoma, age-related macular degeneration (AMD), trachoma, corneal opacity and diabetic retinopathy.

The demands of independent living of blind individuals have motivated researchers to develop assistive devices that can aid the blind in navigation. Electronic travel aids (ETAs) are devices that aim at conveying information on the immediate physical environment to visually impaired individuals so that they can exploit a part of the information that sighted people normally use to experience the world and navigate. The present day requirement of a blind individual (herein after referred as the blind) is to perceive and enjoy the world around, just as a sighted individual would. There are practically no effective travel aids currently available which would enable a visually impaired person to perform tasks of sighted people. This chapter has been prepared mainly on the development of certain ETAs that can detect and identify obstacles in terms of shape, location and distance from and on the blind to navigate collision free.

16.3 THE BLIND POPULATION

The WHO has reported that due to lack of epidemiological data, especially from the developing and under developed countries, the exact number of blind persons in the world is not known. Globally, in 2002, more than 161 million people were visually impaired, of those 37 million were totally blind and the rest had low vision. The totally blind population is estimated to increase to 60 million by 2010 and 75 million by 2020. The WHO also estimates that about 80% of global blindness could have been prevented, controlled or successfully treated and sight restored if the available knowledge and interventions had been applied in a timely manner.

Visual impairment is distributed across age groups. More than 82% of all people who are blind are 50 years or older, although they represent only 19% of the world's population. Childhood blindness remains as a major problem. According to the WHO, in 2003, an estimated 1.4 million blind children were below the age of 15. In every region of the world, and at all ages, females have a significantly higher risk of being visually impaired than males. Except for the most developed countries, cataract remains the leading cause of blindness in all regions of the world; 47% of the total blind are due to cataract. Associated with ageing, it is even more significant as a cause of low vision. Glaucoma is the second leading cause of blindness on the global scale (12.3% of total blind), with AMD (8.7% of total blind) ranking third. However, in developed countries, AMD is the leading cause of blindness, due to the growing number of people over 70 years of age.

More than 90% of the world's blind population live in developing countries. In the US, approximately 4.3 million were visually impaired and of them 82% were over 55 years old. In the UK, nearly one million people were entitled to register as visually impaired persons, in addition to 1.7 million with vision difficulty. This represents over 3% of the UK population. It is also estimated that 32% of the world's blind live in the countries of the South East Asia region. About 25% of the world's blind population is in the Western Pacific region.

In 2002, ORBIS indicated that a survey from the Malaysian Ministry of Health confirmed that there were 60,000 totally blind adults and 450,000 adults with low vision; 5000 totally blind children and 50,000 children with low vision. As per JKM, Malaysia, an alarming increment of 46.9% of visual impairment was noted in Malaysia from 1990 and 1999. These data are officially declared by the Ministry of Health; it is also estimated that almost 75% are still to register with the Ministry.

16.4 BLIND NAVIGATIONAL AIDS

There are several worldwide efforts, effectively from mid-1970s, on the development of ETAs for vision substitution. They range from improvements to the walking cane to more advanced computer-based aids.

The two commonly used aids from early days for blind navigation have been the walking cane and the guide dog. The walking cane is just a mechanical non-intelligent device that relies on the person's skills to interpret the acoustical reflections which result from continuous tapping on the ground. The cane's range is only a few feet and is limited by its length and the person's hand reach extent. Canes neither provide the spatial information on the objects that are away from the blind nor on the environment changes such as moving of objects. The guide dog, on the other hand, alleviates some of the limitations of the cane, but only limited information regarding orientation and dynamics of objects farther away is conveyed to the blind traveller. In addition, dogs require constant care. An extensive training is required for the dog and also for the blind user for a successful collision-free navigation.

The era following World War II stimulated the developments and applications of sensors such as radar, sonar and other electronic technologies. The advent of microelectronics and semiconductor devices has led to the development of systems that were small enough for a pedestrian to carry. By the 1960s and 1970s, a plethora of devices intended to assist blind pedestrians were developed at least to the prototype stage. The early ETA technology has employed ultrasonic sensors to detect the obstacles in their path and the information is provided in terms of auditory and tactile vibratory signals to the blind. Ultrasonic sensors are susceptible to stray echo signals which corrupt the environmental information offered to the blind. Subsequent developments in high-speed computers, microelectronics and varieties of sensors have created the emergence of sophisticated ETAs.

16.5 CLASSIFICATION OF ETA

During the past three decades, several researchers have developed varieties of ETAs. ETAs are varied in terms of size, type of sensors used and also on the feedback signal. However, the main objective of these devices is unique, that is, enhancing the mobility of blind people by providing them the spatial information about the environment. ETAs can be generally classified based on the sensor devices used for scanning the environment. They are as follows:

(a) Electronic sensor-based navigational aids
(b) Single camera-based navigational aids
(c) Stereo camera-based navigational aids

We shall briefly discuss the ETAs classified in (a) and classifications (b) and (c) are the main aspects of this chapter.

16.6 ELECTRONIC SENSOR-BASED NAVIGATIONAL AIDS

The early ETAs have used four types of sensors, namely ultrasonic sensors, sonar sensors, laser sensors and infrared sensors for environment scanning and obstacle detection. In ultrasonic-based ETAs, ultrasonic waves are transmitted and the reflections are decoded to sense the presence of any object and sometimes the distance range of the object in the travel path. The information is provided in terms of vibration or sound.

Some of the noteworthy ETAs using the ultrasonic, sonar, laser and infrared sensors are presented as follows:

The guide cane is a cane attached to a small mobile robot platform with an array of ultrasonic sensors, an odometer, a compass and gyroscope sensors (Figure 16.1). The mobile robot steers around obstacles if detected by the sensors. The blind lightly pushes the device to roll and feels the steering command; the mobile robot senses obstacles present on the path and detours the obstacle simultaneously giving the blind steering information. The blind then follows the device's path without any conscious effort. Unfortunately, the robot is a wheeled platform and therefore restricted to travel along relatively flat surfaces.

An improvement over the guide cane is the Polaron as shown in Figure 16.2. It is a chest-mounted sonic obstacle finder. When an obstacle is in front of the blind, the Polaron gives a vibratory or sound signal. The vibration or the sound intensity increases as the blind gets closer to the obstacle. The detection range is selectable as 1.2, 2.4 and 4.8 m. It is very efficient in wall following, and detecting and going through doors.

The NavBelt is designed by the University of Michigan. This system has an array of ultrasonic sensors that are mounted on a belt around the abdomen of the user. These sensors provide sound information or a tactile signal to the user when he goes near the obstacle.

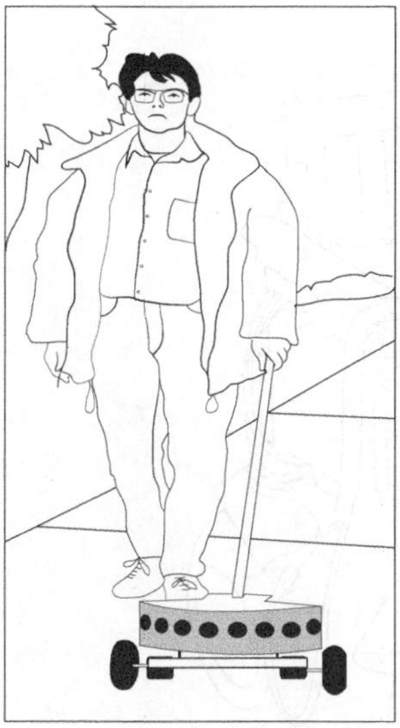

Figure 16.1 Guide Cane

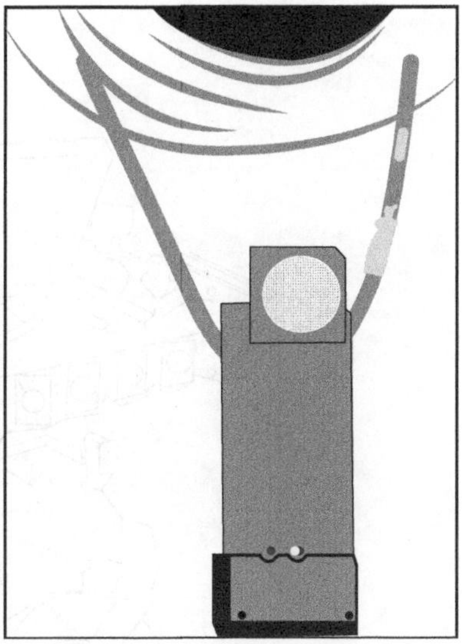

Figure 16.2 Polaron

Figure 16.3 displays a person wearing the NavBelt system. The user has to undergo an intense training since the acoustic signal or tactile vibration used in NavBelt is too complex for the blind to comprehend, even at normal walking speed. In addition, the NavBelt is too heavy and complex to wear.

The sonic guide, as shown in Figure 16.4, consists of an ultrasonic transmitter and receiver mounted on spectacle lenses. Signals reflected from the environment are received and processed so that the user gets an audio signal. Thus, the blind user gets information on the presence of the obstacle and its approximate distance from the user.

Figure 16.3 NavBelt

Figure 16.4 Sonic Guide

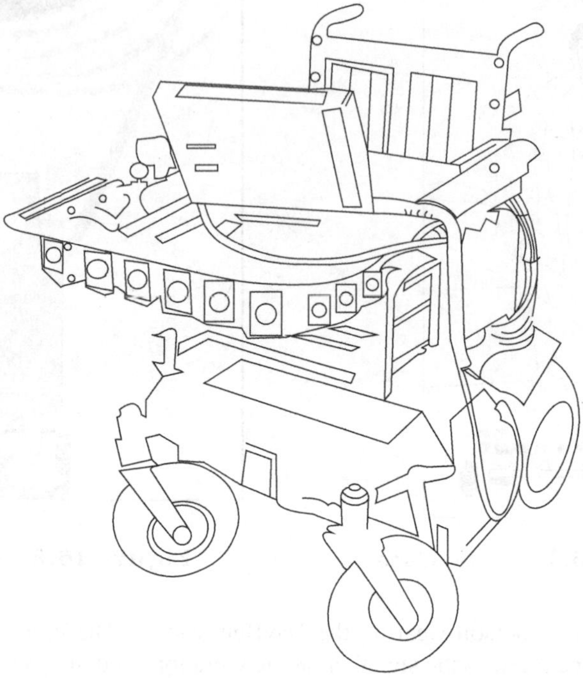

Figure 16.5 NavChair

The NavChair, as shown in Figure 16.5, has a number of ultrasonic sensors to detect obstacles in front. It has other facilities such as a pneumatic bumper to take up any accidental collision and a set of infrared sensors so that the chair can follow the reflective stickers pasted on the floor. This feature enables the wheelchair to go through doorways and other places within a building. The main disadvantage is that it is too expensive due to the cost of wheelchair and its servomechanism.

Several other ETAs were designed by varied research groups around the world at different times. The Sonic Torch, the Russell Pathsounder, the Mowat Sensor and the Nottingham Obstacle Detector are some of the developments. All of these are based mainly on ultrasonic sensors even though a few other sensors are also used to avoid collision with nearby objects. These ETAs are usually heavy and huge to wear and require longer battery life and complex signal processing efforts. During the 1990s, when high-speed digital video cameras became a reality in terms of precisely describing the environment, the design of ETA became entirely dependent on cameras.

The vOICe system, as shown in Figure 16.6, is a significant development in camera-based ETAs. It has a single video camera and a set of earphones embedded in a headgear. The image captured by the camera is converted to a sequence of sounds that are sent to the set of earphones. The blind can recognize the presence of obstacles in front of them by hearing the sounds through the earphones. There is no signal processing effort embedded in to the vOICe system for eliminating the background from the object images. The absence of a powerful software to eliminate insignificant background portion of the image from the

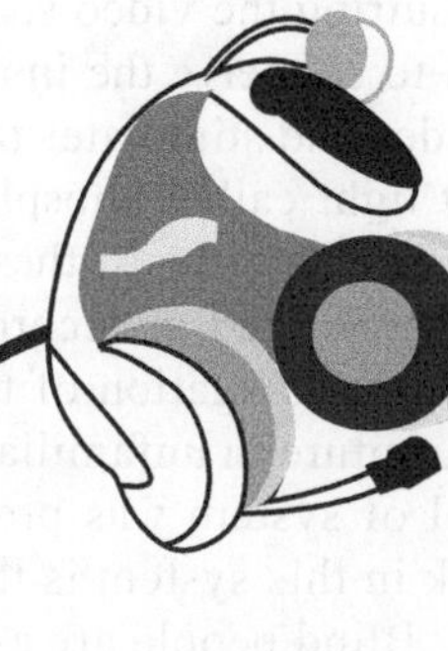

Figure 16.6 The vOICe

object image may be the reason for the blind user having problems in distinguishing the object from the background through sounds. Another setback is the inability of this system to detect the distances of objects from the user.

16.7 DOBELLE EYE

An interesting effort in developing eye-substitution systems is through invasion into the human body. The only example is the Dobelle Eye tested in 2000. The Dobelle Eye, as shown in Figure 16.7, is made up of a sub-miniature television camera for capturing the image of the environment and an ultrasonic sensor for distance determination. They are mounted on a spectacle-shaped wear. A grid of electrodes is strategically

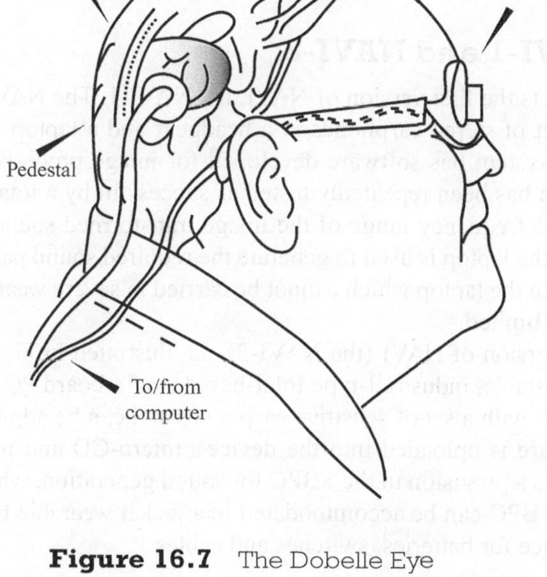

Figure 16.7 The Dobelle Eye

implanted on the surface of the brain's visual cortex and these electrodes are controlled by a wearable computer. After acquiring the video signals through the television camera, the computer processes them to conceive the image. The computer then sends an electrical impulse to these electrodes and stimulates the visual cortex of the brain. The result is the sensation of a bright light called phosphene in the blind eyes which are normally blackened visual fields. The patterns of these phosphenes can convey visual information to the user. They are generated in accordance to what the head-mounted camera 'sees'. These patterns help in navigation of the blind among the surrounding objects such as chairs and other furniture in unfamiliar indoor territories. The research was started in 1968, and a model of system was produced in 2000. This is a highly invasive method and the main risk in this system is the chance of infection due to the implant of electrodes in the brain. Blind people are as yet reluctant to accept this type of navigational aid.

16.8 NAVI

The NAVI (Navigational Assistance to Visually Impaired) has been a research project performed by the author and his associates since 2001. It is based on a process of transforming vision information to a set of stereo sound patterns.

The research on NAVI has two phases – one on image processing towards object recognition and the other on converting the recognized object images to carefully structured sounds. There are two versions of NAVI, NAVI-1 and NAVI-2. The NAVI-1 has a single digital video camera and a set of earphones mounted on a specially designed headgear. The camera continuously acquires the images. The images are sampled at specified time instants and processed to achieve a mono-vision-to-stereo-sound transformation. NAVI-2, the second version, has an improved headgear to suit any blind, with faster and more powerful software for image processing and sound production. The frequency of sound has to be carefully decided to match the blind. There is a provision in the NAVI-2 systems to tune to a frequency that matches the hearing response of the blind person.

16.8.1 NAVI-1 and NAVI-2

Figure 16.8 depicts the first version of NAVI, the NAVI-1. The NAVI-1 has a digital video camera with a set of stereo earphones in a headgear and a laptop – all interconnected to each other. The system has software developed for image processing and sound generation. This version has been repeatedly tested as successful by a totally blind man who has recommended the frequency range of the image transformed sound patterns. The facility of sound card in the laptop is used to generate the required sound patterns. NAVI-1 is obviously heavy due to the laptop which cannot be carried in severe weather conditions; the life of battery is also limited.

The second version of NAVI (the NAVI-2), as illustrated in Figure 16.9, uses a small, rugged, easily portable, industrial-type Intel-based single-board PC (SBPC). The headgear is also redesigned with a set of sensitive earphones and can be adjusted to suit a particular head. The software is uploaded into the device's micro-CD and is interconnected to the headgear. There is a provision in the SBPC for sound generation, which activates the stereo earphones. The SBPC can be accommodated in a jacket wearable by the blind. The jacket also provides space for batteries, switches and cables.

Figure 16.8 A Blind with NAVI-1

16.8.2 Image to Sound Conversion

The camera in the headgear repeatedly acquires images from the environment in front of the blind. A specially developed set of software processes the image data towards enhancing the object in the image and removing all unwanted background information.

The enhanced object image is divided into two halves so that the sound generated in the right half of the image can be heard by the right ear and that in the left half of the image by the left ear. Each image half has a number of pixels (picture elements) and each pixel has its position in the image area. Depending on its position, a sound with a specific frequency and amplitude is produced by the computer and sent to the ears. The computer scans the processed image from right to the centre and simultaneously from left to the centre. Each column of image produces a specific sound at both ears. This results in a stereo sound, which the blind person can hear. Figure 16.10 illustrates the way of stereo sound generation in a (8 × 8) pixel-based image.

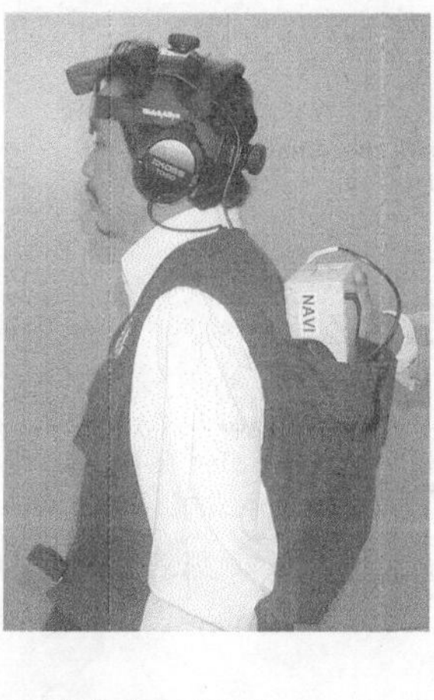

Figure 16.9 A Blind with NAVI-2

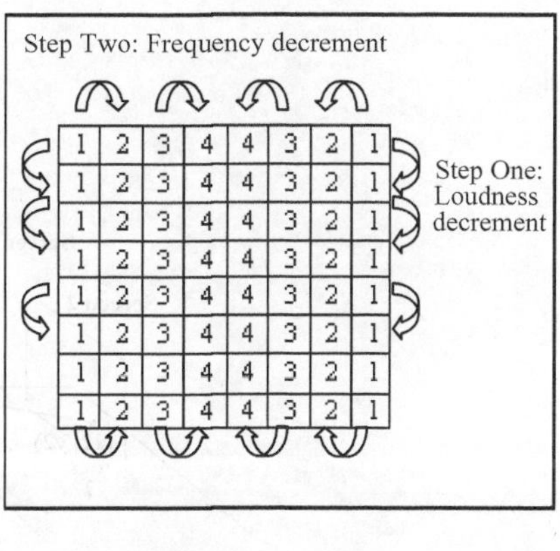

Figure 16.10 Image to Sound Transformation

A well-designed training scheme has been constructed so that the blind undergoes a systematic in-house training before he is allowed to go on his own way inside or outside the rooms. It has been found that the blind can easily understand the sound patterns, identify the objects nearby, and even detect slowly moving objects approaching him or objects that he is approaching. However, the distance between the blind and the object (the *z*-axis distance or depth) is a required aspect in blind navigation and this is not considered in this chapter. There are proposals for determining the distance with single-camera images. However, the distance computed by a single-camera system does not have the expected accuracy.

16.9 SVETA

A natural extension of NAVI is the stereo vision-based ETA (SVETA). SVETA substitutes both the eyes of the blind through a set of two digital video cameras. Our eyes not only offer the vision to identify the objects far and near but also provide an estimation of distance of objects from us. We are accustomed with one eye even for estimating the distances of objects from us. If the object is far away from us, the estimation of distance using one eye vision can be erroneous.

This can be illustrated by a simple exercise of repeatedly touching two needle points held in each hand fully extended and with one eye closed. We tend to miss at least a few times.

Stereo vision is provided by two cameras that are kept side by side at a constant distance. Any object in front of the cameras produces two images on the screen of each camera. The images will not be identical in their positions with respect to a vertical reference centre line in each of the screens. The images are off from the identical reference lines. This off-set is predominant if the object is closer to the cameras. This off-set is known as the disparity. This disparity is inversely proportional to the distance of the object.

16.9.1 Computation of Object Distance

Distance computation using stereo vision is illustrated in Figure 16.11. The two cameras have identical focal lengths (λ cm). The camera lenses are placed side by side with a

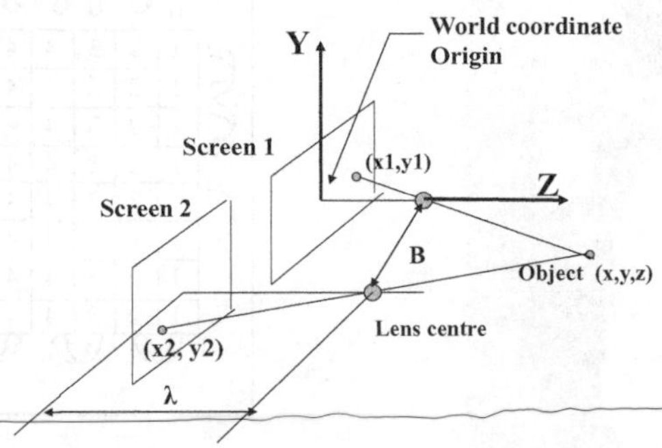

Figure 16.11 A Stereo Camera Experiment

distance of B cm between them. There is a world coordinate frame (X, Y, Z) in which the object is placed at a point D cm away from the lenses. Two images of object are captured by the cameras, at $(x_1, y_1, 0)$ in one screen and at $(x_2, y_2, 0)$ in the other screen, as represented in the world coordinate system. The disparity is computed as $(x_2 - x_1)$. Then, the distance D cm is then represented by a simple relation as

$$D = \lambda - \frac{\lambda B}{x_2 - x_1}. \qquad (16.1)$$

The relation between distance and disparity is shown in Figure 16.12.

16.9.2 SVETA-1

The version 1 of SVETA (SVETA-1) has a hard plastic headgear that houses the set of stereo cameras and stereo earphones. Figures 16.13 and 16.14 show the SVETA system and a blind wearing the headgear. The images captured from the cameras are processed in a computer (not shown in the figures) to produce the stereo sounds. The stereo sound not only gives the information on the size of the object but also indicates the distance of object from the blind through a verbal sound. If the object is at one side (left or right side) of the blind, he will be able to turn his head towards the object after hearing the stereo sound, thereby getting the exact distance of object. It was possible to attain an accuracy of 95% in the estimation of distance through stereo vision. This level of accuracy offers the ability to the blind to go confidently near an object and even touch and grasp it.

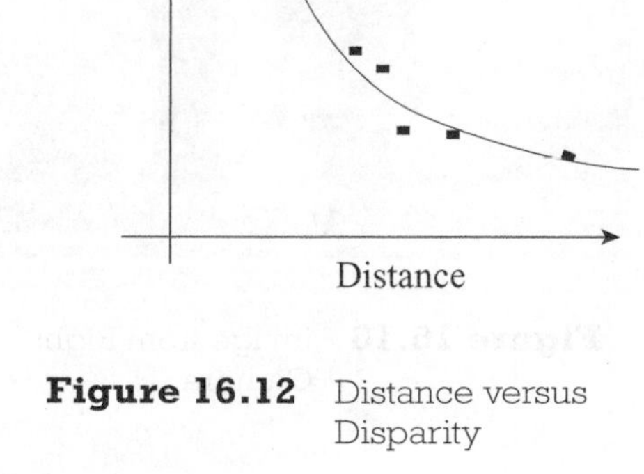

Figure 16.12 Distance versus Disparity

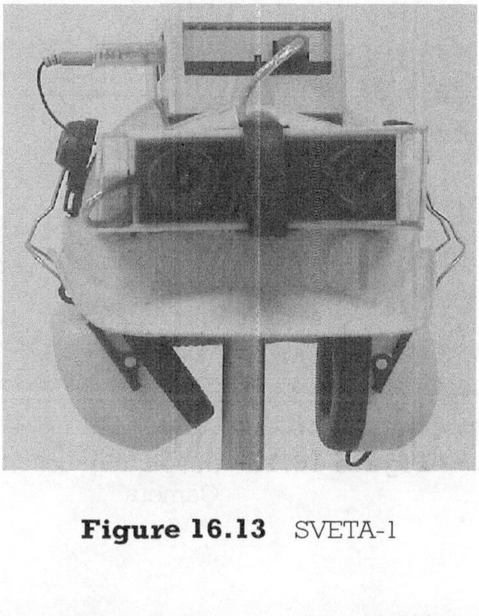

Figure 16.13 SVETA-1

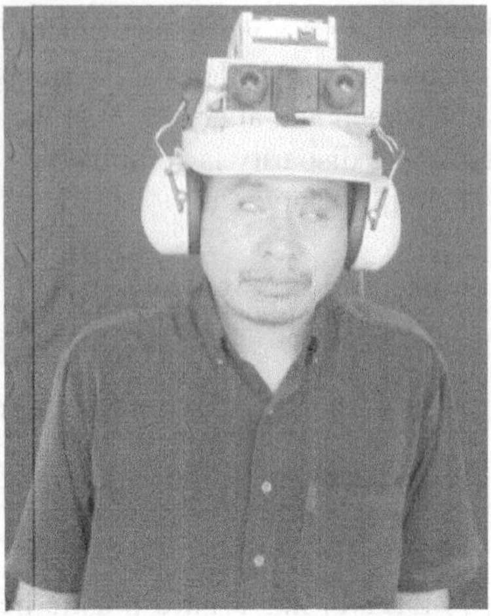

Figure 16.14 A Blind Wearing SVETA-1

16.9.3 Stereo Image to Stereo Sound Conversion

The stereo image to stereo sound processing is similar to the one discussed for NAVI except that the distance information is an additional effort in SVETA. The distance information can be verbal or a specially designed modulated sound transform. The distance in terms of verbal can be, for example, '70 cm'. A total blind cannot conceive distance in terms of a measurement as 'cm'. However, the distance can be perceived in terms of his arm length. This requires brief training.

In NAVI, a process of modulating the sound with frequency and amplitude has been incorporated as depicted in Figure 16.10, but in SVETA, an additional modulation is incorporated to indicate the distance of objects. This requires a complex functional operation using artificial intelligence components such as fuzzy logic and neural network. However, a simple example is given below to illustrate this aspect of modulation:

Figure 16.15 Image from Left Camera

Figure 16.16 Image from Right Camera

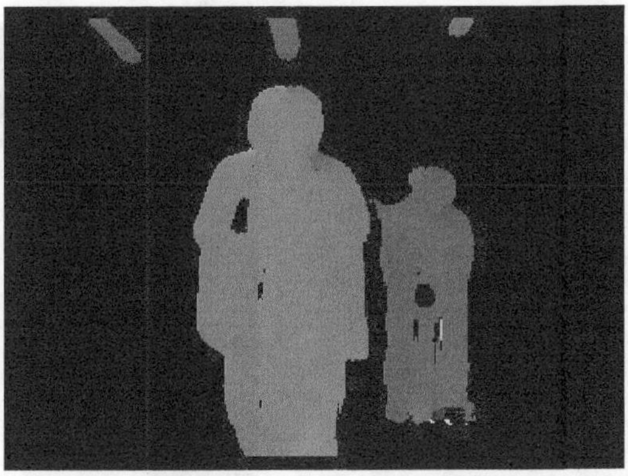

Figure 16.17 The Disparity Image

Figures 16.15 and 16.16 show images acquired by the left and right side cameras, respectively. If closely observed, the images of the first person (nearer to the camera) are not identical in their position with respect to those of second person; a disparity and hence a distance of the first person from the camera can be computed using Equation (16.1). Both images are processed in a specially designed software to bring out a disparity image (Figure 16.17). This disparity image has information on relative distances of objects in such a way that the object farther away from the blind is darker than the other object. The relative distances can be brought in the sound as intensity variations. The intensity of sound can be made higher when the object is nearer to the blind.

16.9.4 The Use of Musical Board

A blind person, when using the NAVI or SVETA, has to hear the sound repeatedly to conceive the presence, the size and the distance of objects as he navigates. Conceiving these is through the selection of a specific set of frequencies. The sound has to be repeatedly heard to recognise the current position of object when the object or the blind or both are moving. Selecting an arbitrary set of frequencies will make the blind tired of repeatedly hearing the sounds.

Two blind persons were involved in this research to test the systems. They were to choose the frequency of sounds that are comfortable to hear repeatedly. It is certain that this selection is person dependent. As an alternative approach to have a comfortable hearing, musical keyboard (or piano key) sounds with selected octaves were introduced and tested by the blinds. This effort has offered a higher percentage of hearing pleasantness than that of selecting some comfortable frequency.

The NAVI and SVETA systems are designed such that the blind repeatedly hears the sound at the rate of 0.8 s. This duration covers all processing times such as image acquisition, image processing, image to sound transformation and sending the sound to earphones. This duration is found to be optimum, since a lower duration makes the blind confused and a higher duration does not indicate precisely the current position of object when the object and/or the person are moving.

(a) Stereo Cameras

(b) Compact Computing Device

(c) A Blind with SVETA-2

Figure 16.18 SVETA-2

16.9.5 SVETA-2

SVETA-1 is designed and constructed basically for testing by the blind people and for selecting correct frequencies; it is heavy and bulky in construction. The hardware such as processing computer, the stereo headphones and the vision cameras are not fully matched in their interfacing.

SVETA version 2 (SVETA-2) is a commercial prototype, lightweight and compact. Figure 16.18 shows the main components of SVETA-2 and a blind using this system. Cameras are of 1.3 Megapixels, focal length as 6 mm and with an adjustable base distance. The cameras are fixed to a specially made headgear. The compact computing device (CCD) has an IEEE 1394 USB interface with a dimension of (38 × 9 × 5) cm including its batteries and 500 g in weight. The blind can use the SVETA-2 system for about 2 hours before recharging the batteries. The CCD is kept in a pouch and attached to blind's waist belt.

16.10 SUMMARY

The research on vision substitution expands due to the continuous emergence of new techniques in image processing, sound generation, microelectronics, energy conservation and sensors. The blind wish to live independently in a dignified way without the sympathetic assistance from others and strive to compete with sighted people when required. Every country has its own regulations to protect the ambitions of blind. There is no doubt that the blind are relatively intelligent compared to several sighted people, perhaps as a compensation for their vision loss. Their opportunities in getting education, training and jobs are increasing. This research is aimed at continually expanding and targeting all the requirements of the blind, including categorizing colours, recognizing the persons in a group, identifying moving objects and enjoying natural sceneries.

REVIEW QUESTIONS

1. Briefly describe two aspects of human navigation.
2. What are the main restrictions of (a) blind cane, (b) blind robots and (c) dogs?
3. What is Dobelle Eye? What is the drawback of using this eye? Explain briefly.
4. What is NAVI? How many cameras are required for NAVI? What is the functionality of using so many cameras?
5. How is image extracted from vision? Explain with a diagram.
6. What experiment will you adopt to compute distance of obstacle? Explain with diagrams.
7. What is the main difference beween SVETA-1 and SVETA-2?
8. Explain the main concept of this chapter?

APPENDIX

MATRICES AND VECTORS

Definitions

A matrix of dimensions $(m \times n)$ with $m, n > 0$ is known as an aray of elements having m rows and n columns, and is denoted by

$$A = \begin{bmatrix} a_{11} & a_{12} & \dots & a_{1n} \\ a_{21} & a_{22} & \dots & a_{2n} \\ . & . & \dots & . \\ . & . & \dots & . \\ a_{m1} & a_{m2} & \dots & a_{mn} \end{bmatrix} \tag{A.1}$$

When $m = n$, the matrix is said to be square. When $m < n$, the matrix has more columns than rows; when $m > n$, the matrix has more rows than columns.

A square matrix A of dimensions $(n \times n)$ is said to be upper triangular if $a_{ij} = 0$ for all $i > j$,

$$A = \begin{bmatrix} a_{11} & a_{12} & \dots & a_{1n} \\ 0 & a_{22} & \dots & a_{2n} \\ . & . & \dots & . \\ . & . & \dots & . \\ 0 & 0 & \dots & a_{nn} \end{bmatrix} \tag{A.2}$$

and the matrix is known as lower triangular when $a_{ij} = 0$ for all $i < j$.

If the elements $a_{ii} > 0$ and other elements are 0 then the matrix is said to be diagonal.

If the elements a_{ii} are 1 and others are 0, then the matrix is called identity matrix. If all elements are 0, the matrix is called null matrix.

An $(n \times n)$ sqare matrix A is said to be skew-symmetrix if $A^T = -A$ for all $i \neq j$ and $a_{ii} = 0$ leading to

$$A = \begin{bmatrix} 0 & a_{12} & \dots & a_{1n} \\ -a_{12} & 0 & \dots & a_{2n} \\ . & . & \dots & . \\ . & . & \dots & . \\ -a_{1n} & -a_{2n} & \dots & 0 \end{bmatrix} \tag{A.3}$$

Matrix Operations

Trace of an $(n \times n)$ square matrix is

$$\text{Tr}(A) = \sum_{i=1}^{n} a_{ii} \tag{A.4}$$

Two $(n \times n)$ matrices are A and B. A and B are added to get C with elements as (c_{ij}) having $i, j = 1, 2, \ldots, n$ as

$$C = A + B. \tag{A.5}$$

If O is null matrix then

$$A + O = A \tag{A.6}$$

Also

$$A + B = B + A \tag{A.7}$$

$$(A + B) + C = A + (B + C). \tag{A.8}$$

When A_s is symmetric matrix and A_n is a non-symmetric matrix, then

$$A = A_s + A_n \tag{A.9}$$

where

$$A_s = ½ (A + A^T) \tag{A.10}$$

and

$$A_n = ½ (A - A^T). \tag{A.11}$$

When $(n \times n)$ matrix D is a product of A and B then

$$D = A\,B \tag{A.12}$$

whose elements are given by $d_{ij} = \sum_{k=1}^{n} a_{ik} b_{kj}$.

The following properties hold:

$$A(BC) = (AB)C \tag{A.13}$$
$$A(B + C) = AB + AC \tag{A.14}$$
$$(A + B)C = AC + BC \tag{A.15}$$
$$(AB)^T = B^T + A^T \tag{A.16}$$

For an $(n \times n)$ square matrix A, the determinent of A is a scalar quantity and is given by

$$\det(A) = \sum_{j=1}^{n} a_{ij} (-1)^{i+j} \det(a_{(ij)}). \tag{A.17}$$

The following property holds:

$$\det(A) = \det(A^T). \tag{A.18}$$

If A and B are square matrices, then

$$\det(A\,B) = \det(A)\det(B). \tag{A.19}$$

The rank $R(A)$ of a matrix A of dimension $(m \times n)$ is the maximum integer r so that at least a non-null minor of order r exists. The following properties hold:

$$R(A) \leq \min(\text{rows, columns}), \tag{A.20}$$
$$R(A) = R(A^T), \tag{A.21}$$
$$R(A^TA) = R(A). \tag{A.22}$$

If B is any matrix with rank equal to $R(B)$, then

$$R(AB) \leq \min(R(A), R(B)). \tag{A.23}$$

The adjoint of a square matrix A is

$$\text{Adj } A = [(-1)^{i+j} \det(A_{i+j})]^{T};\ i = 1, 2, ..., n \text{ and } j = 1, 2, ..., n. \tag{A.24}$$

An $(n \times n)$ square matrix A is said to be invertible if a matrix A^{-1} exists, termed inverse of A, so that

$$A^{-1} A = A A^{-1} = I. \tag{A.25}$$

Since $R(A) = n$, an $(n\times n)$ square matrix A is invertible, that is

$$A^{-1} = (1/\det(A)) \text{ Adj}(A). \tag{A.26}$$

The following properties hold:

$$(A^{-1})^{-1} = A. \tag{A.27}$$

$$(A^{T})^{-1} = (A^{-1})^{T}. \tag{A.28}$$

If the inverse of a square matrix is equal to its transpose, that is, $A^{T} = A^{-1}$ then, the matrix is said to be orthogonal

$$AA^{T} = A^{T}A = I. \tag{A.29}$$

If A and B are invertible square matrices of same dimensions, then

$$(AB)^{-1} = B^{-1} A^{-1}. \tag{A.30}$$

The derivative of an $(m \times n)$ matrix $A(t)$ whose elements $a_{ij}(t)$ are differentiable functions is the matrix

$$(\mathrm{d}A(t)/\mathrm{d}t) = (A \text{ matrix with } (\mathrm{d}a_{ij}(t)/\mathrm{d}t),\ i = 1, 2, ..., n \text{ and } j = 1, 2,..., m). \tag{A.31}$$

Vector Operations

All x_i are vectors of dimension $(m \times 1)$, then the following expression is linearly independent

$$k_1 x_1 + k_2 x_2 + ... + k_n x_n = 0 \tag{A.32}$$

and holds good only when all the constants k_i vanish.

A necessary and sufficient condition is that the matrix

$$A = [x_1 \mid x_2 \mid ... \mid x_n] \tag{A.33}$$

has rank n.

The scalar product $<x, y>$ of two vectors x and y of dimension $(m \times 1)$ is the scalar that is obtained by summing the products of the respective components in a given basis:

$$<x, y> = \sum_{k=1}^{m} (x_k, y_k) = x^T y = y^T x. \tag{A.34}$$

Two vectors are said to be orthogonal when their scalar product is null:

$$x^{T} y = 0. \tag{A.35}$$

The norm of a vector is defined as

$$\|x\| = \sqrt{(x^{T} x)}. \tag{A.36}$$

It is possible to show that both the triangular inequalities

$$\|x + y\| \leq \|x\| + \|y\| \tag{A.37}$$

and the Schwarz inequality is

$$\|x^{T}y\| \leq \|x\| \, \|y\|. \tag{A.38}$$

An unit vector $\bar{y}$ is denoted by

$$\bar{y} = (1/\|y\|)\, y. \tag{A.39}$$

The following holds (χ is a symbol of cross product):

$$x \,\chi\, x = 0, \tag{A.40}$$
$$x \,\chi\, y = -y \,\chi\, x, \tag{A.41}$$
$$x \,\chi\, (y + z) = x \,\chi\, y + x \,\chi\, z. \tag{A.42}$$

Given three vectors x, y and z in Euclidean space, the following expressions hold for scalar tripple products:

$$x^{\mathrm{T}} (y \,\chi\, z) = y^{\mathrm{T}} (z \,\chi\, x) = z^{\mathrm{T}} (x \,\chi\, y). \tag{A.43}$$

If any two vectors of (x, y) are equal, then the scalar product is null, that is

$$x^{\mathrm{T}} (x \,\chi\, y) = 0. \tag{A.44}$$

EXERCISES

1. In ancient China (1028–957 BC), Yan Shi, an engineer, presented a life-size human-shaped figure (a robot) to King Mu of Zhou. The robot walked with rapid strides, moving its head up and down and winking its eyes.

   In 2006, NASA developed a concept of a personal robot (fondly called Red Ball) going round an astronaut in space. The Red Ball is expected to be very useful in protecting the astronaut who comes out of spacecraft for a mission of possible repair or maintenance.

   Such documents are massively available with immense knowledge in the literature to be listed in the history of robot. Expand Table 1.1 from the dates of BC till date and include them in chronological order.

2. Over the years, several varieties of robots with varied mobilities have been developed and implemented by a number of research institutions around the world. They can be having articulations and mobility such as picking, placing, rolling, walking, climbing, swimming, flying, crawling and hopping.

   It will be interesting to collect at least five high-quality video clips which demonstrate their mobilities, articulations and applications.

3. Typical values of annual sales (within a country and through export) of manufactured robot in per unit currency and the number of installations of robots in the country are given in Table Q.1. Determine suitable curve fitting equations for both of these cases and predict the annual sales and number of installations during the year 2014.

Table Q.1 Robot Statistics

Year	Annual Sales (× \$1000)	No. of Robot Installations (Units)
1995	1400	5000
1996	1700	6300
1997	2000	7800
1998	2500	9700
1999	3100	12100
2000	3900	15100
2001	4900	18900

(Continued)

Table Q.1 Continued

Year	Annual Sales (× $1000)	No. of Robot Installations (Units)
2002	6800	24300
2003	9600	32200
2004	13400	43600
2005	18800	59900
2006	26422	83200

4. Prepare detailed reports with illustrative diagrams, photographs and video clips for the following varieties of robots:
   (a) Robots in home
   (b) Military robots
   (c) Hospital robots
   (d) Robots for elder care
   (e) Space robots
5. Name any four basic components of a robot system.
6. Give all possible classifications in a robot system.
7. For the following tasks state whether a gripper or a tool is appropriate, please provide reasons for your answer.
   (a) Welding
   (b) Scraping
   (c) Assembly of two parts
   (d) Drilling a hole
   (e) Tightening a nut of an engine
   (f) To solve hole–peg problems
8. Prepare a state-of-art report on robotics in India.
9. Prepare a state-of-art report on industrial robot and project demand for future.
10. Find out application of robot in space explorations.
11. What are the issues in a robotic control? Explain.
12. Describe a few methods of teaching a robot.
13. Explain the anatomy of human wrist joint and analyse it for type of joint, number of degrees of freedom, number of joints, etc.
14. Initially point G is set at $[3\ \ 0\ \ -1\ \ 1]^T$:
   (i) When it is rotated by 180° about the Z-axis and then translated by 3 units along the Y-axis.

   or

   (ii) When it is translated 3 units along the Y-axis and then rotated by 180° about the Z-axis.

   Explain why the two locations in two cases are same or different.

15. Two frames A and B are initially coincident. Frame B undergoes the following four operations, in sequence with respect to axes of frame A:

    (a) A rotation of $\theta°$ about the Z-axis
    (b) A translation of d cm along the Z-axis
    (c) A translation of a cm along the X-axis
    (d) A rotation of $\phi°$ about the Z-axis.

    Determine the final homogeneous transformation matrix to describe frame B after the above with respect to frame A.
16. Two coordinate frames are coincident initially. A point Q is located at 6 cm along the X-axis of mobile frame. It is then rotated by 60° about the Z-axis of fixed frame. Determine the coordinates of point Q in a fixed coordinate frame.
17. The end point of a link of a manipulator is at $P = [2\ \ 2\ \ 6\ \ 1]^T$. The link is rotated by 90° about the X-axis. Then it is rotated −180° about its own Y-axis. Finally, by −90° about its own Z-axis. Find the resulting homogeneous transformation matrix and the final location of end point.
18. What do you understand by screw transformations? Where are these transformations useful?
19. Explain why homogeneous transformations and homogeneous coordinates are required in modelling robotic manipulators?
20. One of the axes has a gear ratio as $\eta = \{\eta_1/\eta_2\}$ where η_1 and η_2 are the number of teeth at actuator shaft and the number of teeth at the link shaft, respectively. Compute the torque, velocity and distance moved from link shaft to actuator shaft.
21. The trajectory of a joint's motion between start and goal points in a pick and place operation is determined by dividing the motion in two segments. The interpolating polynomial for each segment is cubic and the acceleration is continuous at intermediate point. Determine the coefficients of two cubic polynomials.
22. The trajectory between two points is divided into five segments Five cubic polynomials are used for interpolation. The boundary conditions that the polynomials satisfy are: (i) position constraints at start, lift off, set down and goal positions and (ii) continuity of velocity and acceleration at all the path points. Determine the polynomial for each segment.
23. The velocity profile of a trajectory is trapezoidal with constant acceleration segment of 0.5°/ s² duration and constant velocity of 10°/s. Determine the parameters of smooth trajectory for interpolating the time sequence of position with this type of trajectory.
24. Consider a circle and an ellipse that might be viewed by a machine vision system. The circle has a radius of 4 cm. The ellipse has major and minor axes as 4 cm and 2 cm, respectively. Apply the two definitions of thinness to both elements and compare the results.
25. Develop PALLET, a VAL program to move sequentially nine objects on a (3×3) pallet matrix to a location called MACH.

26. Explicitly and otherwise determine A^{-1} when A is given by

$$\text{(a)} \begin{bmatrix} 0 & 1 & 0 & 0.2 \\ 0 & 0.2 & 0 & 0.6 \\ 1 & 0.6 & 1 & 0.5 \\ 0 & 0 & 0 & 1 \end{bmatrix} \qquad \text{(b)} \begin{bmatrix} 2 & 1 & 1 & -21 \\ 3 & 2 & 3 & -44 \\ 4 & 3 & 5 & -67 \\ 0 & 0 & 0 & 1 \end{bmatrix}$$

27. A pen is mounted on the robot's end effector that is used to write the first and the last letters of your name approximately large about 10 cm × 10 cm on a sheet of paper attached to the surface of work table. Prepare a software, especially in VAL and V+.
28. Two joints of a SCARA manipulator are to move by 20° and 25° in 3 s and 4 s, respectively. Assume that the trajectory has to be cubic for both joints. Determine the coefficients p_j and q_j for the two cubic polynomials

$$Z_1(t) = p_1 + p_2 t + p_3 t^2 + p_4 t^3$$
$$Z_2(t) = q_1 + q_2 t + q_3 t^2 + q_4 t^3$$

29. A single robot with a rotary joint will smoothly rotate from its initial position $\Theta_o = 5°$ to the final position $\Theta_f = 150°$ with zero initial and final velocities. Design a cubic polynomial trajectory to realize this motion in 4 s.
30. Design a LSBP trajectory for a single rotary joint with its initial position $\Theta_o = 0°$, the final position $\Theta_f = 120°$ and acceleration $\alpha = 100°/s^2$ in the acceleration segment. The motion is assumed to be completed in 3 s. Assume the missing data.
31. Design a Bang-Bang trajectory for a single rotary joint with its initial position $\Theta_o = 0°$, the final position $\Theta_f = 120°$ and the maximum acceleration of $\alpha_{max} = 80°/s^2$.
32. Let the second-order system be as follows:

$$x'_1 = x_2$$
$$x'_2 = -\alpha(1 + x_2)^2 x_2 - x_1 \text{with } \alpha > 0.$$

Determine the stability of system using Lyapunov stability theory.

33. Three machines are organized in a machine cell using one robot. The robot is to load and unload the machines with the unfinished parts. The cycle times of three machines are given as in Table Q.2. Determine the robot's idle time and machine interference.

Table Q.2 Machine Details

Machine	Run time (s)	Service time (s)
Machine 1	30	20
Machine 2	15	10
Machine 3	20	10

34. Each of the three machines in the work cell is identical. They have identical cycle times of 50 s. The cycle time is divided between run time (30 s) and service time (load and unload) by the robot (20 s). Determine the robot idle time and the machine interference.
35. Consider a linear system described by the following state-space equation:

$$x'(t) = Ax + Bu,$$

where $x \in R^{n\times 1}$ state variable vector, $u \in R^{m\times 1}$ control vector, $A \in R^{n\times n}$ and $B \in R^{n\times m}$ system parameter matrices. Investigate the stability of the system by Lyapunov method and list the conditions.

36. Consider the following nonlinear systems:

    (1) $x'_1 = x_2$ (2) $x'_1 = x_2$

    $x'_2 = -100\,x_1 - 100\,x_2^{\,2}$ $x'_2 = -100\,x_1^{\,2} - 100\,x_2$

    Determine the stability of the systems using Lyapunov stability theory.

37. An industrial robot program has facilities to pick-off parts from a conveyor and load them into a pallet of about 25 cm from the pick-off point. A mechanical stop, provided on the conveyor, is used to locate the parts in a known position. The parts are to be arranged in a (10 × 10) mm pattern, 120 mm apart in both directions. The two directions are along the *x* and *y* axes, respectively. Prepare a program for the robot. You may use VAL and V+.

38. A robot installation costs ₹850,000. The expected annual costs for operation and maintenance are evaluated as ₹240,000. The robot project is expected to generate a revenue of ₹600,000 annually for 4 years. After this time the project is terminated. The salvage value of robot and its associated sub-systems at the end of 4 years is ₹400,000.

    (a) Determine the payback period of this project.

    (b) Determine the equivalent uniform annual cost for the project using a 30% rate of return in the interest calculation.

    (c) Determine the expected return on investment to be derived from the investment.

39. Write a program in (VAL, V+ and AML) to pick the blocks off a conveyor belt and place them in a pallet in (3 × 3) mm array position. The array is of (12 × 12) mm. The blocks are precisely positioned on the conveyor stops at the pick-up points. The conveyor is reactivated manually.

40. Consider Figure Q.1. The pendulum mass m_2 is attached to the moving mass m_1. The mass m_1 can take any position along the *x*-axis. The unstretchable string length is *r* and it can be anywhere within the *x*-*y* plane. The pendulum is made to oscillate without any initial torque.

    (i) Determine *L*, the Lagrange function, which is the difference between kinetic energy and potential energy.

    (ii) Write the equation of motion of the pendulum.

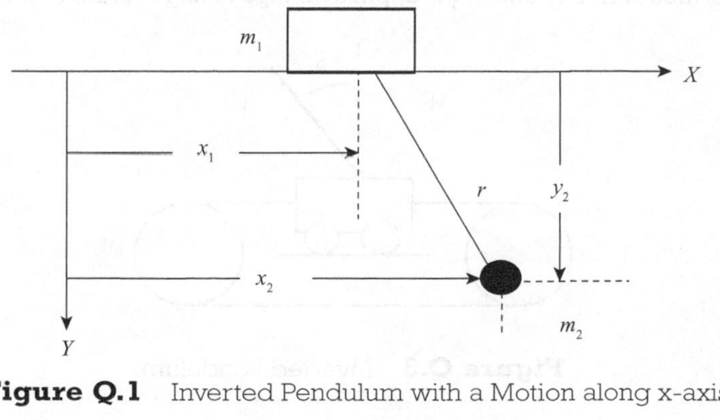

Figure Q.1 Inverted Pendulum with a Motion along x-axis

41. The LISP program is to be written. Assume that you are given locations of two objects in the form of list (x, y, z) for each object. If $x = y$ for both parts and z is greater than the other, then that one is to be kept on the top. This program uses CAR and CDR statements to get individual (x, y, z) values and then performs the comparison.
42. The work cycle layout is shown in Figure Q.2. The work cycle does a simple task in which the robot must move parts weighing 1 kg from Conveyor 1 to Conveyor 2. The Conveyor 1 presents the part in 15 s. The sequence of work cycle is as follows:
    (i) Robot picks up the part from Conveyor 1 which has delivered the part to a known pick-up position.
    (ii) Robot transfers part to Conveyor 2 and releases the part.
    (iii) Robot moves back to start position at Conveyor 1.

    List the elements of the work cycle in a table. Assume different speeds.

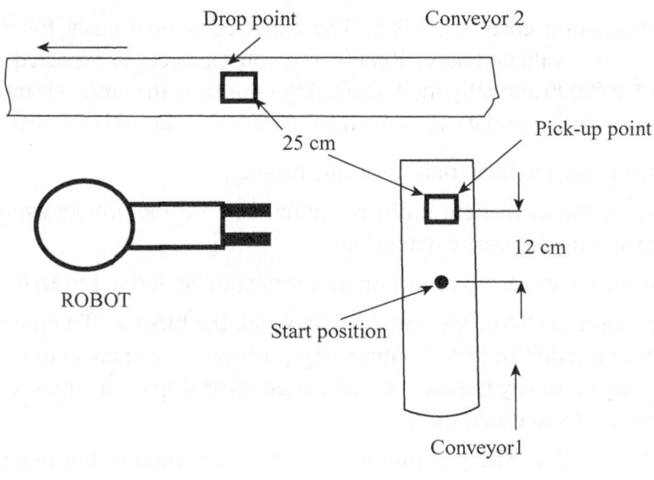

Figure Q.2 Work Cell Layout

43. Figure Q.3 details how the stick L has been maintaining an angle θ and this angle has to be made zero. This is made by moving the cart M right and left synchronously. The electronic control box is not shown in the figure.

    Prepare a list on how to solve this problem. You can decide the appropriate UoDs for position, velocity and V, the applied voltage to larger drum of the system.

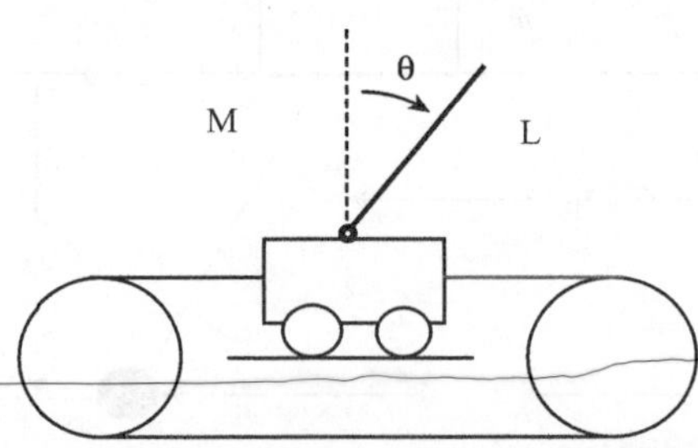

Figure Q.3 Inverted Pendulum

44. Give a set of possible membership functions that are not given in your book.
45. Sort out a combined form of fuzzy sets that are not given in your book.
46. Draw a block diagram of a controlled process which has a fuzzy controller as a logic system. Explain each of block function.
47. Briefly explain three basic functions of medical robots.
48. Which are the main contributors of slow growth of medical robots?
49. What is exoskeleton? Can you prepare an image? In which way does it improve the health of a patient?
50. Explain the main themes of Chapters 15 and 16.
51. What is the statistics of blind and partially blind population in Malaysia and in the world in 2002?

GLOSSARY

INDUSTRIAL ROBOTIC TERMS

Accuracy: After receiving a command from the controller, accuracy is the ability of robot to position and orient its end effector at a specified point over and over again.

Actuator: An electric, hydraulic or pneumatic motor or other automatic device used to produce link motion.

Arm: This is a member of link which is connected by revolute or prismatic joints.

Articulated: It is characterized by one or more joints connected to one another.

Automation: An automatically actuated mechanism capable of autonomous motion.

Bang-Bang Parabolic Blend: Linear segment with parabolic blends (LSPB) trajectory in which the parabolic blends shrink to zero.

Cartesian Space: Generally, x, y, z Cartesian axes in three-dimensional rectangular space.

Compliance: Ability of a robot to move in certain selective directions in which no force is created.

Configuration: The position and orientation of tool or end effector.

Continuous Path: The robot follows a stored trajectory of time-dependent data which are often generated by a form of lead through operation. A paint spraying robot works in this way.

Controlled Path: This is a method of controlling robots that require more sophisticated servo-type path control between individually stored target points.

Coordinate Systems

Base Coordinates: This is a reference in a system. Usually Cartesian with origin is fixed to some non-movable portions of robot base.

Joint Coordinates: A reference frame fixed to each link.

Tool Coordinates: A reference system usually Cartesian coordinates with its origin fixed to a moving object such as a gripper. This has three directions: approach vector (a), sliding vector (s) and normal vector (n), so that $a \times n = s$.

World Coordinates: A reference system with origin fixed to some reference point in the vicinity of one or more robots.

Coordinated Motion: This is the robot movement in which various links are made to move together in a dependent manner.

Coordinate Transformation

Orientation Matrix: Usually, this is called as rotation matrix. It can be associated with a (3 × 3) matrix and named as orientation transformation matrix.

Homogeneous Transformation Matrix: It can be associated with Cartesian coordinate system of rotations and translations in disjoint origin systems.

Denavit–Hartenberg (D-H) Matrix: A special matrix used in robotics; usually indicated by a Cartesian coordinate system fixed to each link, i, relative to previous link, $i - 1$.

D-H Parameters: Four independent parameters, two angles and two linear displacements are computed from D-H matrix

Cubic Segment: A trajectory either in Cartesian space or in joint space specified by a single cubic polynomial function of time.

Degenerate Configuration: Any robot configuration in which freedom of motion is constrained.

Degrees of Freedom: The number of parameters used to specify the configuration of any element in a kinematic chain with respect to any other element.

Direct Drive Robot: A manipulator in which the link axes are directly coupled to rotors of electric motors, thereby eliminating the need for mechanical transmission elements.

Displacement Vector: A vector usually defined between the disjoint origins of different Cartesian coordinate systems.

Dynamics: The study of bodies in motion. Dynamics can be divided into two parts—kinetics and kinematics.

End Effector: It is a gripper or a tool attached to the end of a robot to transport objects.

Euler Angles: Three ordered angles associated with three successive rotations of a Cartesian coordinate system, the first one being made about one of the three major axes.

Flexible Automation: A form of automation that employs reprogrammable and multi-functional mechanisms such as robots.

Free Motion: The most common current form of unconstrained robot motion based on internal sensory information.

Generalized Moment: It is the total moment that directly affects dynamical link behaviour in the direction of permissible link motion. That is, the complete drive terms that are associated with Lagrange's differential equations of motion of a given robot manipulator.

Generalized Torque: A vector composed of torques and forces that are transmitted to links from the tool.

Generalized Force: This is a six vector. This composes of a three vector of force and three vector of moment, both act simultaneously at the tool and/or wrist.

Geometric Configuration: The number and types of joints and links comprising a robot and their positioning relative to one another. For a general purpose three jointed robot is used:

Cartesian: If first three joints are prismatic.

Cylindrical: If first joint is revolute and next two are prismatic.

Spherical: If first two joints are revolute and the third is prismatic.

Revolute: If all joints are revolute.

Guarded Move: This is a constrained motion to locate a surface on which a slow motion is initiated.

Inverse Dynamic: A form of open-loop compensator in which an actuator drive signal is computed from knowledge of differential equations.

Jacobian Matrix: In robotics, a Jacobian matrix represents the transformation between link velocities and Cartesian velocities.

Joint: Location of the physical connection between two adjacent links (or kinematic pairs) in a kinematic chain.

Kinematically Simple: A robot characterized by twist angles equal to 0° or ±90°, with spherical wrist.

Kinematic Chain: A series of rigid bodies (links) connected by joints.

Closed joints: If every link is connected to at least two other links.

Open joints: If certain links are connected to only one other link. Robot is termed as the generation of open joints.

Kinematics: The study of dynamics involving geometry and the time-dependent motion of open-loop kinematic chain. No forces and moments are accounted for.

Kinetics: That part of dynamics involving study of relation between the forces acting on the robot, the masses of various links that comprise robot and link motion.

Knot Points: The intermediate positions and their associated velocities and accelerations used in the specification of robot trajectories.

Lead Through: A method of programming robots by a teach pendant that is employed to move the robot to desired configuration.

Linear Segment with Parabolic Blends: A trajectory that is composed of constant velocity linear segment to which a second-degree (parabolic) polynomial time functions are blended in such a way to provide a continuous velocity at blend times.

Link: A rigid body in a kinematic chain.

Link Space: An *n*-dimensional space that is defined by various displacements.

Manipulator: The articulated portion of robot that is an open kinematic chain as well as any associated actuators and/or sensors.

Manufacturing Cell: An interconnection of robots with their conveyors, feeders, sensors and buffers which are temporarily controlled to perform a specified manufacturing operation.

Measure of Performance: The quantitative and qualitative descriptions of steady-state errors and the transient responses of a dynamical system.

Modular Robot: Robot constructed with capacity of sub-unit (single or multiple link) interchange.

Orientation: Defines a particular axis location of a system fixed to end effector relative to a base coordinate system. Often specified by three Euler angles or their perpendicular projection of the approach, sliding and normal vectors on a Cartesian base frame.

Path: The positional information associated with a time-dependent trajectory.

Personal Robot: Robot designed and built for personal use rather than industrial use.

PID Control: A parallel combination of error-driven compensator consisting of proportional (P), integral (I) and derivative (D) terms. Each term is characterized by an independently adjustable gain set that can be tuned to modify the performance characteristics of the robot system.

Plug-In: A method of programming robots off-line by directly loading and executing another program usually written and debugged at an earlier time.

Point to Point: A non-servo control technique. This method enables a robot to stop at numerous distinct points. These points are often specified by an initial lead-through operation but do not allow any controlled motion between points.

Pole Placement: A control method that outlines conditions under which it is possible to assign arbitrarily all of the closed-loop poles of an nth order system by an appropriate choice of a compensator of order $n - 1$.

Position: The location of a point which is defined for a robot such as a tool usually placed relative to its base coordinates.

Precision: The accuracy, repeatability and resolution associated with a robot.

Prismatic Joint: Connection between a pair of links that allows only relative translation.

Redundant Manipulators: Manipulators that are characterized by more movable links than are necessary to achieve arbitrary configurations.

Repeatability: The ability of robot to position its end effector at the same point during repeated runs.

Resolution: The smallest displacement that a robot can be commanded to move.

Revolute Joint: A connection between a pair of links that allows only relative rotation about a single axis.

Robot: An articulated mechanism together with any auxiliary supporting device that is capable of an autonomous motion.

Sampling Times: The times at which numerical values such as those associated with desired link positions change in a discrete time computer.

Sensor: A transducer whose input is due to physical phenomenon and whose output (usually electric) is a measure of that physical phenomenon.

Singular Values: Any set of link values at which the Jacobian is singular (i.e. the Jacobian has no determinant).

Stable: The system having a property of bounded output for bounded input.

Steady-State Error: This is a performance measure. This reflects the difference between the output of a stable closed-loop system and the external input to system to track after the transient response dies down to zero.

Task Level Programming: A desired form of programming the robots using explicit command statements that specify the desired goals.

Teach Pendant: Hand held box that is used to move the robot to visually determined configuration. The link displacements can be committed to 'taught' points for a subsequent replay during execution.

Teleoperator: This is an articulated mechanism directly, and often remotely, controlled by a human operator.

Trajectory: A path along with appropriate position, velocity and acceleration information.

Trajectory Planner: A special computer program that generally produces a time-varying signal representing the desired trajectory that the robot is to follow in either Cartesian or joint space.

Transient Response: The short-term response of a stable dynamical system that normally converges to zero in some finite time.

Unity Feedback System: This system has a closed-loop configuration; the output of this controlled system is directly fed back for comparison with the input, thus producing an error signal.

Walk-Through: This is a method of programming robots by physically moving the robot by hand through desired trajectory and recoding the entire motion for a subsequent replay. Examples are spray painting, welding and other path-oriented tasks.

Working Envelope: The boundary of the working range (volume).

Working Range: The set of all points that can be reached by any part of robot when articulated through all possible configurations.

Wrist: Robot end effector which can carry a tool or a gripper.

REFERENCES

An, C. H., C. G. Atkeson, and J. M. Hollerbach. *Model Based Control of a Robot Manipulator*. Cambridge, MA: MIT Press, 1988.

Andreff, N., R. Horaud, and B. Espiau. "Robot Hand Eye Calibration Using Structure-Form-Motion." *International Journal of Robotics Research* 20, no. 3 (2001): 228–248.

Andrew, Jr. C. S. *Robotics and AI – An Introduction to Applied Machine Intelligence.* NJ: Prentice-Hall, 1987.

Ardayfio, D. D., R. Kapur, S. B. Yang, and W. A. Watson. "MICRAS: Microcomputer Interactive Codes for Robot Analysis and Simulation." *Mechanism and Machine Theory,* 20, no. 4 (1985): 271–284.

Asada, H., and K. Youcef-Toumi. *Design of Direct Drive Manipulators.* Cambridge, MA: MIT Press, 1987.

Ashley, S. "Magnetostrictive Actuators." *Mechanical Engineering* (1998): 68–70; "Push Pull Magnetostrictive Linear Actuators." *NASA Technical Briefs* (1999): 47–48.

Auslander, D., and C. J. Kempf. *Mechatronics: Mechanical System Interfacing.* Upper Saddle River, NJ: Prentice-Hall, 1998.

Baker, D., and C. Wampler. "On The Inverse Kinematics of Redundant Manipulators." *The International Journal of Robotics Research,* 7, no. 2 (1988), 218–219.

Baker, K. "Improved Robot Joint Calculations." *NASA Technical Briefs,* 8, no. 11 (1988), 79–80.

Balakrishnan, G., G. Sainarayanan, R. Nagarajan, and Sazali Yaacob. "Stereopsis Method for Visually Impaired to identify Obstacles Based on Distance." Proceedings of International Conference on Image and Graphics, China 580–583, 2004.

Balakrishnan, G., G. Sainarayanan, R. Nagarajan, and Sazali Yaacob. "Object Discrimination Using Stereo Vision for Blind Through Stereo Sonification." Proceedings of Indian Conference on Computer Vision, Graphics and Image Processing, India 234–239, 2004.

Balakrishnan, G., G. Sainarayanan, R. Nagarajan, and Sazali Yaacob. "Blind Navigation Using Stereopsis with Fuzzy Rule Base." Proceedings of International Conference on Artificial Intelligence in Engineering and Technology, Kota Kinabalu (Malaysia) 238–243, August 3-5, 2004.

Balakrishnan, G., G. Sainarayanan, R. Nagarajan, and Sazali Yaacob. "NAVI — Applying Stereo Vision Technology For Visual Rehabilitation." International Conference on Biomedical Engineering, Kuala Lumpur (Malaysia) 341–344, September 2004.

Balakrishnan, G., G. Sainarayanan, R. Nagarajan, and Sazali Yaacob. "A Support System for Visually Impaired Persons to Understand Environmental Information Using Acoustic Interface." 1st International Workshop on Artificial Life and Robotics, (AROB), KUKUM, Perlis, 31–36, May 14-15, 2005.

Balakrishnan, G, G. Sainarayanan, R. Nagarajan, and Sazali Yaacob. "Stereo Image Processing Procedure for Vision Rehabilitation." *International Journal on Machine Vision and Applications (USA)*, 22, no. 2 (2005): 10–16.

Bolton, W. *Mechatronics: Electronic Control System in Mechanical and Electrical Engineering.* Longman, NY: Addison-Wesley, 1999.

Bone, G. M., and M. A. Elbestavi. "Dual Sensor Based Robotic Deburring." *ASME Journal of Manufacturing Science and Engineering*, 118, no. 3 (1996): 439–441.

Brooks, R. A. "New Approaches to Robotics." *Science,* 1227, (1991), 400-401.

Burns, J. B., A. R. Hanson, and E. M. Riseman. "Extracting Straight Lines." *IEEE Transactions on Pattern Analysis Machine Intelligence,* 8, no. 4 (1986): 425–455.

Chand, S., and K. Doty. "On Line Polynomial Trajectories for Robot Manipulators." *The International Journal of Robotics Research,* 4, (1985).

Craig, J. J. *Adaptive Control of Mechanical Manipulators.* NJ: Addison-Wesley, 1988.

Craig, J. J. *Introduction to Robotics – Mechanics and Control,* 2nd ed. NJ: Addison-Wesley, 1989.

Craig, J. J. *Introduction to Industrial Robotics – Mechanics and Control,* 3rd ed. New Jersey: Pearson Prentice-Hall, 2005.

Deb, S. R., and S. Deb. *Robotics Technology and Flexible Automation.* New Delhi: Tata McGraw Hill, 2012.

Denavit, J., and R. S. Hartenberg. "A Kinematic Notation for Lower-Pair Mechanism Based on Matrices." *ASME Journal of Applied Mechanics,* 22, no. 10 (1955): 215–221.

Dorf, R. C. *Concise International Encyclopedia Robotics Applications and Automation.* Wiley–Interscience, 1990.

Eman, K. F., Soon Hun Lee, and J. C. Cesarone. *Trajectories, International Encyclopedia of Robotics: Applications and Automation,* Edited by Richard C. Dorf. New York: Wiley, 1988.

Engelberger, J. F. "Robotics in Practice – Management and Applications of Industrial Robots." *Forwarded by Isaac Asimov, American Management Associations*, 1982.

Eppinger, S. D., and W. P. Seering. "Introduction of Dynamic Models Robot Force Control." *IEEE Control Systems Magazine,* 7, no. 2 (1987): 48–52.

Fu, K. S., R. C. Gonzalez, and C. S. G. Lee. *Robotics: Control, Sensing, Vision and Intelligence.* New York: McGraw Hill, 1987.

Ghosal, A. *Robotics – Fundamental Concepts and Analysis.* Oxford University Press, 2006.

Gonzalez, R. C., and Richard Woods. *Digital Image Processing.* Addison-Wesley, 1992.

Groover, M. P., M. Weiss, R. N. Nagal, and N. G. Odrey. *Industrial Robotics – Technology, Programming, and Applications.* McGraw-Hill, 1987.

Hariharan, K. "Design of Force/Position Control Laws for Constrained Robots, Including Joint Flexibility and Actuator Dynamics." *Robotica,* 17, no. 1 (1999): 41–48.

Holland, S. W. "The Next Big Thing in Automotive Robotics." *Robotics World Magazine,* 19, no. 1 (2001).

Hollerbach, J. M. "A Recursive Lagrange Formulation of Manipulator Dynamics and a Comparative Study of Dynamic Formulation Complexity." *IEEE Transactions on Systems, Man, and Cybernetics,* 10, (1980): 730–736.

Horn, B. K. P. *Robot Vision.* New York: The MIT Press, McGraw Hill, 1986.

Hsia, T. C. "Simple Robust Scheme for Cartesian Space Control of Robotic Manipulators." *International Journal of Robotics and Automation* (1994): 167–174.

Janabi-Sherifi, F., and W. J. Wilson. "Automatic Selection of Image Features for Visual Servoing." *IEEE Transactions on Robotics and Automation,* 13, no. 6 (1997): 890–903.

Johnson, G. Colin, and Duncon Marsh. "Modeling Robot Manipulators with Multivariate B-Splines." *Robotica,* 17, no. 3 (1999):239–247.

Jung, S., T. C. Hsia, and R. G. Bonitz. "Force Tracking Impedance Control for Robot Manipulators with an Unknown Environment: Theory, Simulation, and Experiment." *International Journal of Robotic Research,* 20, no. 9 (2001): 765–774.

Kanede, T. "Survey; Region Segmentation: Signals vs Semantics." *Computer Graphics and Image Processing,* 13, (1980): 279–297.

Khalad, R. H., and M. P. Cartmell. "A New Methodology for Designing PD Controllers." *Robotica,* 19, no. 3 (2001): 267–274.

Khatib, O. "A Unified Approach to Motion and Force Control of Robot Manipulators: The Operational Space Formulation." *IEEE Journal of Robotics and Automation,* 3, no. 1 (1987): 43–53.

Lee, C. S. G. "Robot Arm Kinematics." *Dynamics and Control, Computer,* 15, no. 12 (1982): 62–80.

Lee, C. S. G. "Robotics Arm Kinematics, Dynamics and Control." *IEEE Computer,* 15, no. 12 (1982): 62–80.

Lee, C. S. G. "A Geometric Approach Inverse Kinematics of PUMA Robots." *IEEE Transactions on Aerospace and Electronic Systems,* 20, no. 6 (1984): 696–706.

Lee, M. H. "Tactile Sensing: New Directions, New Challenge." *International Journal of Robotics Research,* 19, no. 7 (2000): 636–643.

Lin, C. S., and P. R. Chang. "Joint Trajectories of Mechanical Manipulators for Cartesian Path Approximations." *IEEE Transactions on Systems, Man, and Cybernetics,* 13, (1983): 203–210.

Lin, C. S., P. R. Chang, and J. Y. S. Luh. "Formulations and Optimization of Cubic Polynomial Joint Trajectories for Industrial Robots." *IEEE Transactions on Automatic Control,* 28, (1983): 1066–1073.

Lozano-Perez, T. "A simple Motion Planning Algorithm for General Robot Manipulators." *IEEE Journal of Robotics and Automation,* 3, no. 3 (1987): 224–238.

Luh, J. Y. S., and C. S. Lin. "Optimum Path Planning for Mechanical Manipulators." *ASME Journal of Dynamic Systems, Measurement, and Control,* 102, (1981): 142–151.

Mair, G. M. *Industrial Robotics.* NJ: Prentice-Hall International, 1988.

Manocha, D., and J. F. Canny. "Efficient Inverse Kinematics for a General 6R Manipulators." *IEEE Transactions on Robotics and Automation* 10, no. 5 (1994): 648–657.

Marzurkiewicz, J. "From Dead Stop to 2000 rpm in One Millisecond." *Motion Control* (1999): 41–44.

Mason, M. T. "Compliance and Force Control for Computer Controlled Manipulators." *IEEE Transactions on Systems, Man, and Cybernetics,* 6, no. 1 (1981): 418–432.

McKerrow, P. J. *Introduction to Robotics.* Addison-Wesley, 1991.

Mittal, R. K., and I. J. Nagrath. *Robotics and Control.* New Delhi: Tata McGraw Hill, 2003.

Muralindran, M., R. Nagarajan, Sazali Yaacob, Mohammad Yunus, and Renee Chin. "A Preliminary Study of Patient Lifting Robot for Nurse Assistance." First International Workshop on Artificial Life and Robotics (AROB), KUKUM, Perlis, 2005.

Nagarajan, R., G. Sainarayanan, and Sazali Yaacob. "Fuzzy Based Human Vision Properties in Stereo Sonification Systems for Visually Impaired." International Conference on Intelligent Systems in Design and Manufacturing IV, Boston, MA, (USA) October 28-31, 2001.

Nagarajan, R., G. Sainarayanan, Sazali Yaacob, and Rosalyn Porle. "Fuzzy Based Color Identification for Human Blind." Proceedings of International Conference on Artificial Intelligence in Engineering and Technology, Kota Kinabalu (Malaysia) 238–243, August 3-5, 2004.

Nagarajan, R., G. Sainarayanan, Sazali Yaacob, and Rosalyn R Porle. "Object Identification and Color Recognition for Human Blind." Proceedings of Indian Conference on Computer Vision, Graphics and Image Processing, India 210–215, 2004.

Nagarajan, R., G. Sainarayanan, Sazali Yaacob, and Rosalyn Porle. "Object Preference Assignment in NAVI Using Neural Network." Proceedings of International Conference on Artificial Intelligence in Engineering and Technology, Kota Kinabalu (Malaysia) 232–237, August 3-5, 2004.

Nagarajan, R., Sazali Yaacob, and G. Sainarayanan. "Realisation of Fuzzy Based Image Processing in Sonification System Applied in Blind Navigation." Second International Symposium on Advanced Intelligent Systems, Taejon (Korea) August 24-25, 2001.

Nagarajan, R., Sazali Yaacob, and G. Sainarayanan. "Implementation of AI Image Processing Methodology for Vision Substitution." International Symposium on Artificial Intelligence, Kohlapur (India) December 18-20, 2001.

Nagarajan, R., Sazali Yaacob, and G. Sainarayanan. "A Neuro Fuzzy Scheme for Image Processing Applied in Stereo Sonification System." Proceedings of International Conference on Artificial Intelligence in Engineering and Technology ICAIET-2002, Universiti Malaysia Sabah, (Malaysia) 73–79, June 17-18, 2002.

Nagarajan, R., Sazali Yaacob, and G. Sainarayanan. "Fuzzy Clustering in Vision Recognition Applied in NAVI." Presented at NAFIPS-FLINT 2002 International Conference, New Orleans (USA), June 27-29, 2002.

Nagarajan, Sazali Yaacob, and G. Sainarayanan. "Application of Artificial Intelligence in Image Segmentation Towards Blind Navigation." Paper No: IC51, Presented at the Second World Engineering Congress, Kuching (Malaysia) July 22-25, 2002.

Nagarajan, R., Sazali Yaacob, and G. Sainarayanan. "Role of Object Identification in Sonification System for Vision Impaired." IEEE TENCON 2003, Region 10, Bangalore (India) October 15-17, 2003.

Nagaranjan, R., Sazali Yaacob, M. Paulraj, Amin Shamsudin, Khalid Marzuki, and C. R. Hema. "Edge Detection of Occluded Objects for Bin Picking Applications." Third Seminar on Science and Technology Universiti Malaysia Sabah, Kota Kinabalu, 2004.

Nagarajan, R., Sazali Yaacob, and G. Sainarayanan. "Computer Aided Vision Assistance for Human Blind." *Integrated Computer-Aided Engineering-an International Journal,* 11, no. 1 (2004): 15–24.

Nagarajan, R., G. Sainarayanan, Sazali Yaacob, and Rosalyn Porle. "Fuzzy Rule Based object Identification method for NAVI System." *EURASIP Journal; Spl Issue on Advances in Intelligence Vision System Methods and Applications Part II* 2005, no. 14 (2005): 2260–2267 (USA).

Nagarajan, R., Sazali Yaacob, and G. Sainarayanan. "Fuzzy Learning Vector Quantization in Intelligent Vision Recognition for Blind navigation." *Journal of Institution of Engineers,* 4, (2005): 1215–1218 (India) (Awarded gold medal of honor: The Sir Thomas Ward Memorial Prize as Best Subject Paper).

Nagrath, I. J., and M. Gopal. *Control System Engineering,* 3rd ed. New Age International Ltd., 1999.

Nakamura, Y., and H. Hanafusa. "Inverse Kinematic Solutions with Singularity Robustness for Robot Manipulator Control." *Journal of Dynamic Systems Measurement and Control,* 108, no. 3 (1986): 163–171.

Nakamura, Y., H. Hanafusa, and T. Yoshikawa. "Task-Priority Based Redundancy Control of Robotic Manipulators." *The International Journal of Robotics Research,* 6, no. 2 (1987): 3–15.

Nikku, S. B. "Active Distance Measurement and Mapping Using Non-stereo Vision Systems." Proceedings of Automation '94 Conference, Taipei, Taiwan, ROC, 5, 147–150, 1994.

Nikku, S. B. "Scheme for Active Positional Corrections of Robotic Arms." Proceedings of Fifth International Conference on CAD/CAM; Robotics and Factors of Future, Springer Verlag, 590–593, 1991.

Parker, H. *Step Motors and Servomotors Control Catalog*, Servo-pnuematic Positioning System 2000 Festo, A. G. & Co., 7, 1996.

Paul, R. P. *Robot Manipulators: Mathematics, Programming, and Control.* Cambridge, MA: MIT Press, 1981.

Paul, R. P., and C. N. Stevenson. "Kinematics of Robot Wrists." *The International Journal of Robotics Research,* 2, no. 1 (1983): 31–38.

Rehg, James A. *Introduction to Robotics in CIM Systems,* 4th ed. Prentice-Hall, 2000.

Rivin, E. I. *Mechanical Design of Robots.* NY: McGraw Hill, 1988.

Roger, D. "On Chip Correction for Defective Pixels in an Image Sensor." *NASA Technical Brief,* 34, (2000).

Rout, B. K., and R. K. Mittal. "A Review on Minimizing Variability in Product Performance with the Help of Computational Tools by Controlling Active Noise Factors for Robotic Design." Proceedings of National Seminar on Current Applications of Computers in Design Engineering, Jodhpur, India, 81–90, March 2001.

Saha, S. K. *Introduction to Robotics*. New Delhi: Tata McGraw Hill, 2008.

Sahoo, N. C., S. K. Panda, and P. K. Dash. "A Current Modulation Scheme for Direct Torque Control of Switched Reluctance Motor Using Fuzzy Logic." *Mechatronics, The Science of Intelligent Machines,* 10, no. 3 (2000): 353–370.

Sainarayanan, G., R. Nagarajan, and Sazali Yaacob. "Vision Substitution for Visually Impaired by Stereo Acoustic Transform of Images Using Neural Network." International Conference of EANN 2001, Italy 281–287, July 16-18, 2001.

Sainarayanan, G., R. Nagarajan, and Sazali Yaacob. "Incorporating Certain Human Vision Properties in Vision Substitution by Stereo Acoustic Transform." Invited Paper in Special Session on Enhanced User Interfaces for Non-visual Image Display, Sixth International Symposium on Signal Processing and its Applications, ISSPA-2001, IEEE, 1 Kuala Lumpur (Malaysia) August 3-16, 2001.

Sainarayanan, G., R. Nagarajan, Sazali Yaacob, Marzuki Khalid, and Shamsuddin Amin. "On Clustering Methodology for Image Classification." IMAGE 2003, First Regional Malaysia–France Workshops on Image Processing in Vision Systems and Multimedia Communications, Miri (Malaysia), April 21-22, 2003.

Sainarayanan, G., Farrah Wong, R. Nagarajan, and Sazali Yaacob. "Image Segmentation Applied in Vision Aid for Human Blind." *International Journal of Engineering, Science and Technology,* 2, no. 1 (2002): 81–93.

Schilling, R. J. *Fundamental of Robotics – Analysis and Control.* NJ: Prentice-Hall, 1990.

Schlling, R. J., and R. B. White. *Robotic Manipulation: Programming and Simulation Studies* International Editions. NJ: Prentice-Hall, 1990.

Shahinpoor, M. *A Robot Engineering Text Book.* New York: Harper and Row, 1987.

Shahinpoor, M. *Dynamics, International Encyclopedia of Robotics: Application and Automation;* Edited by Richard C. Dorf, New York: Wiley, 1988.

Sharma, N. N., and R. K. Mittal. "Performance Analysis of Rule Based Fuzzy Control of Robotic Manipulators Using ANOVA." Proceedings of National Seminar on Current Applications of Computers in Design Engineering, Jodhpur, India, 23–30, 2001.

Shetty, D., and R. Kolk. *Mechatronics System Design.* Massachusetts: PWS Publishing, 1997.

Siciliano, B. "Kinematic Control of Redundant Robot Manipulators – A Tutorial." *Journal of Intelligent and Robotic System,* 3 (1990): 201–212.

Snyder, W. *Industrial Robots: Computer Interfacing and Control.* Englewood Cliffs, NJ: Prentice-Hall, 1985.

Spencer, R. "Ten Years of Evolution of Robotics Industry." *Robotics World Magazine,* 19, no. 5 (2001).

Spong, M. W., and M. Vidyasagar. *Robot Dynamics and Control.* Wiley, 1989.

Stadler, W. *Analytical Robotics and Mechatronics* International Edition Wiley & Sons, 1995.

Tan, K. K., S. Y. Lim, T. H. Lee, and H. Dou. "High Precision Control of Linear Actuators Incorporating Acceleration Sensing." *Journal of Robotics and Computer Integrated Manufacturing,* 16, no. 5 (2000): 295–305.

Veloso, M. M. "Entertainment Robotics." *Communication of ACM,* 45, no. 3 (2002): 50–63.

Wamplet, C. W. "Manipulator Inverse Kinematic Solutions Based on Damped Least Square Solutions." *IEEE Transactions on Systems, Man, and Cybernetics,* 16 (1986): 93–101.

Whitney, D. E. "The Mathematics of Coordinated Control of Prosthesis Arm and Manipulators." *ASME Journal of Dynamics and Systems, Measurements and Control,* 94, (1972): 303–309.

Williams, O. R., J. Angeles, and F. Bulca. "Design Philosophy of an Isotropic Six Axis Serial Manipulator." *Robotics and Computer Integrated Manufacturing,* 10, no. 4 (1993): 275–286.

Wise, E. *Applied Robotics.* Prompt Publications, 1999.

Wise, E. *Applied Robotics.* Thomson Delmar Learning, 2003.

Wolovich, W. A. *Robotics: Basic Analysis and Design.* Holt, Rinehart and Winston, 1986.

Wong, A. K. C., S. W. Lu, and M. Rioux. "Recognition and Shape Synthesis of a 3D Object Based on Attributes Hypergraphs." *IEEE Transactions on Pattern Analysis and Machine Intelligence* (1989): 279–290.

Wong Farrah, R. Nagarajan, and Sazali Yaacob. "Fuzzy Segmentation via Thresholding." *The Institution of Engineers* (*India*), 86, (2006): 68–72.

Wong Farrah, R. Nagarajan, and Wong Yaacob. "Neural Network in Stereo Vision Calibration for Blind Navigation." Proceedings of International Conference on Artificial Intelligence in Engineering and Technology, Kota Kinabalu (Malaysia), 244–247, August 3-5, 2004.

Wong Farrah, R. Nagarajan, Sazali Yaacob, Ali Chekima, and Nour-Eddine Belkhamza. "Electronic Travel Aids for Visually Impaired – A Guided Tour." Proceedings of International Conference in Engineering in Sarawak, Malaysia 377–382, May 19-20, 2000.

Wong Farrah, R. Nagarajan, Sazali Yaacob, Ali Chekima, and Nour-Eddine Belkhamza. "A National Conference, KL (Malaysia) 377–382, ParStereo Auditory Display for Visually Impaired." Proceedings of IEEE TENCON Intert II, September 24-27, 2000.

Wong Farrah, R. Nagarajan, Sazali Yaacob, Ali Chekima, and Nour-Eddine Belkhamza. "A Stereo Based Sonification System for Visually Impaired." SCORed, Malaysia, Feburary 2001.

Wong Farrah, R. Nagarajan, Sazali Yaacob, Ali Chekima, and Nour-Eddine Belkhamza. "An Image Segmentation Method Using Fuzzy-Based Threshold." Presented at Sixth International Symposium on Signal Processing and Its Applications, ISSPA-2001, IEEE, Kuala Lumpur (Malaysia), August 13-16, 2001.

Wong Farrah, R. Nagarajan, Sazali Yaacob, Ali Chekima, and Nour-Eddine Belkhamza. "Application of Fuzzy Image Processing Methodology in Blind Navigation." Proceedings of International Conference on Artificial Intelligence in Engineering and

Technology ICAIET-2002, Universiti Malaysia Sabah (Malaysia) 67–72, June 17-18, 2002.

Xu, W. L., and J. D. Han. "Joint Acceleration Feedback Control for Robots: Analysis, Sensing, and Experiments." *Journal of Computers and Integrated Manufacturing,* 16, no. 5 (2000): 307–320.

Yang, C. C., and M. M. Marefat. "Spatial Quantization Errors in Active Vision Inspection." IEEE Proceedings of International Conference on Systems, Man, and Cybernetics, San Antonio, TX, 1994.

Yoshikawa T. "Manipulability in Robot Mechanisms." *International Journal of Robotics Research,* 4, no. 2 (1985): 3–9.

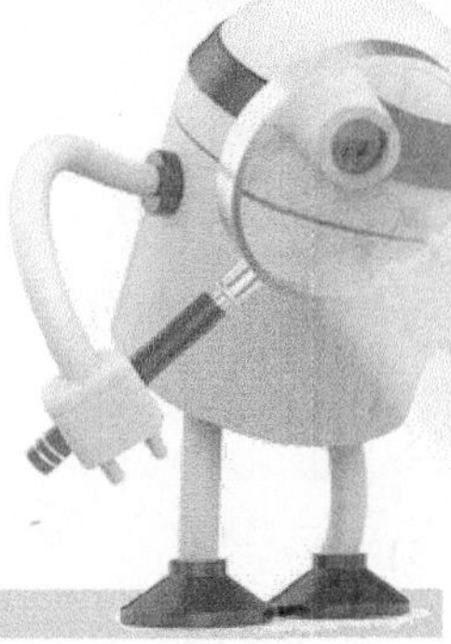

INDEX